JET PROPULSION

This book is a self-contained introduction to the aerodynamic and thermodynamic design of modern civil and military jet engines. Through two engine design projects, for a large passenger aircraft and a new fighter aircraft, the text explains modern engine design. Individual sections cover aircraft requirements, aerodynamics, principles of gas turbines and jet engines, elementary compressible fluid mechanics, bypass pressure ratio selection, scaling and dimensional analysis, turbine and compressor design and characteristics, design optimization and off-design performance.

The civil aircraft, which was the core of Part 1 of earlier editions, has now been in service for several years as the Airbus A380. Attention in the aircraft industry has now shifted to two-engine aircraft with a greater emphasis on reduction of fuel burn. The model created for Part 1 is therefore a hypothetical new two-engine aircraft aimed at high efficiency.

Nicholas Cumpsty is Professor Emeritus at Imperial College London. He was Professor of Aerothermal Technology in the University of Cambridge until 2000, when he became Chief Technologist of Rolls-Royce. On retiring from Rolls-Royce, he became a professor of mechanical engineering at Imperial College until retiring in 2008. He is a visiting professor at MIT and a Fellow of the Royal Academy of Engineering, the ASME and the AIAA.

Andrew Heyes is Professor of Mechanical Engineering and Head of Department in Mechanical and Aerospace Engineering at the University of Strathclyde. He has previously held positions at the University of Leeds and Imperial College London, where he spent a number of years teaching engine design based on the second edition of *Jet Propulsion*. Before Imperial, he worked with Rolls-Royce and British Aerospace (Military Aircraft Division, Warton). He is a Chartered Engineer and Fellow of the Institution of Mechanical Engineers.

JET PROPULSION

A Simple Guide to the Aerodynamics and Thermodynamic Design and Performance of Jet Engines

NICHOLAS CUMPSTY
Imperial College London

ANDREW HEYES
University of Strathclyde

CAMBRIDGE
UNIVERSITY PRESS

CAMBRIDGE
UNIVERSITY PRESS

32 Avenue of the Americas, New York, NY 10013-2473, USA

Cambridge University Press is part of the University of Cambridge.

It furthers the University's mission by disseminating knowledge in the pursuit of education, learning and research at the highest international levels of excellence.

www.cambridge.org
Information on this title: www.cambridge.org/9781107511224

First edition published 1998
Second edition 2003
Third edition 2015

Printed in the United States of America

A catalog record for this publication is available from the British Library.

Library of Congress Cataloging in Publication Data
Cumpsty, N. A.
Jet propulsion : a simple guide to the aerodynamics and thermodynamic design and performance of jet engines / Nicholas Cumpsty, Imperial College, Andrew Heyes, Imperial College.
pages cm
Includes bibliographical references and index.
ISBN 978-1-107-51122-4 (pbk.)
1. Airplanes – Turbojet engines. 2. Airplanes – Jet propulsion. I. Heyes, Andrew (Andrew L.) II. Title.
TL709.3.T83C85 2015
629.134′3533–dc23 2015010596

ISBN 978-1-107-51122-4 Paperback

Additional resources for this publication at www.cambridge.org/Cumpsty

CONTENTS

PREFACE

PREFACE TO THE THIRD EDITION

The civil aircraft, the Airbus A380, which resembles the New Large Aircraft of Part 1 of earlier editions, has been in service for several years. Attention in the aircraft industry has now shifted to two-engine aircraft with a greater emphasis on reduction of fuel burn, so the model created for Part 1 is the New Efficient Aircraft, a twin aimed at high efficiency. There is another change to highlight, which is the switch to using fan pressure ratio as the independent design variable rather than bypass ratio. In the time since the first edition, the typical fan pressure ratios have been reduced, and this has necessarily led to a considerable increase in complexity. The changes relating to military combat engines are relatively small.

Another major change is the inclusion of a co-author. Andy Heyes had used the second edition in teaching a course in Imperial College and was well placed in terms of knowledge and experience to work on the third edition.

For the third edition, we would like to acknowledge additional help from friends and colleagues. From Rolls-Royce, we would mention Conrad Banks, John Bolger, Simon Gallimore, Peter Hopkins, Glen Knight, Paul Madden, Alan Newby, Ian Rainbow and Joe Walsh; from Pratt and Whitney, Yuan Dong, Alan Epstein and Jayant Sabnis; from Stanford University, Juan Alonso and Anil Variyar; from the University of Cambridge, Chez Hall and John Young; from Imperial College, Aaron Costall, Ricardo Martinez-Botas and Peter Newton; from Ohio State, Mike Dunn, retired from NASA, Tony Strazisar; retired from General Electric, Meyer Benzakein; and retired from Airbus, Jeff Jupp.

PREFACE TO THE FIRST EDITION

This book arose from an elementary course taught to undergraduates, which forms the first ten chapters, concerned with the design of the engines for a new 600-seat long-range airliner. Introductory undergraduate courses in thermodynamics and fluid mechanics would provide the reader with the required background, but the material is also presented in a way to be accessible to any graduate in engineering or physical sciences with a little background reading. The coverage is deliberately restricted almost entirely to the thermodynamic and aerodynamic aspects of jet propulsion, a large topic in itself. The still larger area associated with mechanical aspects of engines is not covered, except that empirical information for such quantities as maximum tip speed are used, based on experience. To cover the mechanical design of engines would have required a much bigger book than this and would have required a mass of knowledge which I do not possess.

In preparing the course, it was necessary for me to learn new material, and for this, I obtained help from many friends and colleagues in industry, in particular at Rolls-Royce. This brought me to realise how specialised the knowledge has become, with relatively few people having a firm grasp outside their own speciality. Furthermore, a high proportion of those with the wide grasp are nearing retirement age, and a body of knowledge and experience is being lost. The idea therefore took hold that there is scope for a book which will have wider appeal than a book for students – it is intended to appeal to people in the aircraft engine industry who would like to understand more about the overall design of engines than they might normally have had the opportunity to master. My ambition is that many people in the industry will find it useful to have this book for reference, even if not displayed on bookshelves.

The original course, Chapters 1–10, was closely focused on an elementary design of an engine for a possible (even likely) new large civil aircraft. Because the intention was to get the important ideas across with the least complication, a number of simplifications were adopted, such as taking equal and constant specific heat capacity for air and for the gas leaving the combustor, as well as neglecting the effect of cooling air to the turbines.

Having decided that a book could be produced, the scope was widened to cover component performance in Chapter 11 and off-design matching of the civil engine in Chapter 12. Chapters 13–18 look at various aspects of military engines; this is modelled on the treatment in Chapters 1–10 of the civil engine, postulating the design requirement for a possible new fighter aircraft. In dealing with the military engine, some of the simplifications deliberately adopted in the early chapters are removed; Chapter 19 therefore takes some of these improvements from Chapters 13–18 to look again at the civil engine.

Throughout the book, the emphasis is on being as simple as possible, consistent with a realistic description of what is going on. This allows the treatment to move quickly and the book to be brief. But more important, it means that someone who has mastered the simple formulation can make reasonably accurate estimates for performance of an engine and can estimate changes in performance with alteration in operating conditions or component behaviour. Earlier books become complicated because of the use of algebra; furthermore, to make the algebra tractable frequently forces approximations which are unsatisfactory. The present book uses arithmetic much more – by taking advantage of the computer and the calculator, the numerical operations are almost trivial. The book contains a substantial number of exercises which are directed towards the design of the civil engine in the early chapters and the military engine in the later chapters. The exercises form an integral part of the book and follow, as far as possible, logical steps in the design of first the civil engine and then the military combat engine. Many of the insights are drawn from the exercises.

Because Chapters 1–10 were directed at undergraduates, there are elementary treatments of some topics (most conspicuously, the thermodynamics of gas turbines, compressible fluid mechanics and turbomachinery), but only that amount needed for understanding the remainder of the treatment. I decided to leave this elementary material in, having in mind that some readers might be specialists in areas sufficiently far from aerodynamics and thermodynamics that a brief but relevant treatment would be helpful.

ACKNOWLEDGEMENTS

It is my pleasure to acknowledge the help I have had with this book from my friends and colleagues. The largest number are employed by Rolls-Royce (or were until their retirement) and include Alec Collins, Derek Cook, Chris Freeman, Simon Gallimore, Keith Garwood, John Hawkins, Geoffery Hodges, Dave Hope, Tony Jackson, Brian Lowrie, Sandy Mitchell, James Place, Paul Simkin, Terry Thake and Darrell Williams. Amongst this group I would like to record my special gratitude to Tim Camp, who worked through all the exercises and made many suggestions for improving the text. I would also like to acknowledge the late Mike Paramour of the Ministry of Defence. In the Whittle Laboratory, I would like to record my particular debt to John Young and also to my students Peter Seitz and Rajesh Khan. I am also grateful for the help from other students in checking late drafts of the text. In North America, I would like to mention Ed Greitzer and Jack Kerrebrock (of the Gas Turbine Laboratory of MIT), Bill Heiser (of the Air Force Academy), Phil Hill (of the University of British Columbia), Bill Steenken and Dave Wisler (of GE Aircraft Engines) and Robert Shaw. Above all, I would like to express my gratitude to Ian Waitz of the Gas Turbine Laboratory of MIT, who did a very thorough job of assessing and weighing the ideas and presentation – the book would have been very much the worse without him. In addition to all these people, I must acknowledge the help and stimulus from the students who took the course and the people who have added to my knowledge and interest in the field over many years.

THE EXERCISES

An important part of the book are exercises related to engine design. To make these possible, it is necessary to assume numerical values for many of the parameters, and appropriate values are therefore assumed to make the exercises realistic. These values are necessarily approximate, and in some cases so, too, is the model in which they are used. The answers to the exercises, however, are given to a higher level of precision than the approximations deserve. This is done to assist the reader in checking solutions to the exercises and to ensure some measure of consistency. The wise reader will keep in mind that the solutions are in reality less accurate than the number of significant figures seems to imply.

The usefulness of the book will be greatly increased if the exercises are undertaken. In some cases, one exercise leads to another, and a few simple calculations on a handheld calculator suffice. In others, it is desirable to carry out several calculations with altered parameters, and such cases call out for a computer and spreadsheet. Those exercises marked with an asterisk provide results to be entered onto the design sheets for the New Efficient Aircraft and the New Fighter Aircraft.

SOLUTIONS TO THE EXERCISES

Solutions to all the exercises may be obtained from the publisher by e-mailing
solutions@cambridge.org.

GLOSSARY

afterburner   a device common in military engines where fuel is burned downstream of the turbine and upstream of the final propulsive nozzle. Also known as an augmentor or as reheat.

aspect ratio   the ratio of one length to another to define shape, usually the ratio of span to chord.

blades   the name normally given to the aerofoils in a turbomachine (compressor or turbine). Sometimes stationary blades are called stator vanes (or just vanes) and rotor blades are called buckets.

booster   a name given to compressor stages on the LP shaft in two-shaft engines. The booster stages only affect the core flow.

bypass engine   an engine in which some of the air (the bypass stream) passes around the core of the engine. The bypass stream is compressed by the fan and then accelerated in the bypass stream nozzle. These are sometimes called turbofan engines or fan engines.

bypass ratio   the ratio of the mass flow rate in the bypass stream to the mass flow rate through the core of the engine.

chord   the length of a wing or a turbomachine blade in the direction of flow.

combustor   also known as a combustion chamber. The component where the fuel is mixed with the air and burned.

compressor   the part of the engine which compresses the air, a turbomachine consisting of stages, each with a stator and rotor row.

core   the compressor, combustion chamber and turbine at the centre of the engine. The core turbine drives only the core compressor. A given core can be put to many different applications, with only minor modifications, so it could form part of a high-bypass ratio engine, a turbojet (with zero bypass ratio) or part of a land-based power generation system. The core is sometimes called the gas generator.

drag   the force D created by the wings, fuselage, etc. in the direction opposite to the direction of travel.

fan   the compressor operating on the bypass stream; normally the pressure ratio of the fan is small, not more than about 1.8 for a modern high-bypass civil engine (in one stage with no inlet guide vanes) and not more than about 4.5 in a military engine in two or more stages.

gross thrust   the thrust F_G created by the exhaust stream without allowing for the drag created by the engine inlet flow; for a stationary engine, the gross thrust is equal to the net thrust.

HP   the high-pressure compressor and turbine are part of the engine core. They are mounted on either end of the HP shaft. In a two-shaft engine, they form the core spool.

incidence   sometimes called angle of attack, the angle at which the wing is inclined to the direction of travel or the angle at which the incoming flow direction is inclined to a wing or to a compressor or turbine blade.

IP	the intermediate-pressure compressor or turbine, mounted on the IP shaft. There is only an IP shaft in a three-shaft engine.
jetpipe	the duct or pipe downstream of the LP turbine and upstream of the final propulsive nozzle.
LCV	the lower calorific value of the fuel; the energy released per unit mass of fuel in complete combustion when the products are cooled down to the inlet temperature but none of the water vapour is allowed to condense.
lift	the force L created, mainly by the wings, perpendicular to the direction of travel.
LP	the low-pressure compressor and turbine are mounted on either end of the LP shaft. Combined, they form the LP spool.
nacelle	the surfaces enclosing the engine, including the intake and the nozzle.
net thrust	the thrust F_N created by the engine available to propel the aircraft after allowing for the drag created by the inlet flow to the engine. (Net thrust is equal to gross thrust minus the ram drag.)
ngv	the nozzle guide vane, another name for the stator row in a turbine.
nozzle	a contracting duct used to accelerate the stream to produce a jet. In some cases, for high-performance military engines, a convergent-divergent nozzle may be used.
payload	the part of the aircraft weight which is capable of earning revenue to the operator (can be freight or passengers).
pylon	the strut which connects the engine to the wing.
ram drag	the momentum of the relative flow entering an engine.
sfc	specific fuel consumption (actually the *thrust* specific fuel consumption) equal to the mass flow rate of fuel divided by net thrust. The units should be in the form $(kg/s)/kN$ but the units are often given as $lb\,h^{-1}lb^{-1}$ or $kg\,h^{-1}kg^{-1}$.
specific thrust	the net thrust per unit mass flow through the engine, units m/s.
spool	used to refer to the compressor and turbine mounted on a single shaft, so a two-spool engine is synonymous with a two-shaft engine.
stagnation	stagnation temperature is the temperature which a fluid would have if brought to rest adiabatically. The stagnation pressure is the pressure if the fluid were brought *isentropically* to rest. Stagnation quantities depend on the frame of reference and are discussed in Chapter 6.
static	static temperature and pressure are the actual temperature and pressure of the fluid, in contrast to the stagnation quantities defined earlier.
turbine	a component which extracts work from a flow. It consists of rotating and stationary blades. The rotating blades are called rotor blades, and the stationary blades are called stator blades or nozzle guide vanes.
turbofan	a jet engine with a bypass stream.
turbojet	a jet engine with no bypass stream – these were the earliest types of jet engines and are still used for very-high-speed propulsion.

NOMENCLATURE

a	speed of sound $\sqrt{\gamma R T}$
A	area
A_R	Aspect ratio
bpr	bypass ratio
c	chord of wing or blade
c_p	specific heat at constant pressure
C_D	drag coefficient
C_L	lift coefficient
D	drag (force opposing motion)
d_r, D	diameter
E	energy state $m(gh + V^2/2)$
E_s	specific energy state $gh + V^2/2$
F_G	gross thrust
F_N	net thrust
g	acceleration due to gravity
h	static enthalpy
h_0	stagnation enthalpy
h, H	altitude
h	blade height (i.e. span)
H	range factor
i	incidence
L	lift (force perpendicular to direction of motion)
LCV	lower calorific value of fuel
m	mass
$\dot{m}$	mass flow rate
$\overline{m}$	non-dimensional mass flow rate, $\dot{m}\sqrt{c_p T_0}/A p_0$
M	Mach number
n	load factor
N	shaft rotational speed
opr	overall pressure ratio
p	static pressure
p_0	stagnation pressure
q	dynamic pressure $\frac{1}{2}\rho V^2$
Q	heat transfer
$\dot{Q}$	heat transfer rate
r	radius (Chapters 9 and 18)
r	pressure ratio
R	gas constant
s	entropy

SAR	specific air range
SEP	specific excess power
sfc	specific fuel consumption
T	static temperature
T_0	stagnation temperature
U	blade speed
V	velocity
V_j	jet velocity
V^{rel}	velocity relative to moving blade
w	weight
W	work
$\dot{W}$	work rate, power
α	flow direction (measured from axial)
α^{rel}	flow direction relative to moving blades
β	blade direction measured from axial
γ	ratio of specific heats c_p/c_v
δ	flow deviation (Chapters 9 and 18)
δ	p_0/p_{0ref}
η	efficiency
θ	$\sqrt{T_0/T_{0ref}}$
ρ	density
ε	cooling effectiveness (Chapter 5)
Ω	angular velocity

Subscripts

a	ambient
ab	afterburner
air	air
b	bypass
c	core
dry	no afterburner in use
e	combustion products (c_p and γ)
f	fuel
$isen$	isentropic (efficiency)
m	mean
p	polytropic (efficiency)
sl	sea level
$therm$	thermal (efficiency)

A NOTE ON NOMENCLATURE

The various stations or positions throughout an engine are given numbers, and different companies have different conventions for the many positions along the flow path of a multi-spool engine. An internationally recommended numbering scheme applies to some of the major stations, and of these the most important station numbers to remember are the following:

2   engine inlet face
3   compressor exit and combustion chamber inlet
4   combustion chamber exit and turbine inlet

The preceding brief list shows the one superficial snag: the inlet face of the engine is station 2, whereas most teaching courses call it station 1. The reason for this discrepancy is that for some engine installations, particularly in high-speed aircraft, there can be a substantial reduction in stagnation pressure along the inlet; station 2 is after this loss has taken place. In this book, the international standard will be used, where appropriate, with 2 at the inlet to the engine, and a simplified guide is shown in Figure 7.1. For more detailed treatment of the engine, the scheme in Figure 12.7 or Figure 15.1 should be consulted.

Subscript zero is used to denote stagnation conditions, for example stagnation pressure, p_0, and stagnation temperature, T_0. (See Chapter 6 for an explanation of the terms *stagnation pressure* and *stagnation temperature.* Some people use the word *total* in place of *stagnation.*) The stagnation pressure at engine inlet is therefore written p_{02} and temperature at turbine entry as T_{04}.

TERMINOLOGY

There are differences between British and American usage, but usually these are small – *aeroplane* and *airplane*, for example. It may be noted that in Britain, it is normal to use the word *civil* when referring to aviation, aircraft and air transport, where in the United States the word *commercial* would normally be used. In the book, the British usage *civil* is adopted. However, although it is still quite common in Britain to refer to *reheat*, the corresponding American term *afterburner* is used throughout the book.

Part 1

Design of Engines for a
New Efficient Aircraft

CHAPTER 1   THE NEW EFFICIENT AIRCRAFT: REQUIREMENTS AND BACKGROUND

1.0 INTRODUCTION

This chapter sets out the background to the new airliner which is to form the basis of the first part of this book. The aircraft, to be called the New Efficient Aircraft (NEA), will be a large wide-body aircraft designed to give low fuel burn, in anticipation of the likely rise of fuel price and pressure to reduce CO_2 emissions. The aircraft will have two engines.

The costs and risks of a new aircraft or engine project are huge, but the profits might be large too. Some background is first discussed concerning the history and business of jet propelled aircraft and the impact of concerns for the environment. In explaining the requirements some of the units of measurement used are discussed. Design calculations in a company are likely to assume that the aircraft flies in the International Standard Atmosphere (or something very similar) and this assumption will be adopted throughout this book. The standard atmosphere is introduced and discussed towards the end of the chapter.

1.1 SOME BACKGROUND

The age of jet travel really got started when the Boeing 707 entered service in 1958. By the time this aircraft was initiated, Boeing had already acquired considerable experience of large multi-engine jet aircraft, bombers and tankers, so it was in a strong position to make good design choices. The 707 was conceived as a long-range aircraft, which in those days meant it was capable of flying across the Atlantic non-stop with a full load of passengers, typically 110 in a two-class cabin. The range with maximum payload was only 2800 nautical miles (nm), but the shortest distance between London and New York is 2991 nm and going west there are normally headwinds that increase the effective distance. Such flights would therefore operate with less than the maximum payload, which would mean less than maximum freight on board, if all seats were taken.

The Boeing 707 had four engines and very much set the trend for aircraft which followed it. The fuselage was circular for most of its length, the wings were swept and the turbojet (no bypass flow) engines were mounted on pylons to hang under and forward of the wing. Hanging the engines under the wing has many advantages, in particular their significant weight is close to where the lift is generated, which makes the structure lighter. Furthermore, in the case of an uncontained engine failure, the risk of catastrophic damage to the airframe or other engines is reduced. The under-wing engine arrangement has subsequently become the preferred for commercial aircraft. Other things about the 707 also came to affect later designs. The take-off field length was 11,000 feet (3353 m), which affected not only later aircraft specifications but also airport design around the world. The initial cruising altitude was 31,000 ft, and the cruise Mach number was 0.85, both of which strongly affected later designs. In the same year that the 707 entered service the Douglas DC8 had its first flight and the following year the Convair 880 flew for the first time; both these aircraft had similar features to the 707, but never achieved a similar market success.

Only twelve years after the Boeing 707, the Boeing 747 entered service. The 747 had a two-aisle layout in the cabin and initially could carry up to 366 passengers in three classes. The full payload range was under 3000 nm. The four engines were carried under the swept wings, but an important difference with the 747 is that now the engines were of the high bypass ratio type. For the initial 747 the engines were only made by Pratt & Whitney, who had also made the engines for the 707. The 747, like the 707, was conceived as a long-range aircraft and for that reason four engines seemed the rational choice. For the early engines achieving enough thrust for take-off was hard and the margins were small. At take-off the engines must be sized so that if one fails at the worst possible moment, just when the aircraft rotates prior to leaving the ground, the take-off can be continued and the plane can land safely. For a four-engine plane it means that take-off thrust is 4/3 times the minimum required, whereas for a two-engine plane (a "twin") the thrust must be twice the minimum. For a turbojet or a low bypass engine sizing the engines to meet this double thrust requirement implies extra engine weight and excess capacity at cruise. The extra weight of this excess capacity would reduce the range and efficiency of the aircraft. As we will see, the situation changed considerably with the advent of modern high bypass ratio engines and the twin has now become the norm even for long range. In addition, in the early days of over-ocean travel the added security during cruise of four engines was deemed prudent and the use of four engines for long range persisted so that both the A340, entering service in 1993, and the A380, in 2007, had four engines. In the case of the A380, however, it is now size which makes four engines more attractive than two. Were the Airbus A380 to have only two engines they would be so large that it would be necessary to raise the wing to fit them under. Raising the wing is not considered practical since, if the wing were higher off the ground than current large aircraft, it would raise the cabin, and if the cabin were raised higher the existing passenger handling facilities at airports would be unusable. Raising the wing would also make the undercarriage much bigger and heavier. Furthermore, if the Airbus A380 were to have only two engines, these would be too large to be moved conveniently as air freight, which is occasionally necessary when an engine needs replacement or overhaul.

Shortly after entry into service of the four-engine Boeing 747, two "tri-jets" with three high bypass engines came into service: the Douglas DC10 and the Lockheed L1011. The engines were made by General Electric and Rolls-Royce respectively. These were conceived as intermediate-range aircraft (2350 nm for the DC10 and 2324 nm for the L1011) for which three engines gave a good mix of performance and added security. Like the 747 they were wide-body aircraft with two aisles in the cabin. There was also a general acceptance that shorter range would imply a smaller aircraft with fewer passengers, a view which still carries some weight today.

The mould was broken by a new international company with headquarters in France: Airbus. As the name "Airbus" implies, they conceived of an aircraft to move many people a relatively short distance. The aircraft they designed was the Airbus A300, a wide-body aircraft with only two engines, often referred to as a twin, which entered service in 1974. Not only was it a wide-body aircraft with only two engines, the A300 was aimed at short range and initially had a full payload range of only about 900nm (see Table 1.1). The A300 could carry 220 passengers in a three-class layout and as many as 375 passengers in a one-class cabin. The Boeing 767, a wide-body two-engine plane, entered service eight years later with a full payload range of 2200 nm for the 767-200 but almost 5000 nm for the 767-200ER. Extensive use was made of the 767 for long-range flights and special regulations, referred to as Extended-range Twin-engine Operational Standards (ETOPS), were introduced to allow operation over oceans; ETOPS is briefly explained below. The 767-200ER established that the twin could be a long-range aircraft and meanwhile the original Airbus A300 range had also greatly increased. About a decade after the Boeing 767 Airbus introduced the A330 and Boeing the 777, both wide-body twins. Initial full payload ranges were relatively short (3700 and 3300 nm, respectively), and Airbus had little incentive to push the range because it almost simultaneously released the four-engine A340 for long ranges. However, range increased very rapidly with the Boeing 777, so that the 777-200LR has a full-payload range of about 7600 nm. The success of the A330 and 777 then form the background to the launch of the Boeing 787 programme in 2000 and the launch of the Airbus A350 in 2006. Problems with the 787 delayed entry into service until late in 2011 and the A350 entered service late in 2014.

Both Airbus and Boeing have embarked on a process of improving existing twin jets, starting with the single-aisle A320 and B737, but more recently including the large twin-aisle aircraft. The A330neo (new engine option) was announced in 2014 and is due to be delivered to airlines at the end of 2017. This will have a new, more efficient engine with a larger fan. The wings will be modified and lengthened, and together this is claimed to offer a fuel-burn savings of 10% at full payload and maximum range. (The engine improvements are estimated to save 10% fuel and aircraft aerodynamics another 3%, but the drag of the larger engine penalises the aircraft by 1% and the increased weight by 2%.) Airbus claims that the performance will match the B787, but the aircraft will be significantly cheaper. In 2013 Boeing launched the B777X, which is to enter service in 2019. It will also offer a new engine and improved wings. The wings will use the carbon-composite technology of the 787 and their span will be increased by 6.5 m, requiring the use of folding wing tips. The heavier version, the B777-9X, will have a capacity of 407 passengers, putting it where only the B747 would have been only a few years ago. These are very long range aircraft with the range for the B777-8X offered as 9300 nm and for the larger-payload B777-9X as 8200 nm.

The full payload ranges for both the 787 and A350, in the initial versions, are in excess of 5000 nm. The performance of both are shown in Table 1.1, together with some of the key two-engine aircraft that preceded them. Many of the parameters in the table are self-evident and others will be explained later. In the table a number of weights (or masses) are referred to. The take-off weight m_{TO} is made up of the empty weight m_E, the payload (which is the weight of passengers

Table 1.1   Salient characteristics of some large two-engine aircraft

	Airbus A300-B2	Airbus A300-600	Boeing 767-200	Airbus A330-300	Boeing 777-200	Boeing 777-200LR	Boeing 787-8	Airbus A350-900
Entry into service	1974	1983	1982	1993	1995	2006	2011	2014
Normal max. passengers	269	274	216	335	375	301	242	315
Range at max. passenger (nm)	1500	3100	4000	5400	4800	9100	7650	8100
Maximum range at max. payload, R1	900	1950	2200	3700	3300	7600	5500	5900e
Max. payload	30.6	43.3	33.2	55.2	54.6	64.0	40.2e	47.6
Max. take-off weight	137	165	143	233	243	348	228	268
Empty weight	86	87	80	120	136	145	122e	133e
Fuel burn at R1 (see Section 2.1)	15	29	24	49	42	124	60e	65
Cruise Mach no.	0.78	0.78	0.80	0.82	0.84	0.84	0.85	0.85
Cruise L/D (estimates)	16	16	18	20	20	20	21	21
Estimated engine sfc $(kg\,h^{-1}kg^{-1})$	0.64	0.63	0.62	0.58	0.57	0.54	0.52	0.51
Wing span (m)	45	45	48	60	61	65	60	65
Wing area (m^2)	260	260	283	362			325	443

Note. All weights in tonne. Weights for the 787-8 are those given in November 2011 by Lyssis, (http://www.lissys.demon.co.uk/boeing787-2011) with empty weight significantly higher than the original design specification given by Boeing. All the values should be treated as indicative of the trend rather than exact, and 'e' indicates that this is estimated rather than from a published source. R1 is the range with maximum payload and maximum take-off weight. Fuel burn is fuel weight at take-off less reserves estimated at 4% of maximum take-off weight.
Note that numbers with 'e' indicates that they are only estimates.

and freight) m_{PL} and the fuel m_F. The fuel weight can be divided into fuel which is expected to be burned, m_{FB}, and the fuel which is held in reserve for contingencies, m_{FR}, which is not expected to be burned. As a reasonable approximation, $m_{FR} = 0.04 m_{TO}$, and the reserve may be treated as an addition to the empty weight.

As the table shows, the cruise Mach number and the range have increased over time, so the latest twins are designed to cruise at the same speed as the long-range four-engine planes, $M = 0.85$. It may also be noted that a full complement of passengers does not provide maximum payload: each passenger may be taken to weigh 95 kg with baggage. Typically the freight carried in the hold ("belly freight") is comparable in weight to a full load of passengers. In quoting range it is not uncommon to give the value when only a full load of passengers is carried, with no freight in the hold.

The growth in twin aircraft for long ranges could not have happened without several things. One is that the engines have become very much more reliable, so that long flights of two-engine aircraft can be allowed over the oceans with acceptable safety. Long-range flights of two-engine aircraft are now governed by the Extended-range Twin-engine Operational Standards (ETOPS), referred to above, which specify how much flight time (flying with only one engine in operation) they may be away from a suitable airport for diversion in case of a problem. The amount of time depends on the testing procedure they have been through, and experience of reliability in testing and operations to date. For the 787, for example, the ETOPS limit is 330 minutes and the A350 is intended to be 350 minutes. ETOPS has led to increased attention to reliability. Another reason for the use of twins for long range is the recognition that high bypass engines are quite well matched at take-off and cruise, meaning that, unlike the old engines, they operate at full capacity in both cases and are not carrying excess capacity (and weight) to ensure safe take-off capability. A key consequence of this is that long range flight with twins becomes efficient. This satisfactory match is a topic which will be discussed in more detail in later chapters. A further reason for the increase in twins is that engines are expensive items requiring maintenance and having two rather than four engines reduces operating cost. The New Efficient Aircraft (NEA) which is the object of the first ten chapters of this book, will therefore have two engines.

As already noted, large aircraft are usually intended for long-range operation, whilst small aircraft are for short range. The smaller aircraft have a single-aisle configuration in the passenger cabin; this category includes regional jets but is most conspicuous in the Boeing 737 and the Airbus A320 series. There is some variation in full-payload range between types of single-aisle aircraft, but most can carry a full load of passengers around 3000 nm, even though the majority of flights are much less than 1000 nm. Using relatively small aircraft for short flights allows the traffic to be handled with frequent flights and this is generally attractive to passengers. There are also efficiencies associated with short range aircraft due to their increased ratio of passenger to fuel weight. As noted above, the original Airbus A300 was a short-range, wide-body aircraft, but the majority of wide-body aircraft are primarily for long range. The aerodynamic performance of the single-aisle aircraft is perhaps 10–15% worse than the larger aircraft, measured in terms of drag per unit of weight. Likewise, the engine specific fuel consumption may be 20% worse than that of the best larger engines. The New Efficient Aircraft (NEA) will therefore be a wide-body aircraft capable of carrying about 300 passengers with a design range more like that of the current single-aisle aircraft, about 3000 nm. The aircraft aerodynamics and structural efficiency will be comparable to the latest twin aircraft, the 787 and A350, and engine performance will be comparable to those engines installed on these aircraft.

1.2 ENVIRONMENTAL ISSUES

When jet propelled passenger transport was initiated, little or no thought was given to the environment, either near the airports or in the upper atmosphere. By the late 1960s the situation near airports was becoming intolerable, mainly because of the noise, but also because of pollution. The pollution involved unburned hydrocarbons, smoke (i.e. small particles of soot, which is unburned carbon) and oxides of nitrogen (NO_x). Gradually steps have been taken to rein in the nuisance near airports by international agreement, with regulations both for combustion produced emissions near airports and for noise during take-off and landing.

The international agreements are reached so that the interests of various parts of the industry (from manufacturers of engines through to the airlines which operate rather old aircraft) are addressed. The net result is that the international agreements have lagged behind public pressure for amelioration so that local regulations at important airports around the world have tended to be more challenging for the makers of new engines to meet. The international limits on noise are so far above the noise produced by new aircraft with modern engines that the international limit serves merely as the benchmark from which the margin of lower noise is set. For noise the airport which tends to determine the level which new large aircraft have to achieve is London Heathrow. For products of combustion an airport which sets the level is Zurich, where charges are varied depending on the amount of pollution released in a standard landing and take-off operation. The issues and rules for emissions of pollutants are addressed briefly in Section 11.5. Noise is considered in an appendix at the end of the book.

Concerns about noise, particulates (small soot particles) and NO_x remain, but these have more recently been overshadowed by concerns about the impact of aviation on global warming. The principal motivation behind the introduction of the early jet aircraft, such as the Boeing 707, was higher speed and comfort and in terms of fuel burn, or CO_2 production, they were far worse than piston engines they replaced; indeed it is only recently that jet-propelled aircraft have achieved fuel efficiency as good as the later piston engine airliners. According to the Air Transport Action Group,[1] the global aviation industry currently produces 2% of the global CO_2 emissions, which is about 12% of the CO_2 from transport. In addition, oxides of nitrogen do make a contribution to global warming, as do the condensation trails left by aircraft when flying through an atmosphere which is supersaturated relative to ice. There is large uncertainty associated with both these factors, and they are sometimes described, quite inappropriately, as giving a multiplier to the effect of the carbon

---

[1] http://www.atag.org/facts-and-figures.html.

dioxide generated. Certainly carbon dioxide has a well-recognised effect and persists much longer in the atmosphere, for times on the order of 100 years. In this book we direct attention at reducing the emission of carbon dioxide, whilst recognising that this is not the whole impact of aviation. The reduction in the amount of fuel burned automatically leads to a reduction in the emission of CO_2, and with this in mind the International Air Transport Association and the International Civil Aviation Organisation have set goals for fuel efficiency improvement of 1.5% per year to 2020 and 2% per year from then to 2050. It is hard to maintain year-on-year improvements like this. Indeed, the historical trend going back to the introduction of the Boeing 707 is about 0.7% per year, and it is widely thought that improvements are becoming harder to achieve.

Reducing fuel burn is an obvious saving in cost to any airline. At the time when *all* aircraft in service in 2015 were conceived, and designed, the cost of fuel was a relatively small fraction of the overall direct operating cost, typically no more than 20%. This meant that considerable inefficiency in the use of fuel could be allowed if the aircraft were to be more flexible in operation. So, in particular, a plane capable of long range could be attractive to an airline even if the requirement for long range was infrequent. This has driven the trend for range so apparent in Table 1.1. Likewise, the ability to carry significant belly freight was attractive, even if the amount carried on the majority of flights was well below the maximum possible, so the fuel burn penalty of the increase in aircraft structural weight and larger engine thrust to allow this freight capacity was accepted. Fuel is now more expensive and is expected to become more so, either because of a rise in the cost of unrefined petroleum or because of taxes or charges added to the fuel. Possible charges on emitted CO_2 are equivalent to a rise in the cost of fuel. Moreover, there are plans to introduce certification levels for CO_2 or fuel burn, broadly analogous with those currently for noise and NO_X, which will further drive aircraft towards lower fuel burn. This is the background for the proposed New Efficient Aircraft, a wide-body aircraft embodying the latest technology optimised for low fuel burn in the majority of operations.

1.3 COMMERCIAL ASPECTS OF NEW LARGE AIRCRAFT

It takes several years to design, develop and certify (i.e., test so that the aircraft is approved as safe to enter service) a new aircraft. It seems to take even longer to develop the engines, but until the specifications of the aircraft are settled, it is not clear what engine is needed. There are three major engine manufacturers (Rolls-Royce in Britain, Pratt & Whitney and General Electric in the United States). The costs of developing a wholly new engine are so high that it has been the ambition of each manufacturer to use, whenever possible, an existing engine, perhaps with some adaptation or uprating. On the Boeing 777 all three major manufacturers offered an engine, and the competition was fierce. Pratt & Whitney and Rolls-Royce offered developments of existing large engines, but in this case General Electric developed a wholly new engine, the GE90. Rarely does it serve the makers of new engines or new aircraft to give much information on the costs they have incurred. Nevertheless, the *Economist* of 18 September 1999 reported that the GE90 had cost General Electric $1 million per day for four and a half years, in total about $1.6 billion; it is not clear how much extra was spent by risk-sharing partner companies. This huge sum can be made more understandable if an average wage for an employee, with the appropriate overheads, is taken to be \$150,000 per annum – the \$1.6 billion cost then translates into over 10,000 man-years of work.

If engines are expensive, the airframe is even more so. The *Seattle Times* of 5 February 2011 quoted a Wall Street analyst who says that the original Boeing estimate for the cost of the new 787 was \$5 billion, but that the problems Boeing encountered raised the cost by between \$12 billion and \$18 billion. The *Seattle Times* of 24 September 2011 gave a 'conservative estimate' that the cost to Boeing of producing the 787 was in excess of \$32 billion. A spokesman, David Strauss, from the investment bank UBS, is quoted, who believes Boeing must sell between 1100 and 1900 787s to break even. An estimate by Reuters news agency on 22 October 2012 refers to Airbus having to spend \$15 billion producing the A350, but it is not known how much has actually been spent in bringing the aircraft into service.

This is a business in which large financial risks are taken, which, in turn, leads to caution and conservatism. On the other hand, the developments evident in Table 1.1 point to a continuing increase in design range, which is now significantly greater than the overwhelming majority of flights; airlines have a liking for the flexibility that long design range brings, even if this comes at the costs of higher fuel consumption. So long as airlines want increased range in their new aircraft, it is difficult to get change in the style or operation of aircraft. It is even harder to get any radically new configuration, such as blended wing bodies or open rotors (two rows of propellers), and this book will be concerned with more conventional engine and aircraft configurations propelled by a turbofan engine. The New Efficient Aircraft considered here will be the conventional fuselage with wing (sometimes called tube and wing) and the engines will be mounted under the wings. Engines are large and heavy; for example, a Rolls-Royce Trent 800, which is the lightest engine to power the Boeing 777, weighs about 8 tonne when installed on the aircraft. Because most of the lift is generated by the wings, hanging the comparatively massive engines where they can most easily be carried makes good structural sense. This reduces the wing root bending moment and makes possible a reduction in the strength and weight of the whole aircraft.

Whilst discussions are going on between aircraft manufacturers and airlines, they are also going on between aircraft manufacturers and the engine manufacturers. As specifications for the 'paper' aircrafts alter, the 'paper' engines designed to power them will also change; many potential engines will be tried to meet a large number of proposals for the new aircraft before any company finally commits itself. The first ten chapters of the book will attempt, in a very superficial way, to take a specification for an aircraft and design the engines to propel it – this is analogous, in a simplified way, to what would happen inside an engine company.

1.4 Specification of the New Efficient Aircraft

The New Efficient Aircraft (NEA) is to have a clear design objective of moving passengers with the least environmental impact, consistent with meeting the needs of passengers and airlines. The aircraft will be designed to have a shorter range, about 3000 nm, than the latest two aisle types (Boeing 787 and Airbus A350), which have a full-payload range of 5500 nm or more. The NEA will have the advanced aircraft aerodynamics and engine fuel consumption consistent with what is being offered for the latest new long-range large aircraft. The NEA will adopt the lower cruise speed of short-range planes, $M = 0.78$, in place of 0.85 for recent new large aircraft – this is a reversion to the speed of the A300 and close to that of the B767. It is also the cruising speed of the A320 and the Next Generation B737. The shorter range and lower speed, it will be shown, give substantially lower fuel burn than that which has been offered for the smaller single-aisle types available for shorter ranges.

The anticipated fuel-efficiency benefits of a wide-body aircraft compared with the latest single-aisle aircraft derive from the observation that in the past the wide-body aircraft, most of which were aimed at long range, have been significantly more efficient in terms of aircraft aerodynamics and engine performance. There is another factor which leads to large increases in fuel burn and that is congestion of air space. This leads to inefficient routing and delays during which aircraft circle whilst waiting for permission to land. Congestion is likely to get worse and is already a problem, not only in North America and Europe but also in China. In terms of congestion the relatively small single-aisle aircraft is as much a problem as a larger twin-aisle, such as the NEA. So there is the possibility of containing, or even reducing, the problem of congestion by reducing the number of flights by replacing small aircraft with larger ones. It is noteworthy that recently Airbus has agreed to the production of an A330 designed specifically for shorter-range operation, with China particularly in mind.

As noted in the previous paragraph, the design of a new aircraft involves many iterations and changes. The starting point for a passenger aircraft appears to be a decision on the number of passengers to be carried and the range with a full load of passengers. This is somewhat arbitrary, since different airlines have different seating arrangements and therefore different densities, but some typical value is chosen. The assumed maximum number of passengers is considerably smaller than the maximum allowed by regulations for emergency exits, for with such close packing the level of comfort would be unacceptable for most passengers. Planes aimed at long-range operation typically have less dense seating than those for shorter range. Having fixed on the number of passengers and the effective floor area per passenger, the floor area inside the fuselage is fixed. This then offers a volume under the cabin floor potentially available for cargo and, knowing the typical density of cargo, the payload of freight which could be carried is calculable. For many twin-aisle aircraft, like the Boeing 787, the maximum total payload is about double the mass of the full complement of passengers alone. The design range may be specified in terms of a full load of passengers only or in terms of maximum payload. Design range essentially fixes the weight of fuel at take-off and, with the stipulation of maximum payload, essentially determines size and maximum take-off weight of the aircraft.

Some key parameters for the New Efficient Aircraft are given in Table 1.2, where they are compared with the corresponding figures for the 787-8 and A350-900. The aircraft lift/drag

Table 1.2   A comparison of the New Efficient Aircraft with two modern, large, two-engine aircraft

	NEA	Boeing 787-8	Airbus A350-900
Entry into service	2020?	2011	2014
Normal max. passengers	280	242	315
Range at max. payload, R1 (nm)	3000	5500	5900
Max. payload	40.2	40.2	47.6e
Max. take-off weight	175	228	268
Empty weight	100	122e	133
Fuel burn at R1	27.8	60	65
Fuel burn per nm at R1	9.2	10.9	11.0
Cruise Mach no.	0.78	0.85	0.85
Cruise L/D	21.6	21e	21e
Engine sfc ($\mathrm{kg\,h^{-1}kg^{-1}}$)	0.50	0.52e	0.51e
Wing span (m)	60	60	65
Wing area ($\mathrm{m^2}$)	304	325	443
Wing sweep	25°	32°	32°

Weights in tonne (10^3 kg), range in nm; 'e' indicates estimate.
R1 is maximum range at maximum payload.

ratio, L/D (discussed in Chapter 2), and the engine specific fuel consumption, sfc, discussed in the Section 1.5, are at the same technology level as the 787 and A350, though the values are altered by the different flight Mach numbers.

Because the NEA is derived from the 787-8, it is useful to consider the improvement in fuel burn by comparing these two aircraft at the R1 range, i.e. maximum range at full payload. Reducing the design range full payload of the 787-8 from 5500 nm to the 3000 nm of the NEA has been computed to be worth about 10% in fuel burn expressed as kg-fuel per unit payload transported one kilometre (written kg-fuel/available tonne-kilometre). This can be understood when one notes that for the long-range 787-8 about 26% of the maximum take-off weight (MTOW) is fuel for a full-payload flight at range R1, which is substantially more than the payload fraction of MTOW. Fuel must be burned to keep this weight of fuel aloft. On a shorter flight less fuel needs to be carried, there is less weight overall and more of the fuel is used to transport the payload – which is its intended purpose – rather than to transport fuel to be burned later in the flight. Reducing the cruise Mach number to $M = 0.78$ from 0.85 for the 787-8 is computed to give a further reduction in fuel burn due to speed worth about 7%. Reducing the flight Mach number allows a wing with less sweep to be used without incurring strong shock waves and this improves the aerodynamics of the aircraft (i.e., improves its lift-drag ratio, L/D). Higher L/D means a reduced thrust is required for a given aircraft weight. Reduced sweep also makes the structure lighter, reducing the aircraft empty weight. Reduced empty weight also means less thrust is required and results in more of the fuel being used to transport the payload rather than the weight of the aircraft. Combining reduced range and reduced Mach number is predicted by a detailed design analysis that is beyond the scope of this book to give an improvement of fuel efficiency of approximately 15%. With the historical experience of 0.7% improvement per year, 15% is equivalent to about 23 years of aircraft evolution relying largely on technology improvement.

1.5 THE UNITS USED

In Table 1.1 most of the quantities are in SI units, but this is not entirely usual in the aviation industry because it is dominated by the United States, which has been rather slow to see the advantages of SI units. It is helpful to remember that

$$
\begin{aligned}
&\text{1 lb mass (lbm)} &&= 0.4536 \, \text{kg}, \ 1000 \, \text{kg} = 1 \, \text{tonne}, \\
&\text{1 lb force (lbf)} &&= 0.4536 \, \text{kgf} = 4.448 \, \text{N}, \\
&\text{1 foot (ft)} &&= 0.3048 \, \text{m}, \\
&\text{1 nautical mile (nm)} &&= 1.852 \, \text{km}, \\
&\text{1 knot} &&= 1 \, \text{nm/hour} = 0.5144 \, \text{m/s}.
\end{aligned}
$$

Altitude is frequently given in feet. The nautical mile is *not* arbitrary in the way other units are, but is the approximate distance around the surface of the earth on the equator corresponding to 1 minute of longitude (east–west). (The circumference of the earth around the equator is therefore $360 \times 60 = 21{,}600$ nautical miles, which is close to the accepted figure of 21,639 nm.) If the earth were a true sphere, 1 nautical mile would also correspond to 1 minute of latitude.

The data in Table 1.1 also give the cruising speed as a Mach number, defined as V/a, the ratio of the flight speed V to the local speed of sound a. Wherever possible aerodynamicists use non-dimensional numbers and Mach number is one of the most important in determining the performance of the aircraft. The speed of sound is given by

$$
a = \sqrt{\gamma R T},
$$

where T is the local atmospheric temperature (i.e., the *static* temperature), γ is the ratio of the specific heats c_p/c_v (which is taken here to be 1.40 for air) and R is the gas constant ($0.287 \, \text{kJ kg}^{-1}\text{k}^{-1}$ for air). Since $c_p = \gamma R/(\gamma - 1)$, this leads to $c_p = 1.005 \, \text{kJ kg}^{-1}\text{k}^{-1}$. These values will be used for the atmosphere and in Part 1 (Chapters 1–10) for the gas in the engine. These values would *not* be accurate enough for use in a real design either for the products of combustion or for pure air at elevated temperatures. Although the simplification $\gamma = 1.4$ suffices for the treatment in Part 1 of the book, something better will be used in later parts for combustion products.

Sometimes non-dimensional parameters cannot be used. For example, there is a need to give a value to the weight of a passenger and this is frequently taken to be 95kg including baggage.

Engine fuel consumption is usually expressed in terms of specific fuel consumption, *sfc* (more precisely, thrust specific fuel consumption); specific fuel consumption is the fuel flow rate (in kg/s) divided by the engine thrust (in kN). The *sfc* is therefore correctly given as $\mathrm{kg\,s^{-1}kN^{-1}}$, but it is often expressed as $\mathrm{kgm\,h^{-1}kgf^{-1}}$, which is numerically equal to $\mathrm{lbm\,h^{-1}lbf^{-1}}$ to separate mass and force.

For the whole aircraft it is important to have a metric for assessing the fuel burn. A convenient one is to divide the fuel burned for a given journey m_{FB} (including taxi, take-off, climb, cruise, descent and taxi, but neglecting the reserves) by the product of payload m_{PL} and distance or range flown R to give $m_{FB}/(m_{PL}R)$ with units $\mathrm{kg\,kg^{-1}km^{-1}}$. More usually it is expressed as kg of fuel burned per available tonne-kilometre, which is sometimes written as kg-fuel/ATK. It is this parameter which is predicted to be 15% better for the New Efficient Aircraft than the 787.

Exercises[2]

1.1 The shortest distance between two places on the surface of the earth is the *Great Circle Distance*, which, for a perfectly spherical earth, would be equal to the radius R_e of the earth times the angle A subtended between vectors from the centre of the earth to the points on the surface.

Express the positions of points 1 and 2 on the surface of the earth in terms of Cartesian vectors about the centre of the earth, using θ_1 and ϕ_1 to denote the latitude and longitude respectively for point 1 and likewise θ_2 and ϕ_2 for point 2. Then take the dot product of the vectors to show that the cosine of the angle A is given by $\cos A = \cos\theta_1\,\cos\theta_2\,\cos(\phi_1 - \phi_2) + \sin\theta_1\,\sin\theta_2$.

Find the shortest distance in nautical miles between London (latitude 51.5° N, longitude 0) and Sydney in Australia (latitude 33.9° S, longitude 151.3° E). **(Ans: 9170 nm)**

1.2 Confirm that $sfc = 1.00\,\mathrm{lbm\,hr^{-1}lbf^{-1}} = 1.00\,\mathrm{kgm\,hr^{-1}kgf^{-1}}$ is equivalent to $sfc = 0.0283 \times 10^{-3}\,\mathrm{kg\,s^{-1}N^{-1}}$.

1.3 For a Boeing 787-8 it is plausible to assume that the fuel burn metric with full payload is about 0.149 kg-fuel/ATK (which is kg of fuel burned per available tonne-kilometre) for maximum payload and maximum take-off weight (so range is R1). Use data from Table 1.1 to convert the fuel metric into passenger-miles per US and per Imperial gallon. (Take 1 US gallon to be 3.79 litre and 1 Imperial gallon to be 1.201 US gallon, with 1 litre of jet fuel assumed to weigh 0.808 kg, and 1 statute mile = 1.609 km.)
(Ans: 78.1 mile/US.gal; 93.8 mile/imp.gal)

1.4 a The maximum take of weight (MTOW) of the New Efficient Aircraft is 175 tonne. The range with maximum payload is 3000 nm and cruise begins at 35,000 ft. Express MTOW in pounds (the units that much of the airline industry uses), altitude in km and range in km.
(Ans: MTOW = 386 10^3 lb; alt. = 10.67 km; range = 5556 km)

b** Calculate the speed of sound and the flight speed at 35,000 ft when $M = 0.78$ and then make a rough estimate of the flight time for a range of 3000 nm if cruise were all at the initial altitude and constant Mach number. (Obtain the speed of sound using temperature from Table 1.3.)
(Ans: $a = 296.5$ m/s. V = 231 m/s; time of flight ≈ 6.67 hours)

---

[2] Exercises with an asterisk produce solutions which should, for convenience, be entered on the Design Sheets at the back of the book.

Note that to maintain consistency and to make checking of solutions easier, answers are given to a precision which is much greater than the accuracy of the assumptions warrants.

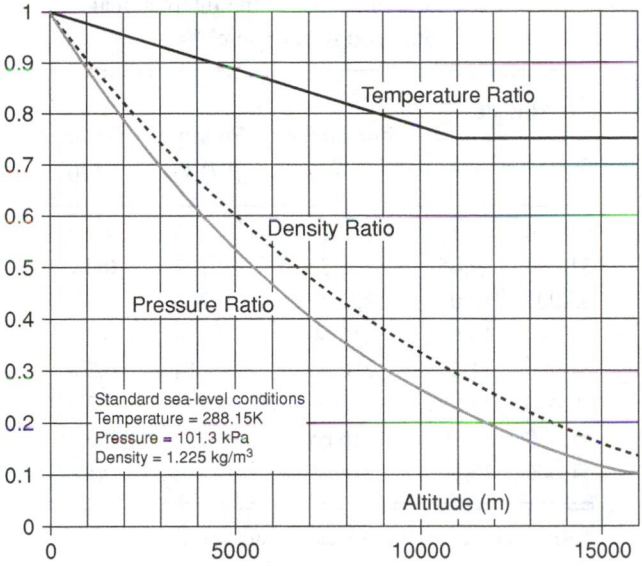

Figure 1.1. The International Standard Atmosphere. Note the tropopause at 11 km altitude.

1.6 THE STANDARD ATMOSPHERE

The atmosphere through which the aircraft flies depends on the altitude, with the pressure, temperature and density falling as altitude increases. The temperature profile with height is determined primarily by the absorption of solar radiation by water vapour and subsequent radiation back into space. At high altitude the variation with season, location and time of day is much less than at ground level and it is normal to use a standard atmosphere in considering aircraft and engine performance. Temperature, pressure and density are plotted in Figure 1.1 according to the *International Standard Atmosphere* (ISA). Standard sea-level atmospheric conditions are defined as $T_{sl} = 288.15$ K, $p_{sl} = 101.3$ kPa and $\rho_{sl} = 1.225$ kg/m^3. In the standard atmosphere, temperature decreases linearly with altitude at 6.5 K per 1000 m below the *tropopause* (in the standard atmosphere is assumed to be at 11,000 m, or 36,089 feet), but to remain constant above at 216.65 K. (The discontinuity in temperature gradient must give a discontinuity in the pressure and density gradients too, but this is small and the curve fitting programme has smoothed it out.) In the exercises sea-level temperature is taken to be 288.15 K, but in the text this and other temperatures are rounded to whole numbers.

As noted already, non-SI units are common in aviation, and air traffic control over much of the world assigns aircraft to corridors at altitudes defined in feet. Cruise now normally begins at a minimum altitude of 35,000 ft and increases in steps of 2000 ft up to 41,000 ft. Although this book will be based on SI units, altitudes for the civil aircraft will be given in feet and Table 1.3 may be

Table 1.3   Useful values of the International
Standard Atmosphere*

Altitude		Temperature	Pressure	Density
feet	km	K	105 Pa	kg/m^3
0	0	288.15	1.013	1.225
31,000	9.45	226.73	0.287	0.442
33,000	10.05	222.82	0.260	0.410
35,000	10.67	218.80	0.238	0.380
37,000	11.28	216.65	0.214	0.344
39,000	11.88	216.65	0.197	0.316
41,000	12.50	216.65	0.179	0.287
51,000	15.54	216.65	0.110	0.179

* Also known as the ICAO Standard Atmosphere.

helpful. For the purpose of this book the conditions of the standard atmosphere will be assumed to apply exactly – this makes for consistency in the numbers and facilitates checking the exercises. It will be clear, however, that the standard atmosphere is at best an approximation to conditions averaged over location and season.

The temperature in a real atmosphere deviates more than the pressure from ISA and this variation is greatest close to the ground. It is not uncommon, for example, for the temperature at an airport in continental North America to be as low as −40°C in winter and as high as +40°C in summer.

It is normal to refer to conditions relative to the standard atmosphere, so that if at 35,000 ft altitude the temperature were 228.8 K, it could, by reference to Table 1.3, be described as ISA+10°C. The corrections from standard conditions are often large for high-altitude airports. Johannesburg airport, for example, is 5557 feet above sea level and the ISA temperature for this altitude is 4.0°C: suppose on a hot day that the temperature at Johannesburg airport were 35°C – in this case the conditions would be described as ISA+31°C. Atmospheric air is not dry, and for saturated air the rate of temperature drop is given as 4.9 K per km, compared with 6.5 K per km in the International Standard Atmosphere. Different 'standard' atmospheres are sometimes used to model situations more closely; for example, over Mumbai in the monsoon season, the atmosphere is very different from over Saudi Arabia in summer or northern Russia or Canada in winter. As already noted, the International Standard Atmosphere will be used throughout this book. As Exercise 1.6 shows, it is not necessary to use tables for the ISA, but the values of temperature, pressure and density can be calculated directly. Calculations of ambient air here always assume that it is dry.

Exercises

1.5 The pressure change with altitude h due to hydrostatic effects is given by $dp = -\rho g\, dh$.

a For the idealised International Standard Atmosphere the temperature falls linearly with altitude at a constant rate so that $\partial T/\partial h = -k$, where k is a constant given as 6.5 K/km. Show that the pressure p at altitude H can be written

$$p = p_{sl}\left\{1 - kH/T_{sl}\right\}^{g/Rk} = p_{sl}\left(T/T_{sl}\right)^{g/Rk}, \tag{1.1a}$$

where p_{sl} and T_{sl} are the static pressure and temperature at sea level, 101.3 kPa and 288.15 K.

For the International Standard Atmosphere the rate of change of temperature with altitude is taken to be 6.5 K per 1000 m up to the tropopause at 11 km. Show that when $g = 9.81$ m/s^2 and $R = 287$ J kg^{-1}K^{-1}, the pressure at altitude H, in metres, is given by

$$p = p_{sl}\left(T/T_{sl}\right)^{5.26} = p_{sl}\left\{1 - 2.26 \times 10^{-5}H\right\}^{5.26} \tag{1.1b}$$

up to the tropopause, above which the pressure is given by

$$p = pT \exp\left\{-1.58 \times 10^{-4}\left(H - 11.10^3\right)\right\}, \tag{1.1c}$$

where pT is the pressure at the tropopause. Confirm that these values agree with Table 1.3.

Note: With Equations 1.1b and 1.1c there is no longer any need for tables of pressure of the standard atmosphere since it may be calculated directly.

b If the relationship between pressure and density were that for isentropic changes (i.e. reversible and adiabatic) $p/\rho\gamma = $ constant, show that the pressure at altitude H can then be written as

$$p = p_{sl}\left[1 - \frac{\gamma - 1}{\gamma}\frac{gH}{RT_{sl}}\right]^{\gamma/(\gamma-1)}.$$

Plot a few values of pressure, density and temperature on Figure 1.1, the International Standard Atmosphere.

Note: The temperature of the standard atmosphere is always higher than the temperature for the same pressure achieved by isentropic expansion from sea level.

1.6 To maintain the cabin of an aircraft habitable it is necessary to supply pressurised air. Most aircraft are maintained at 75 kPa, corresponding to about 8000 ft altitude, and 0.5 kg/min is supplied for each passenger at about 15°C to the cabin where it mixes with recirculated air. In this exercise consider a reversible (i.e. an ideal and loss-free) environmental control system (ECS) which takes in ambient air and delivers it to the cabin. (Real systems, it should be noted, are well removed from this ideal state.)

Find the ideal power required per passenger at 35,000 ft and the heat required to deliver air at 75 kPa 288 K. **(Ans:** power in 710 W, heat removed 131 W)

As noted at the end of Exercise 1.5, the standard atmosphere assumes a slower reduction in temperature with altitude even below the tropopause than that which would follow from an isentropic relation between pressure and temperature; the standard atmosphere is therefore stable. To understand why it is stable, imagine the atmosphere perturbed so that a packet of air is made to rise slowly; the pressure will immediately equal that of the surrounding air, but the temperature away from the edges of the packet will not be equal to those of the surrounding air. As a reasonable approximation the ascending air well inside the packet has an isentropic relation between temperature and pressure so that $p/T^{\gamma/\gamma-1} = $ constant. If this expanded air were slightly cooler than its surroundings it would be more dense than the surrounding air and would tend to drop down again; such an atmosphere is stable. If, on the other hand, the ascending packet of air which has expanded isentropically has a temperature higher than that of its immediate surroundings it would be less dense than the surrounding air and continue to rise; such an atmosphere would be unstable. In the standard atmosphere the temperature is always higher than that corresponding to the isentropically expanding packet of air and the atmosphere is stable. However, in the first few hundred metres above the ground convection frequently tends to make the atmosphere locally unstable, which is useful because it helps disperse pollutants. Stable atmospheres can occur near ground level, and frequently do at night under windless conditions when radiation leads to the ground cooling more rapidly than the air above it. Under stable conditions near the ground the natural mixing of the atmosphere is suppressed and the conditions for fog and pollution build-up are liable to occur.

SUMMARY OF CHAPTER 1

The combined effect of rising fuel price and possible regulation to reduce fuel burn makes it appropriate to consider a hypothetical New Efficient Aircraft (NEA) as the basis of the first ten chapters of this book. This is a wide-body aircraft with two large and efficient engines slung under the wings, and it is designed for a maximum of 280 passengers. The full-payload range is comparatively short for a modern wide-body aircraft, around 3000 nm, to achieve fuel savings on the overwhelming majority of flights. The cruise Mach number is equal to 0.78, the value currently common for single-aisle aircraft designed for shorter range, and this too leads to an improvement in aircraft fuel efficiency. The engine specific fuel consumption and aircraft lift/drag ratio are assumed to correspond to a level of technology comparable to those used in the Boeing 787 or Airbus A350, but the reduced design range and cruise Mach number are predicted to give an overall efficiency improvement of about 15%.

New engines are expensive to develop and the risk of designing an engine for an aircraft which does not get built is a serious concern. Unfortunately the time to design and develop the engines has in the past been greater than for the airframe. The expense of engines has encouraged the use of two engines where possible and engines are now sufficiently reliable that aircraft with two engines may operate with adequate safety at large distances from any airport. Furthermore, as is explained in later chapters, high bypass engines are relatively well matched between take-off and cruise, so that the two-engine plane is efficient for long flights. For very large aircraft the engines would be so large in diameter if there were only two that the wing would have to be higher and this is another reason why the largest aircraft still have four engines.

There is an International Standard Atmosphere used for calculating aircraft performance, which gives temperature, pressure and density as a function of altitude. Temperature is assumed to fall linearly with altitude (at 6.5 K per km) until 11 km, beyond which it is constant to 20 km; using Equations 1.1 the pressure and density for the standard atmosphere may be found directly. Subsonic civil air transport does not normally fly above 41,000 ft (12.5 km), though business jets fly at up to 51,000 ft (15.5 km). The atmospheric temperature given by the International Standard Atmosphere falls more slowly with altitude than is implied by an isentropic variation between temperature and pressure, meaning the distribution is stable.

Whenever possible non-dimensional variables are used, such as Mach number. When non-dimensional variables cannot be used, SI units are used throughout the book unless there is a clear reason otherwise (e.g. feet for altitude and nautical miles for range). For fuel consumption of the engine the specific fuel consumption *sfc* is used and for aircraft fuel burn the parameter kg-fuel/ATK (that is, kg of fuel burned per available tonne-kilometre) is employed.

Environmental issues apart from CO_2 are important, with the emphasis in regulations currently being around the airport. The potentially more serious effects of emissions in the upper atmosphere will probably be the subject of future regulation. Limiting noise during take-off and landing will remain important and may well preclude use of open rotors.

Chapter 1 has set out to define the needs, the operating environment and the broad specification of the aircraft. Chapter 2 moves to the next stage, which is to consider the aircraft itself.

APPENDIX

The NEA aircraft has been designed[3] with the Program for Aircraft Synthesis Studies (PASS) developed in the Aircraft Design Group in Stanford University to give shorter range and lower speed than current large aircraft. PASS is a conceptual design tool capable of analysing the performance of existing or future aircraft that are described in terms of their mission, geometry, aerodynamics, propulsion system, structural design and a detailed set of weights and performance constraints. It is able to describe many aspects of the resulting performance of the airplane, including fuel burn for a mission and other environmental impacts. PASS is typically used to study the performance of future aircraft under a wide variety of technology and mission assumptions. For the present choice of a New Efficient Aircraft the Boeing 787-8 was used as the basis, and PASS was matched to this so the fuselage was the same size and shape. The total payload for the NEA is chosen to be 40.2 tonne, the same as the 787-8, but the number of passengers is assumed to be increased to 280, reflecting the more cramped conditions tolerated on shorter flights. A typical assumed weight of a passenger with baggage is 95 kg per person, so 280 passengers comes to 26.6 tonne and the structure would allow freight in the hold to weigh up to 13.2 tonne.

---

[3] The authors are most grateful to Mr Anil Variyar of the Department of Aeronautics and Astronautics of Stanford University for carrying out these designs and to Stanford University for making PASS available.

CHAPTER 2 THE AERODYNAMICS OF THE AIRCRAFT

2.0 INTRODUCTION

The engine requirements for an aircraft depend upon the size, range and speed selected, but they also depend on the aerodynamic behaviour of the aircraft and the way in which it is operated. In this chapter some very elementary aspects of civil aircraft aerodynamic performance are described (if further explanation is needed the reader is referred to Anderson (2011)). These lead to a brief description of the conditions which are most critical for the engine: take-off, climb and cruise. It is possible to see why cruising fast and high is desirable, and to calculate the range. We can also develop the New Efficient Aircraft, the NEA, to form the basis of the exercises in Part 1 of the book. Knowing the ratio of lift to drag it is possible from this stipulation of the NEA to estimate the total thrust requirement.

2.1 PAYLOAD VERSUS RANGE

It is common to show aircraft performance by plotting maximum allowable payload against range capability, and an example corresponding to the Airbus A330-300 is shown in Figure 2.1. The presumption is that only enough fuel is carried to enable the mission, though reserves are also carried to enable diversions or other contingencies to be handled safely – it is the intention that these reserves will not be used. As the range is increased the weight of fuel at take-off must increase. The maximum payload allowed by the structure is marked by line OA and, as the range is increased along the line OA, the payload can remain constant at its maximum value, despite the increased weight of fuel, until the point A is reached. At point A the total weight of the aircraft has reached the maximum take-off weight and the range for this is denoted by R1. For the range to be increased beyond R1 requires additional fuel but, since the maximum take-off weight has been reached, for extra fuel to be carried the payload must be reduced. Payload may be substituted by fuel in this way until the tanks are full, denoted by point B in Figure 2.1. To increase range beyond that for B implies the drag, and consequently aircraft weight, has to be reduced in order to reduce fuel consumption, and this leads to a steep drop in payload for a small increase in range. In the extreme, one gets to the ferry range at which condition there is no payload at all, and this might be used to deliver empty aircraft to distant airfields. R1, the maximum payload range when both payload and take-off weights are at their maximum value, is important in defining the aircraft. It should also be noted that the maximum payload is considerably higher than the weight of a full

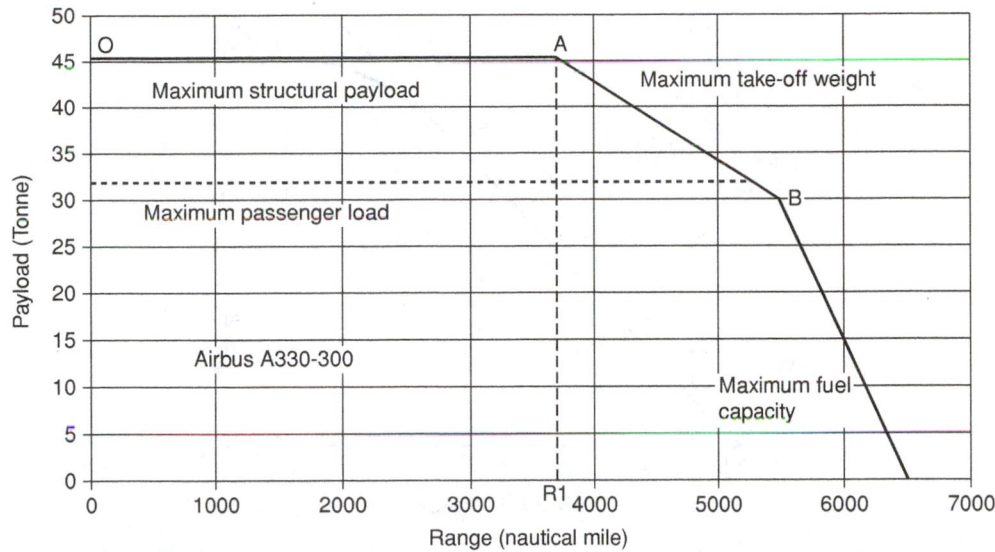

Figure 2.1. Payload versus range for the A330-300. (Maximum passenger load assumes 335 people each of 95 kg.)

load of passengers. Passengers are usually estimated to weigh, with baggage, between about 85 and 95 kg. The maximum number of passengers is not in fact well defined, for it depends on the comfort level; an Airbus A330-300 can carry 295 passengers in a three-class configuration but 440 in an all-economy layout. An upper estimate for the weight of 440 passengers, taking 95 kg per passenger, is about 42 tonne, well below the maximum payload of 49 tonne. The difference is freight carried in the hold under the cabin floor.

2.2 Sizing the Wing

The aircraft has to be a compromise between a machine which can travel fast for cruise and relatively slowly for take-off and landing. Some modification in the wing shape and area does take place for take-off and landing by deploying slats and flaps, but there is a practical limit to how much can be done. As mentioned before, it is normal to work with non-dimensional variables whenever possible. Lift and drag coefficient are defined by

$$C_L = \frac{L}{\frac{1}{2}\rho A V^2}$$

$$C_D = \frac{D}{\frac{1}{2}\rho A V^2}$$

(2.1)

where L is the lift force, which is the force acting in the direction perpendicular to the direction of travel, and D is the drag force acting parallel to and against the direction of travel. In addition, ρ is

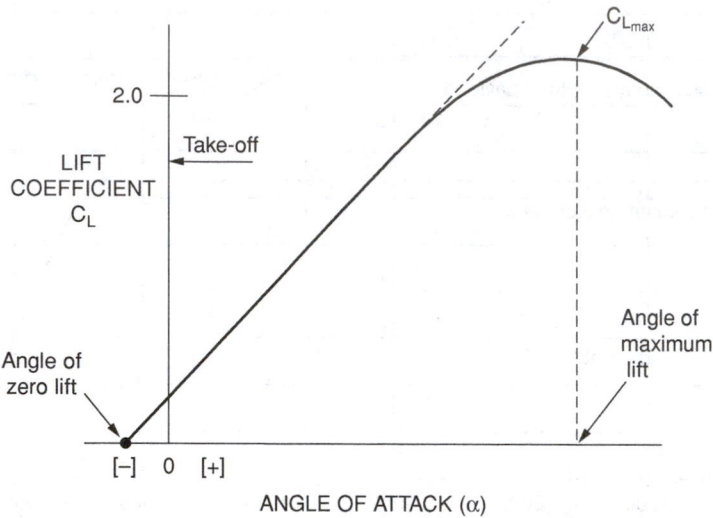

Figure 2.2. A typical curve of lift coefficient versus incidence for a large subsonic civil transport at low speed.

the air density, A the wing area and V the flight speed. In steady level flight the lift is equal to the aircraft weight and drag is equal to the thrust of the engines.

Figure 2.2 shows schematically how the lift and drag coefficients of the aircraft vary with angle of attack (i.e. incidence) at low speeds, such as at take-off. It can be seen that C_L rises almost in proportion to the incidence for small angles, but as incidence increases the curve turns over to give a peak, beyond which it falls rapidly. This rapid fall in lift is referred to as stall and, in simple terms, occurs when the flow ceases to follow the shape of the upper wing surface, that is when the boundary layer separates from the upper surface of the wing. If an aircraft were to stall near the ground it would be in a desperately serious condition and it is important to make sure that this does not occur. To make sure that stalling is avoided, it is essential that the flight speed is high enough for lift to be equal to aircraft weight at a value of the lift coefficient which is well below the stalling value.

The drag coefficient C_D varies in a quite different way. As a good approximation, it is given by

$$C_D = C_{D0} + (K_v / \pi A_R) C_L{}^2$$

where C_{D0} is independent of lift coefficient (and angle of attack) and is due to viscous drag and separation effects. The second term is the so-called induced drag, which increases with the square of the lift coefficient. In the coefficient of the second term A_R is the wing aspect ratio and K_v is the vortex drag factor which is numerically approximately equal to 1. The induced drag is discussed further in connection with combat aircraft in Section 14.2. The magnitude of the friction drag, C_{D0}, varies with the configuration of the wing, notably with the wing flaps. If the flaps are not deployed

C_{D0} may be less than 0.03, but with large deflection of the flaps, as for landing, C_{D0} may be larger than 0.1. C_L increases almost linearly with angle of attack until separation effects become large, at a value of C_L approaching 1. The variation in C_D is therefore approximately parabolic with incidence. As a result, the lift-drag ratio L/D initially rises quite sharply but peaks at an angle of attack and a value of C_L well below the stalling limit. Aircraft operating at the peak value of L/D are thus simultaneously aerodynamically efficient and resistant to stall.

The fully laden aircraft at take-off is heavy and having such a heavy machine moving at high speed along the ground is potentially hazardous. It takes a considerable distance to accelerate the aircraft to its take-off speed, or to decelerate it on landing, and, apart from the cost of making the airfield very long, there is the problem of overheating the tyres if the speed becomes too high or is maintained for too long. For these reasons, the velocity of the aircraft along the runway when the wheels leave the ground is not normally allowed to be more than about 90 m/s (175 knot, 201 mph or 324 km/h). The approach speed, coming in to land, is lower and typically about 72 m/s (140 knot). The lower speed for landing means that it is this condition which normally determines the wing area.

Because the lift is proportional to the density of the air, problems can be encountered when the air is hot and/or the airport is high. Johannesburg is famously difficult, with an elevation of 5557 ft (1694 m). On a hot day the temperature might be as high as 35°C (ISA+31°C), and under such conditions the density would be less than 80% of that at sea level on a standard 15°C day.

Exercise

2.1 a The lift coefficient for the New Efficient Aircraft (NEA) at take-off and for landing is not to exceed 1.56. The flight speed at lift-off (i.e. when the lift is just able to equal the weight of the aircraft) is not to exceed 85 m/s (177 knots), whilst for landing it is to be 70 m/s. Maximum landing weight is the empty weight plus maximum payload plus fuel reserves. (Fuel reserves are taken to be 4% of maximum take-off weight.) Find the wing area needed for take-off and for landing, assuming a standard day at sea level. Data on weights are given in Table 1.2.

(**Ans:** Take-off wing area 249 m², landing wing area 308 m². Landing evidently fixes wing area.)

b Aircraft designers often talk of wing loading defined as lift per unit area. Find this for the wing at the take-off condition. Compare this with an estimate of the weight of a car divided by the area it covers on the ground. (**Ans:** 5.57 kN/m²)

2.3 LIFT, DRAG, FUEL CONSUMPTION AND RANGE

It is an objective of a transport aircraft to lift as much as possible with the smallest drag. Reducing the drag for the same lift allows the aircraft to use less fuel or to travel further – the impact on range is taken up in Section 2.3. For steady level flight at small incidence, as for cruise, two statements can be made on the basis of simple mechanics:

$$\text{lift} = \text{weight} \quad \text{and} \quad \text{drag} = \text{thrust of the engines}$$

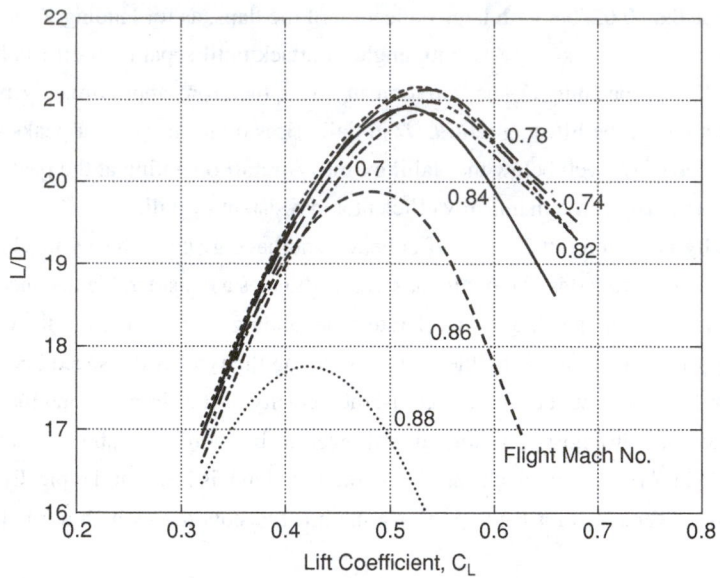

Figure 2.3. Lift-drag ratio versus lift coefficient for various flight Mach numbers.

These assume that the aircraft jet and thrust are aligned in the direction of flight[1] (and therefore in the direction of drag).

The lift is therefore fixed by the weight of the empty aircraft plus its fuel and the payload. The drag is a quantity we wish to minimise, since this has to be matched by the engines by burning fuel. It is rare to find results given for real aircraft, but Figure 2.3 shows the ratio of lift to drag, L/D, versus lift coefficient, C_L, for a Boeing 787-8. These results have been generated for flight at 35,000 ft using the programme PIANO-X produced by the company Lyssys. It may be assumed that similar curves would exist for the proposed New Efficient Aircraft and that similar trends could be observed. The curves in Figure 2.3 are for various Mach numbers. For each Mach number, L/D reaches a maximum value and then falls as C_L, and hence the angle of attack, increases. At low Mach numbers stall will be caused by the separation of the boundary layers towards the rear of the wing, but at the higher Mach numbers the stall is likely to be induced by strong shock waves. Even when the flight speed is subsonic, patches of supersonic flow on the upper surface of the wing are common. When the lift coefficient is raised by increasing incidence the flow is made to curve more and this accelerates the air to higher velocity so strong shocks become more of a problem at higher lift coefficients. For this reason, as the Mach number increases, there is a reduction in both the maximum lift-drag value and C_L at which it occurs, with the reduction becoming more rapid the further one is above the design Mach number. The separation caused by the shocks can be unsteady, leading to fluctuations in lift, a problem referred to as buffeting, which needs to be

---

[1] The assumption that thrust is in the direction of flight is a simplification. In Exercise 14.5 it is shown that the required thrust is minimised if thrust is directed down at an angle θ, where $\tan \theta = (L/D)^{-1}$, so that $\theta = 2.9°$ for $L/D = 20$. The reduction in thrust is approximately equal to $(L/D)^{-2}$ which gives 0.25% reduction for $L/D = 20$.

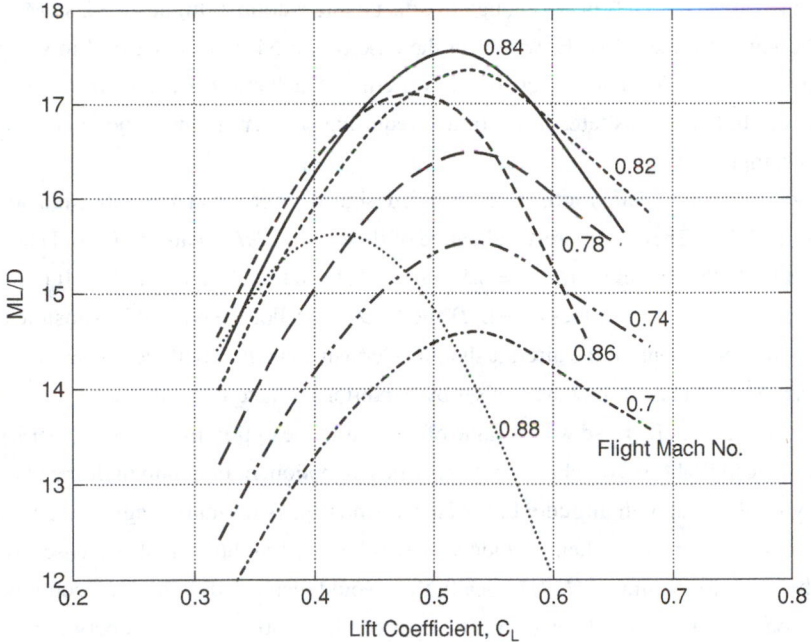

Figure 2.4. Mach number × lift/drag versus lift coefficient for various flight Mach numbers.

avoided. Operation up to $M = 0.86$ is likely to be satisfactory for the Boeing 787, but if maximum L/D were the sole criterion one would operate at a Mach number of about 0.8. Maximising the lift-drag ratio gives the maximum aircraft weight which can be kept aloft for a given engine thrust. Further consideration will reveal that this is not the primary optimum desired for a transport aircraft; instead one wants to travel the maximum possible distance for a given amount of fuel burn. L/D tells us about the rate at which fuel must be burnt to keep the aircraft aloft. If we assume that the aircraft has a fixed fuel capacity, it therefore indicates what the endurance or maximum flight time will be. If we are to achieve the maximum possible range, it is apparent then that the quantity to be optimised is the product of flight speed and lift-drag ratio, VL/D. Most airliners cruise in a band of altitudes over which the temperature and speed of sound do not alter greatly. Hence optimising Mach number times lift-drag ratio, ML/D, is virtually equivalent to VL/D, and plots of this for the Boeing 787-8 are shown in Figure 2.4.

In Figure 2.4, as in Figure 2.3, a maximum occurs for each flight Mach number at a lift coefficient of about 0.5, but Figure 2.4 differs in having the maximum of ML/D for the 787 occur at a flight Mach number of about 0.84, substantially higher than the Mach number for maximum L/D. At still higher Mach numbers ML/D drops precipitously. It is because of the rapid fall at higher Mach numbers that the condition for cruise has normally been set no higher than $M = 0.85$. Thus, for the Boeing 787, L/D has its highest peak at a Mach number of about 0.80 whereas the highest ML/D occurs around the design cruise Mach number of 0.85 and it is ML/D which determines fuel efficiency and range. The NEA is based on the 787 but has been designed for a lower cruise Mach number, $M = 0.78$, and at this condition L/D is predicted (see Section 1.4) to be 21.6. This increase in L/D is not enough to offset the reduction in flight speed, so ML/D would actually be lower for the NEA. However, in the case of the NEA, as discussed in Chapter 1, an additional highly significant advantage of a lower cruise Mach number comes from a reduction in structural weight that necessitates lower thrust, requiring smaller engines, and hence a lower rate of fuel consumption.

Figures 2.3 and 2.4 are for the Boeing 787, designed for cruise at $M = 0.85$, and do not apply directly to the NEA. They are merely indicative of the form of ML/D and L/D, and show that the peak values for both occur close to $C_L = 0.5$. It has also been found for other aircraft that this value of C_L is near where the peak values of ML/D and L/D occur. For the NEA it is plausible to assume that for low fuel consumption the aircraft should be cruising at a lift coefficient near this value, that is $C_L = 0.5$. The wing area, however, is set by conditions at take-off and landing, as in Exercise 2.1a. To make the aircraft cruise with a value of C_L near to the optimum, one can either fly near the ground at reduced flight speed, which is not an attractive option, or maintain high speed and reduce the density by flying at high altitude. Large airliners nowadays normally begin cruising at 35,000 ft or higher and then increase their altitude as fuel is burned and the weight decreases to maintain C_L near the value for optimum ML/D. Ideally they would increase their altitude continuously, but air traffic control places aircraft in 12 bands separated by 1000 ft for flights between 29,000 and 41,000 ft. This means that aircraft normally increase altitude during cruise in 2000 ft steps up to a maximum of 41,000 ft, with a 1000 ft separation between aircraft travelling in opposite directions.

Exercises

2.2 a Find the velocity and Mach number at which it would be necessary to fly to maintain the cruise value of lift coefficient, $C_L = 0.5$, if it were decided to fly at low altitude (so that the conditions may be taken to be those of the standard atmosphere at sea level). Use the aircraft mass and wing area from Exercise 2.1.
(**Ans:** 135 m/s or 283 knots; $M = 0.40$)

b If it were intended to fly at a Mach number of 0.78 at low altitude, what would be the lift coefficient?
(**Ans:** $C_L = 0.129$)

(**Note:** The answers to Exercise 2.2 show that if the optimum lift coefficient is around 0.5, the cruising speed would be low near sea-level. Apart from other practical reasons, this is a good reason for cruising high. Frank Whittle seems to have been one of the first to recognise the benefits for flying fast and high in a final year thesis he wrote in 1928, at the age of 21. He also recognised that at the speeds he envisaged, around 500 miles per hour, a propeller would be very inefficient.)

2.3 The NEA takes off at maximum take-off weight, 175 tonne. Cruise starts at 35,000 ft, by which time fuel has been burned equivalent to 2% of maximum take-off weight. Given the wing area for the NEA derived in Exercise 2.1, find the lift coefficient at start of cruise when $M = 0.78$.   (**Ans:** 0.536)

Find the aircraft mass at 39,000 ft when fuel has been burned such that the lift coefficient is equal to value at start of cruise and Mach number is constant at 0.78. (At 39,000 ft $p_a = 19.7$ kPa and $T_a = 216.7$ K).
(**Ans:** 142 tonne)

2.4*a The NEA is assumed to have a maximum lift-drag ratio of 21.6. Find the engine thrust needed for each engine for steady flight at start of cruise at 35,000 ft when the aircraft weighs 171.5 tonne.

(**Note:** This effectively fixes the thrust and, ultimately, the size of engines needed.)   (**Ans:** 38.9 kN)

b If the take-off thrust for the two engines operating is 0.3 times the maximum take-off weight, what is the required take-off thrust from each engine?
(**Ans:** 257 kN)

2.4 SPECIFIC FUEL CONSUMPTION

One measure of engine performance relates the mass flow rate at which fuel is used to the net thrust produced. The conventional measure of this is the *specific fuel consumption* (sometimes referred to as the thrust-specific fuel consumption), denoted by

$$sfc = \dot{m}_f/thrust$$

In consistent units this is kg/s of fuel per Newton of thrust, that is, $kg\,s^{-1}N^{-1}$; more conveniently, because of the magnitudes, this is often $g\,s^{-1}kN^{-1}$. In the English-speaking world, the usual units used in the industry are pounds of fuel per hour divided by pounds of thrust, written $lb\,hr^{-1}lb^{-1}$, and this is numerically equal to *sfc* expressed in kilogramme, $kg\,hr^{-1}kg^{-1}$. (Sometimes *m* and *f* are added to distinguish mass and force, namely, $lbm\,hr^{-1}lbf^{-1}$ and $kgm\,hr^{-1}kgf^{-1}$.)

The useful work done by the engine in unit time is the product of thrust and flight velocity, which for steady, level flight is drag times velocity. The rate at which burning the fuel releases energy is $\dot{m}_f\,LCV$. One can therefore define an overall efficiency by

$$\eta = \frac{VD}{\dot{m}_f LCV} = \frac{Vw}{\dot{m}_f\,LCV(L/D)}$$

where V is the flight velocity, L is the lift, D is the drag and $w = mg$ is the weight. This may be related to specific fuel consumption by

$$\eta = \frac{V}{sfcLCV}$$

Exercise

2.5 Specific air range (*SAR*) is the distance travelled per unit mass of fuel burned. The New Efficient Aircraft flies at $M = 0.78$ at 35 kft altitude, with $L/D = 21.6$ and $sfc = 0.50$ lb hr^{-1}lb^{-1}. Find the *SAR* at start of cruise when the aircraft mass is 171.5 tonne.   (**Ans:** 210 m/kg)

2.5 BREGUET RANGE EQUATION

Since the drag of an aircraft is proportional to its weight and the weight varies during the flight as fuel is burned, it is necessary to carry out an integration to calculate range in terms of fuel. The Breguet equation is the result of an analysis published in 1923 and is an idealisation applicable to the cruise portion of a flight. On long flights most of the fuel is burned in the cruise portion, but for flights less than, say, 1000 nm this is no longer a good approximation. Although we consider here only the cruise part of the flight and neglect the distance travelled in the initial climb and in the final descent, the non-cruise part can be allowed for with a simple correction.

To estimate the range we need to relate the fuel used to the thrust, and the conventional measure of this is the specific fuel consumption. During cruise the aircraft mass m changes at a rate equal to the mass flow rate at which fuel is burned. For steady level flight the total engine thrust is equal to aircraft drag and drag is aircraft weight divided by L/D. Then, expressing sfc in $\text{kg s}^{-1}\text{N}^{-1}$,

$$\frac{dm}{dt} = -\dot{m}_f = -sfc \times \text{net thrust} = -sfc \times \text{drag}$$
$$= \frac{sfc \times mg}{L/D}.$$

This can be rearranged to get

$$\frac{dm}{m} = \frac{dw}{w} = -\frac{g\,sfc\,dt}{L/D}$$

This equation can then be rewritten in terms of the distance travelled or range R as

$$\frac{dm}{m} = -\frac{g\,sfc\,dR}{V(L/D)} = -\frac{dR}{H} \tag{2.2}$$

where

$$H = \frac{V(L/D)}{g \times sfc}$$

is the so-called range factor (units metre). Range factor can be expressed in terms of overall engine efficiency η_0, defined in Chapter 4, as

$$H = \frac{\eta_0 LCV(L/D)}{g}$$

It can be seen that H is a measure of the quality of the aircraft, since high-flight speed, high lift-drag ratio, high engine overall efficiency and low sfc are all desirable attributes. (It should be noted that if the sfc is given in the units of $\text{kgm s}^{-1}\text{kgf}^{-1}$, the weight w is given in kg and the gravitational constant g is not required; if sfc is given in the more rigorously defined units of $\text{kg s}^{-1}\text{N}^{-1}$, then weight must be given in newtons too.)

An aircraft will obtain maximum range if it flies at a value of H which is close to the maximum. Keeping H constant (which is true if VL/D and sfc are constant), Equation 2.2 can then be integrated to give

$$m_{end}/m_{start} = \exp\{-R/H\} \tag{2.3}$$

where R is the range or distance travelled and m_{start} and m_{end} are the total aircraft masses at the start and end of cruise respectively, with their difference being equal to the mass of fuel used during cruise, m_f. Rearranging one gets the *Breguet's* range formula

$$R = -H \times \ln\left(\frac{m_{end}}{m_{start}}\right) \tag{2.4}$$

As noted above, an aircraft is more likely in practice to fly at constant ML/D than at constant VL/D, but the difference between the two will be small. To maintain either VL/D or ML/D constant would require a continuous very slow climb, whereas, as already noted, air traffic control normally requires steps of 1000 or 2000 ft. The specific fuel consumption is also a function of flight speed, but the changes to hold L/D constant, namely, an increase in altitude at constant Mach number, also maintain the engine at a constant operating condition.

The mass of the aircraft (by convention referred to as the weight) is made up of the empty mass m_{empty}, the payload m_{pl} and the mass of fuel m_f. The empty mass is open to many uncertainties, depending on how the aircraft is configured, but for our purpose, this is a detail. As already noted, some of fuel is carried as a reserve in case of diversion or delay and is not normally burned. The reserve fuel for long-range aircraft is typically about 4% of the take-off mass and for simple consideration may be considered as an addition to the empty weight. The Bregeut equation may be integrated to give the fuel burn parameter

$$\frac{m_{fb}}{Rm_{pl}} = \frac{1}{R}\left[1 + \frac{m_e}{m_{pl}}\right]\left[\exp\{R/H\} - 1\right].$$

Integrated in this way the importance of empty weight is emphasised. As noted above, it is a reduction in empty weight which is responsible for the improved fuel burn of the NEA compared with the B787.

Rather than maximum take-off mass it is common to refer to maximum take-off weight and this is often denoted by MTOW. At the maximum take-off weight the sum of payload and fuel load will be a maximum. Separately payload mass is limited by structural considerations and by internal volume, whilst fuel load is limited by tank capacity. Maximum take-off mass can be limited by structural considerations, aircraft aerodynamic properties and limits on engine thrust. The condition when take-off mass and payload are both at a maximum is the condition at which the range is $R1$.

Exercises

2.6a If the lift-drag ratio of the New Efficient Aircraft is to be 21.6 and the engine *sfc* is to be 0.50 kg m h^{-1} kgf^{-1}, find the range factor in nautical miles and kilometres.   (**Ans:** 35,978 km; 19426 nm)

b Find the value of range factor for the Boeing 787-8 (use Table 1.2)   (**Ans:** 36660 km; 19795 nm)

(**Note:** Although the NEA is to have aerodynamics and engine performance at least as good as the 787-8, the range factor is lower because of the lower flight speed. The benefit from lower flight speed is realised because the empty weight can be lower.)

2.7 Table 1.2 gives weights of the New Efficient Aircraft. The reserve fuel, approximated as 4% of take-off weight, may effectively be added to the empty weight. The mass of fuel actually burned over the entire flight, the block fuel, with maximum payload and maximum take-off weight is $m_{fuel} = 27.6$ tonne. As reasonable approximations, the fuel burned during take-off and climb is 2% of take-off weight, and during climb the aircraft travels 160 nm. Likewise during descent, it may be assumed that fuel equivalent to 0.2% of take-off weight is burned and the aircraft travels 130 nm.

a Find the aircraft mass at beginning and end of cruise for maximum payload and maximum take-off weight (i.e. conditions to give range R1).   (**Ans:** 171.5 tonne; 147.8 tonne)

b Use the Breguet equation to estimate the range R1 for the cruise sector of the flight of the NEA for maximum payload with maximum take-off weight. Then find the total distance travelled.

(**Ans:** 2896 nm; 3186 nm)

c Use the Breguet equation again to estimate the cruise range of the NEA if payload is reduced so that no freight is carried but only the maximum number of passengers, 280, each estimated with baggage to weigh 95 kg. (First find the fuel available to be burned and thence the ratio of mass at beginning and end of cruise.)   **(Ans: 4801 nm; 5091 nm)**

d Estimate the fuel-burn metric, kg-fuel/ATK (i.e. kg-fuel/tonne-kilometre) for cases b and c.
                                              (Ans: b. 0.116 kg/ATK; c. 0.165 kg/ATK)

(Note: The range calculated here for the NEA is some 6% higher than was deduced from the Stanford University code PASS used to design the NEA. Breguet is an idealisation which will always overestimate range if H is correct, so this is not altogether surprising. Applying the same Breguet approach to the 787-8 gives a range of 6082 nm, including distance travelled in climb and descent, about an 11% overestimate.)

2.6 SELECTING ENGINE THRUST FOR CLIMB

Although engine size is usually referred to in terms of thrust when the engine is stationary on the ground, and is therefore available during the acceleration to take-off, this is not the quantity which normally fixes the size of the high bypass ratio engine. The critical condition for sizing the engine is the top-of-climb, when the aircraft is still climbing as it approaches its cruising altitude. For a two-engine aircraft with high bypass ratio engines, such as we are considering here, engines giving adequate thrust at top-of-climb will be shown in Chapter 8 to give ample thrust at take-off under normal conditions.

When the flight condition during cruise is chosen to be at maximum ML/D it follows, for a given aircraft and fixed Mach number, that the drag is proportional to the weight, and is therefore independent of air density. As we show later, the thrust from the engines is roughly proportional to the density and therefore falls rapidly with altitude. We therefore want to size the engine so that it is operating at an efficient condition whilst producing the necessary thrust at the altitude which will set the aircraft at its optimum ML/D.

The aircraft has to climb to its cruising altitude and there are operational advantages if it can do so quickly. It is also an advantage to have some safety margin of climbing ability. The minimum rate of climb is usually given as 300 feet per minute at cruising altitude, usually referred to as top-of-climb, which corresponds to about 1.5 m/s. If the aircraft is climbing at an angle to the horizontal θ, it is easy to show that the lift L (which is perpendicular to the direction of travel through the air) is

$$L = w \cos\theta,$$

where $w = mg$ is the aircraft weight. The difference between the net thrust F_N and the drag D is equal to the component of weight in the direction of travel, that is,

$$F_N - D = w \sin\theta.$$

At altitudes for which cruise normally takes place, that is, greater than 35,000 feet, the magnitude of θ is very small. For a flight Mach number of 0.78 at 35,000 ft the velocity is about 231 m/s and a rate of climb equal to 300 ft/min is a climb rate of 1.5 m/s. The magnitude of θ is therefore about 0.37°, so $\cos\theta \approx 1$, and a reasonable approximation is to take lift equal to weight. It then follows that

$$F_N/w = D/w + \sin\theta = 1/(L/D) + \sin\theta,$$

and it is easy to find the thrust needed at the top-of-climb.

Most of the flight is spent at the cruise condition, which is the engine condition most important in determining the total fuel consumption during the flight and therefore the range. Civil engines are allowed to operate for a continuous period of no more than five minutes at the maximum take-off power, which represents only a very small proportion of the total operation time, and because comparatively little fuel is used at this condition the specific fuel consumption at take-off power is immaterial. (At take-off power the temperatures in the engine are sufficiently high that the engine would deteriorate rapidly if it were allowed to operate for long periods, but at the cruise condition the temperatures are lower and the rate of deterioration is low.) The design point, in the sense of lowest fuel consumption, should therefore correspond to the cruise condition, but the engine must have the capacity to generate some additional thrust to allow the aircraft to climb, and the engine condition when non-dimensional parameters are at their highest value, discussed in later chapters, is usually at top-of-climb.

Take-off thrust is specified to give an acceptable length of runway for safe take-off, certainly no more than 11,000 feet. The aircraft must be able to continue the take-off even if one engine fails, without alteration of the engine throttles. Once the undercarriage is raised the lift-drag ratio is about 14, significantly lower than the cruise value, but when flying with only one engine the asymmetry increases the drag and this condition essentially fixes the minimum required take-off thrust. For a wide range of modern aircraft it is found that the maximum take-off thrust with the aircraft stationary at sea level is equal to about 0.3 times the maximum take-off weight.

Exercise

2.8 Cruise for the NEA will begin at 35,000 ft, and it should arrive at this altitude while flying at $M = 0.78$. Assume $L/D = 21.6$, and suppose that the engines at this top-of-climb condition produce 20% more thrust than that required for steady cruise. What is the inclination of flight to the horizontal and what is the rate of climb?   (**Ans**: 0.53 deg; 422 ft/min)

SUMMARY OF CHAPTER 2

Subsonic airliners tend to have maximum lift-drag ratio L/D and maximum VL/D when the lift coefficient is about 0.5. This requires cruise at high altitude and the altitude is increased during a long flight, as the weight of fuel carried decreases, to keep near to the optimum VL/D. Maximum VL/D and maximum ML/D occur at essentially the same lift coefficient.

The design range R1 is the range achievable when take-off occurs with maximum payload and at maximum take-off weight. This fixes the mass of fuel at take-off. In order to increase range beyond R1 the mass of fuel at take-off must be increased and it is necessary to reduce payload.

The range can be estimated by the Breguet range equation, which shows the dependence of range on VL/D and the inverse dependence on the specific fuel consumption of the engine. Knowing the range and the mass of fuel burned, a useful parameter for the fuel burn can be found: kg-fuel burned per tonne payload flown one kilometre, kg-fuel/ATK.

Large wide-body aircraft with two engines are observed to have higher L/D and lower sfc than the small single-aisle aircraft commonly used for shorter ranges. The current aircraft, the New Efficient Aircraft (NEA), is therefore selected to be a wide-body twin-engine aircraft with technology equivalent to the latest wide-body aircraft designed for long range, the Boeing 787 and Airbus A350. The size and payload of the NEA are those of the 787-8, but the full-payload design range is only 3000 nm. (If the aircraft flies with the payload reduced to just a full complement of passengers, i.e., no freight, the range can increase by about 50%.) Cruise Mach number for the NEA is reduced to 0.78. Together the shorter design range and the lower cruise speed are estimated to reduce the fuel burn metric of the NEA by about 15% relative to the 787.

If the aircraft is to cruise near the maximum of VL/D, since weight = lift, the cruise drag of the aircraft is determined for a given weight. The minimum thrust needed from each of the engines is therefore specified and for the NEA is 38.9 kN at the start of cruise at 35,000 ft altitude. The engines must be capable of increased thrust to allow climb at cruise altitude at a rate of not less than 300 ft/min.

Having considered some aspects of the aircraft specification and performance, we have been able to decide on the thrust needed at the altitude selected for initial cruise. We now need to look at the engine itself to decide how it works and how to specify what form it should take.

APPENDIX: SPECIFIC AIR RANGE AND ICAO

Because specific air range (SAR) has been selected as the basis for the ICAO fuel-burn metric, it is described briefly here. SAR is the aircraft equivalent of miles per gallon or kilometre per litre for a road vehicle. It is defined as the distance travelled per unit mass of fuel consumed or velocity per unit flow rate of fuel:

$$SAR = V/\dot{m}_f.$$

The rate of fuel burn is equal to the specific fuel consumption, sfc, multiplied by the thrust. In steady level flight the thrust is equal to the aircraft weight divided by the lift-drag ratio, so the mass flow rate of fuel is given by

$$\dot{m}_f = sfc\, w/(L/D),$$

where w is the instantaneous weight of the aircraft. The distance travelled per unit mass of fuel, defined as the *SAR*, is therefore given by

$$SAR = V(L/D)/(sfc\,w).$$

The specific range can also be written in terms of the range factor H so that, with instantaneous weight $w = mg$,

$$SAR = H/m = Hg/w.$$

As is to be expected, *SAR* is proportional to the lift-drag ratio (which is a measure of how good the aircraft aerodynamics are) and inversely proportional to engine specific fuel consumption (for which a low value indicates goodness).

SAR is inversely proportional to aircraft weight, so large aircraft have a lower value than small aircraft, just as large trucks go fewer miles per gallon than small cars. *SAR* on its own is therefore a poor way of comparing the performance of aircraft of different sizes. Nevertheless it appears that this will be the metric used for comparing fuel burn and CO_2 emissions in standards being drawn up by the International Civil Aviation Organisation (ICAO). To give some correction for size, the *SAR* is multiplied by a measure of fuselage size called the reference geometric factor (*RGF*), approximately the cabin floor area. *RGF* is raised to a power of 0.24, so the metric adopted is therefore

$$\left(SAR.RGF^{\,0.24}\right)^{-1},$$

where lower values mean better fuel economy. Raising the measure of size to a power less than unity has the effect of making small aircraft, such as business jets and regional jets, appear better relative to large aircraft. *SAR* is inversely proportional to aircraft weight, but does not distinguish between aircraft empty weight and payload. Increasing the ratio of payload to empty weight is a way of increasing the efficiency of an aircraft in terms of what is desired, namely to move payload over a distance, and it is a measure of the quality of an aircraft. Because this is not included in the metric selected by ICAO for fuel burn, this important aspect of aircraft specification and design is neglected.

CHAPTER 3 — THE CREATION OF THRUST IN A JET ENGINE

3.0 INTRODUCTION

This chapter looks at how a jet engine produces thrust, which is a simple consequence of Newton's laws of motion applied to a steady flow. It requires the momentum to be higher for the jet leaving the engine than the flow entering it, and this inevitably results in higher kinetic energy for the jet. The higher energy of the jet requires an energy input, which comes from burning the fuel. This gives rise to the definition of propulsive efficiency (considering only the mechanical aspects) and overall efficiency (considering the energy available from the combustion process).

With few exceptions this book will be concerned with bypass engines. These are engines where some or most of the incoming air passes around and outside the core of the engine: this is the bypass stream. A fraction of the air enters the core and passes through the combustor. The bypass ratio is defined by mass flows of air as

$$bpr = \dot{m}_{bypass}/\dot{m}_{core}.$$

The total mass flow rate is given by

$$\dot{m}_{air} = \dot{m}_{bypass} + \dot{m}_{core}.$$

Early bypass engines had more air going through the core than through the bypass, that is a low bypass ratio, but modern high bypass engines have around ten times as much air in the bypass stream. The jet velocity from the core and bypass need not be equal but they are normally designed to be similar and for the present purposes may be taken to be equal.

3.1 MOMENTUM CHANGE

The creation of thrust is considered briefly here, but a more detailed treatment can be found in other texts, for example Hill and Peterson (1992). Figure 3.1 shows an engine on a pylon under a wing. Surrounding the engine a control surface has been drawn, across which passes the pylon. The only force applied to the engine is applied through the pylon and the component of this force in the direction of travel is the thrust. We assume that the static pressure is uniform around the control surface, which really requires that the pylon is long enough that the surface is only weakly affected by the wing. In fact we assume that the wing lift and drag are unaffected by the engine and the engine is unaffected by the wing; this is not strictly true, but is near enough for our purposes.

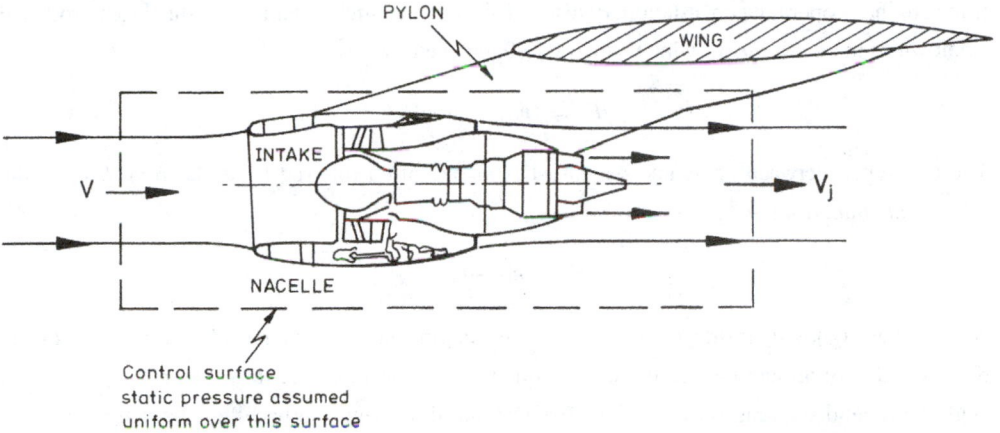

Figure 3.1. A high bypass ratio engine installed under a wing.
(A General Electric CF6 installed in the outboard position on a Boeing 747.)

A flow of fuel $\dot{m}_f$ passes down the pylon, but its velocity is low and it conveys negligible momentum. The mass flow of air $\dot{m}_{air}$ entering the engine is typically at least two orders of magnitude greater than $\dot{m}_f$.

It is assumed for simplicity here that the jet is uniform as it crosses the control surface with velocity Vj. In the case of a bypass engine this requires that any velocity difference between the core and the bypass streams has mixed out, but it would be straightforward to consider separate jets or, in principle, to integrate the momentum of a non-uniform jet.

The thrust is calculated by considering the flux of momentum across the control surface around the engine – since the pressure is assumed uniform over the control surface, the pressure creates no net force. We consider momentum in the frame of reference moving with the engine, so that air enters the control surface with the flight velocity V. Most of the air entering the control surface passes around the engine and of the total only a small part, $\dot{m}_{air}$, passes through the engine. The air that passes around the engine leaves the control surface with the same velocity V as the flight speed and, since the momentum flux of this is equal on entering and leaving the surface, it does not contribute to the thrust. Considering the flow which crosses the control surface and passes through the engine, we can write down what the two fluxes are:

$$\text{Flux of momentum entering the engine} = \dot{m}_{air}V$$
$$\text{Flux of momentum leaving the engine} = (\dot{m}_{air} + \dot{m}_f)V_j.$$

The extra mass flow in the jet of the fuel is included here for completeness, but this represents a small contribution for high bypass ratio engines.

The *net thrust* F_N which is available in flight is given by the difference between the two momentum fluxes, that is

$$F_N = (\dot{m}_{air} + \dot{m}_f)V_j - \dot{m}_{air}V. \tag{3.1}$$

If the engine is operating on a stationary test bed, or with the aircraft stationary, the thrust produced is known as the *gross thrust*, which, since $V = 0$, is given by

$$F_G = (\dot{m}_{air} + \dot{m}_f)V_j. \tag{3.2}$$

The difference between gross and net thrust is $\dot{m}_{air}V$, often referred to as the *ram drag* or the *inlet-momentum drag*, so

$$F_N = F_G - \dot{m}_{air}V.$$

Here we have ignored the drag on the outside of the engine nacelle, which leads to a ring of reduced relative velocity around the jet and a consequent reduction in useful thrust. For the present we neglect this and the drag on the pylon. Because the drag from the nacelle cannot be measured independently, its magnitude can give rise to disputes between the manufacturers of the engine and of the airframe. Unfortunately, nacelle drag is becoming a more significant factor as engines become larger in diameter for the same thrust.

3.2 PROPULSIVE EFFICIENCY

The increase in velocity between the flow entering the engine and that leaving in the jet involves an increase in kinetic energy. This kinetic energy increase is the effect of useful work supplied by the engine to the air, neglecting the work used to overcome losses which just raise the temperature of the jet. Again assuming equal velocity in the core and bypass jets, the rate of change of kinetic energy for the flow through the engine from well ahead to downstream in the jet is given by

$$\Delta \text{KE} = \tfrac{1}{2}\left[(\dot{m}_{air} + \dot{m}_f)V_j^2 - \dot{m}_{air}V^2\right]. \tag{3.3}$$

The power actually associated with propelling the aircraft is given by

$$\text{Power to aircraft} = \text{flight speed} \times \text{net thrust} = V F_N$$
$$= V\left[(\dot{m}_{air} + \dot{m}_f)V_j - \dot{m}_{air}V\right]. \tag{3.4}$$

The *propulsive efficiency* compares the power supplied to the aircraft with the rate of increase in kinetic energy of the air through the engine, that is the power to the jet. Propulsive efficiency η_p is straightforward to define by

$$\eta_p = \frac{\text{Power to aircraft}}{\text{Power to jet}} = \frac{V\left[(\dot{m}_{air} + \dot{m}_f)V_j - \dot{m}_{air}V\right]}{\tfrac{1}{2}\left[(\dot{m}_{air} + \dot{m}_f)V_j^2 - \dot{m}_{air}V^2\right]}. \tag{3.5}$$

Since, as already noted, the mass flow rate of fuel is very much less than that of air, it is possible to write η_p with sufficient accuracy for our present purposes as

$$\eta_p = \frac{\dot{m}_{air}V[V_j - V]}{\tfrac{1}{2}\dot{m}_{air}[V_j^2 - V^2]} = \frac{2V}{V + V_j} \tag{3.6}$$

which is known as the *Froude* equation for propulsive efficiency. Although approximate, since mass flow of fuel is neglected, this equation contains the essential features associated with propulsion. If the jet velocity is nearly equal to the forward speed, the kinetic energy of the jet is used very efficiently and η_p tends to unity. Unfortunately the net thrust is given by $\dot{m}_{air}(V_j - V)$ so that as V_j tends to V the *net* thrust goes to zero. For modern civil engines low fuel consumption tends to be the most important goal, and this requires a high propulsive efficiency; for military combat aircraft the principal requirement is high thrust and for military applications a lower propulsive efficiency is tolerated. Modern civil aircraft engines have bypass ratios up to about 10 and propulsive efficiencies in excess of 80%, whereas military engines typically have bypass ratios less than unity.

3.3 OVERALL EFFICIENCY

The propulsive efficiency relates the rate at which work is done in propelling the aircraft to the rate at which kinetic energy is added to the flow through the engine, but it does not relate the work to the thermal energy made available by burning the fuel. For this we define a *thermal efficiency* by

$$\eta_{th} = \frac{\Delta \mathrm{KE}}{\dot{m}_f LCV}, \tag{3.7}$$

where $\Delta \mathrm{KE}$ is the rate at which kinetic energy is added to the air, which is the work done by the engine on the air going through it. The thermal efficiency is the ratio of gas turbine work output to the energy input from burning fuel. Here LCV is the lower calorific value of the fuel, sometimes called lower heating value. It is the chemical energy converted to thermal energy on complete combustion in air if the water in the products is not condensed but remains as vapour. (The exhaust of gas turbines is invariably hot enough that this is the case.) Again assuming the mass flow rate of fuel is much less than that of air, the *thermal efficiency* can then be written

$$\eta_{th} = \frac{\dot{m}_{air}\left[V_j^2 - V^2\right]/2}{\dot{m}_f LCV}. \tag{3.8}$$

The *overall efficiency* is given by

$$\eta_o = \frac{\text{Useful work}}{\text{Thermal energy from fuel}} = \eta_p \times \eta_{th}, \tag{3.9}$$

the product of propulsive and thermal efficiencies, which can be expanded using the preceding expressions into

$$\eta_o = \frac{\text{Thrust} \times \text{Speed}}{\dot{m}_f LCV} = \frac{\text{Thrust}}{\dot{m}_f} \frac{\text{Speed}}{LCV} = \frac{1}{sfc} \frac{V}{LCV}. \tag{3.10}$$

As one might expect, the overall efficiency is inversely proportional to the product of specific fuel consumption and the heating value of the fuel. Less obviously, η_o is proportional to the flight speed where the dependence of *sfc* on flight speed is overlooked. To take this analysis further it is necessary to understand what determines the thermal efficiency, η_{th}, which will be addressed in Chapter 4.

Exercises

3.1 Find the propulsive efficiency for the following two engines at cruise:

a a Rolls-Royce Trent 1000 or a GE GEnx engine in the Boeing 787 at 35,000 ft, flight Mach number 0.85, with jet velocity approximately equal to 350 m/s   **(Ans: 83.7%)**

b an Olympus 593 (in Concorde) at 51,000 ft ($p_a = 11.0$ kPa, $T_a = 216.7$ K), flight Mach number 2.0, approximate jet velocity 1000 m/s   **(Ans: 74.2%)**

3.2 For the Trent 1000 or GEnx at $M = 0.85$ and 35,000 ft cruise the *sfc* is about 0.50 kg h^{-1}kg^{-1}. For the Olympus 593 (in Concorde) at $M = 2.0$ and 51,000 ft *sfc* was about 1.19 kg h. Use results of Exercise 3.1 to find the overall efficiency and, using Equation 3.9 with Exercise 3.1, the thermal efficiency in each case. Take $LCV = 43$ MJ/kg.

(Ans: Trent or GEnx: $\eta_o = 41.4\%$, $\eta_{th} = 49.4\%$; Olympus: $\eta_o = 40.7\%$, $\eta_{th} = 54.9\%$**)**

Note: (1) The lift-drag ratio for a Boeing 787-8 is estimated to be about 21 at cruise whereas the lift-drag ratio of Concorde was between 6 and 7. Notwithstanding the high efficiency of its engines, Concorde was still an energy-inefficient way to travel!

(2) The effective thermal efficiency appropriate here for the bypass engine also includes the effect of the LP turbine efficiency and the fan efficiency.

3.3 The Trent 1000 and GEnx engines are designed for cruise at $M = 0.85$ at 35,000 ft, at which condition the *sfc* is about 0.50 kg h^{-1}kg^{-1}. Find the *sfc* and the overall efficiency when cruise Mach number is reduced to 0.80 if thermal efficiency stays constant. (Use $\eta_{th} = 49.4\%$ from Exercise 3.2.) The jet velocity should be assumed to remain unchanged at approximately 350 m/s. (In fact the mass flow of air through the engine will reduce by about 5%, as explained in Chapter 6.)

(Ans: *sfc* $= 0.488$ kg h^{-1}kg^{-1}; $\eta_o = 40.0\%$**)**

3.4 Consider two very different engines, one with a jet velocity of 350 m/s and the other 1000 m/s. For simplicity we assume here that the jet velocity does not change with forward speed and in both cases that the jet is uniform.

a For unit mass flow calculate the thrust from each engine when the aircraft is stationary and when it is on the point of taking-off; at the point of take-off the forward speed gives a Mach number of 0.25. The static thrust is the gross thrust. Assuming that the air mass flow through each engine is the same for both conditions, calculate the ratio of the net thrust from each engine for $M = 0$ and 0.25. Take the conditions to be standard atmosphere at sea level and express the results for each engine as a ratio of the static thrust.

(Ans: 0.757 and 0.915)

b For the two engines calculate the ratio of cruise net thrust to the thrust when the aircraft is stationary on the ground. Cruise is at $M = 0.85$ and an altitude of 35,000 ft. Assume that the mass flow through the engine at cruise is 0.44 times that at sea level static conditions.   **(Ans: 0.123 and 0.329)**

Note: (1) As the jet velocity decreases the effect of forward speed on net thrust increases. So for early jet engines, which had high jet velocities, it was relatively hard to get enough thrust for take-off but there was excess thrust capacity at cruise; for modern civil engines, with relatively low jet velocity, take-off is comparatively easy.

(2) The estimation of change in mass flow into the engine between static sea level and cruise can be addressed with the methods introduced in Chapter 6.

3.5 In Exercise 3.1a the propulsive efficiency was estimated to be 83.7% for modern engines at cruise at $M = 0.85$ at 35,000 ft. What would the jet velocity have to be to increase the propulsive efficiency to 90%? By what ratio would the mass flow through the engine need to increase to hold net thrust constant? If the mass flow per unit area through the engine were held constant, estimate the ratio of engine diameter.

(Ans: $V_j = 308$ m/s; mass flow ratio 1.75; diameter ratio 1.32**)**

Note: Raising propulsive efficiency is clearly desirable but, as this exercise shows, a comparatively small increases in η_p, from 83.7% to 90%, leads to a 32% increase in diameter. If a crude estimate is that weight depends on the cube of linear dimension, and the fan is the largest part of the weight, then this would entail the weight more than doubling.

3.6 For steady level cruise the engine thrust must equal the drag, which is given by the aircraft weight divided by the lift-drag ratio of the aircraft. There must be additional engine thrust if the aircraft is to climb. If at 35,000 ft the flight Mach number is 0.78 and the rate of climb is 500 ft/min, find the angle of climb. If $L/D = 21.6$, find the proportional increment in net thrust needed to achieve this climb rate. **(Ans: 23.7%)**

3.7 Adequate take-off performance is achieved for civil jet aircraft with a take-off thrust equal to 0.3 times maximum take-off weight. (This is with all engines operating satisfactorily and leaves margin for take-off to continue safely if thrust from one engine is lost after achieving V1 speed on the runway.) If fuel used for take-off and climb is equal to 2% of maximum take-off mass, find the ratio of take-off thrust to initial cruise thrust when $L/D = 21.6$. Also find the ratio of take-off thrust to climb thrust at top-of-climb when the climb rate is 500 ft/min. **(Ans: 6.61, 5.34)**

Note: It is shown later that the condition which determines the size of a modern engine for a civil transport is the top-of-climb condition, when the aircraft is climbing at the required rate and approaches the initial cruise altitude. This is when the non-dimensional variables in the engine are at their highest values, not at take-off. Nevertheless take-off thrust is still the normal parameter used to describe or characterize engines.

SUMMARY OF CHAPTER 3

Thrust is produced by increasing the momentum of airflow through the engine. The net thrust is that which is actually available in flight; the gross thrust is that which would be produced under the same conditions with the engine stationary:

$$\text{Net thrust} = \text{Gross thrust} - \text{ram drag}$$

$$\text{i.e.} \quad F_N = F_G - \dot{m}_{air}V = \dot{m}_{air}V_j - \dot{m}_{air}V.$$

For high net thrust there must either be a high jet velocity or a large mass flow of air. Here the jet is assumed to have a single uniform velocity, V_j.

Propulsive efficiency compares the rate of work done on the aircraft to the rate of kinetic energy increase of the flow through the engine. It may be approximated for the typical case, when the mass flow of fuel is much smaller than that of air, by

$$\eta_p = \frac{2V}{V + V_j},$$

and this shows that high propulsive efficiency requires the jet velocity V_j to be not much greater than the flight speed V.

Only a fraction of the energy released when the fuel is burned is converted into the kinetic energy rise of the flow; the remainder appears as internal energy in the exhaust. (In common terms, the exhaust is at a higher temperature than it would be if all of the energy had been converted into kinetic energy.) The ratio of the kinetic energy increase to the lower calorific value is denoted by the thermal efficiency η_{th}. The overall efficiency, relating the work done on the aircraft to the energy released in the fuel, is given by

$$\eta_o = \eta_p \eta_{th},$$

and it is easy to show that in terms of specific fuel consumption and calorific value,

$$\eta_o = \frac{1}{sfc}\frac{V}{LCV}.$$

To make more concrete statements and to design the engine it is necessary to consider how a gas turbine works and to find a means of estimating its performance.

Modern civil engines, for which the jet velocity may be less than 1.5 times cruise velocity, experience large variations in net thrust with forward speed and net thrust is much less than gross thrust. Therefore for modern engines, which are sized for cruise or for top-of-climb, achieving take-off is normally no longer the condition which stretches engine capability. Engines with high jet velocity generate net thrust which varies relatively little with forward speed and is a large fraction of gross thrust.

CHAPTER 4 THE GAS TURBINE CYCLE

4.0 INTRODUCTION

The gas turbine has many important applications but it is most widely used as the jet engine. Many of the gas turbines used in land-based and ship-based applications are derived directly from aircraft engines. The gas turbines designed specifically for land use, most often as part of a combined cycle plant, are rather different, but the concepts for the cycle is similar to the jet engine and some of the technology is related to that for aircraft propulsion.

The attraction of the gas turbine for aircraft propulsion is the large power output in relation to the engine weight and size – it was this which led the pre-Second World War pioneers to work on the gas turbine. Most of the pioneers then had in mind a gas turbine driving a propeller, but Whittle and later von Ohain realised that the exhaust from the turbine could be accelerated to form the propulsive jet. Sometimes the gas turbine is still used to drive a propeller to form an efficient engine for relatively low-speed flight, the turbo-prop. However the gas turbine is used for aircraft propulsion, it is the high power output for a given weight is that makes it attractive.

Purists will object to this description of the gas turbine as a *cycle*. Strictly speaking a cycle uses a fixed parcel of fluid which in a gas turbine would be compressed, heated in a heat exchanger, expanded in a turbine and then cooled in a heat exchanger. The ideal gas turbine is sometimes called a Joule or Brayton cycle. The gas turbine 'cycle' we consider here takes in fresh air, burns fuel in it and then discharges it after the turbine: in other words it does not cycle the air. Here we are nevertheless adopting the standard terminology of the industry.

This chapter looks at the operation of simple gas turbines and outlines the method of calculating the power output and efficiency. The treatment is simplified by treating the working fluid as a perfect gas with the properties of air, but later some examples are discussed to assess the effect of adopting more realistic assumptions. It is assumed throughout that there is a working familiarity with thermodynamics – this is not the place to give a thorough treatment of the first and second laws (something covered very fully in many excellent text books, e.g., Moran et al. 2010 and Borgnakke and Sonntag 2008). Nevertheless, in the appendix to this chapter a brief account is given to remind those whose knowledge of engineering thermodynamics is rusty or to familiarise those who have learned thermodynamics in connection with a different field.

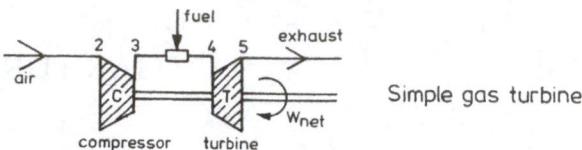

Simple gas turbine

thermodynamically equivalent to :-

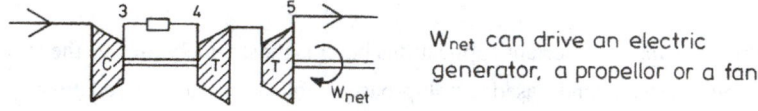

W_{net} can drive an electric
generator, a propellor or a fan

or to :-

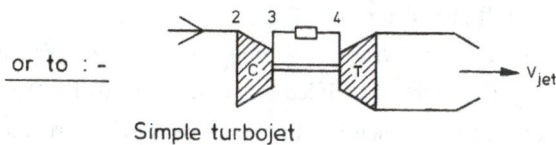

Simple turbojet

The **core** of the engine is shown by dotted lines
and can be essentially identical.

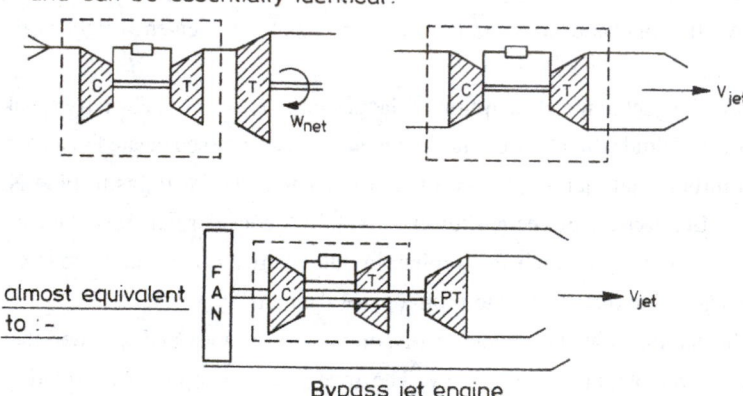

almost equivalent
to :-

Bypass jet engine

Figure 4.1. Gas turbines – variations on a core theme.

4.1 GAS TURBINE PRINCIPLES

The essential parts of a gas turbine are shown schematically in Figure 4.1. The pressure of the air is
raised in the compressor and fuel is burned in this high-pressure air in the combustor (combustion
chamber). The hot, high-pressure gas leaving the combustor enters the turbine. In most cases there
are several turbine stages, one or more to drive the compressor, the others to drive the load. The
turbine driving the load may be on the same shaft as the compressor or it may be on a separate

shaft. The load may be an electric alternator, a ship's propeller or the fan on the front of a high bypass ratio jet engine.

The central part of a gas turbine, the compressor, combustor and turbine driving the compressor, is often referred to as the *core* and the same core can be put to many different applications. In Figure 4.1 the turbine power $\dot{W}_t$ would partly be used to supply the compressor power $\dot{W}_c$ and partly to supply the useful or net power $\dot{W}_{net}$ which is equal to $\dot{W}_t - \dot{W}_c$. At this stage we do not need to consider how $\dot{W}_{net}$ is taken from the core, but it is worth remembering that there is the special case of the pure jet engine when all the net power is used to accelerate the core stream and produce a jet at exit. This was the basis of the engine envisaged and then constructed by Whittle and is still useful for propulsion at supersonic speeds; Concorde, for example, was propelled by a pure jet engine. Pure jets are also used at subsonic speeds when fuel economy is unimportant but first cost and weight do matter, for example to propel missiles.

The first law of thermodynamics can be applied to a steady process through the engine, where air enters the engine at temperature T_2 and exhaust products leave at temperature[1] T_5. (It may seem odd to make the entry condition station 2, but this is chosen here to be compatible with the internationally recommended practice for aircraft engines.) If the effect of combustion is represented by an equivalent net heat transfer to the gas $\dot{Q}_{net}$, the first law may be written

$$\dot{Q}_{net} - \dot{W}_{net} = \dot{m}_{air}\Delta h \tag{4.1}$$

where Δh is the enthalpy difference between the inlet air and the exhaust based on stagnation conditions. The mass flow of fuel is neglected in this equation as a small quantity. If the exhaust gas can be modelled as a perfect gas with the same properties as air, Equation 4.1 reduces to

$$\dot{Q}_{net} - \dot{W}_{net} = \dot{m}_{air}c_p\left(T_5 - T_2\right).$$

The combustion process, which is represented here as an equivalent heat transfer,

$$\dot{Q}_{net} = \dot{m}_{air}c_p\left(T_4 - T_3\right),$$

can in turn be written in terms of the lower calorific value of the fuel

$$\dot{m}_f LCV = \dot{m}_{air}c_p\left(T_4 - T_3\right). \tag{4.2}$$

For kerosene, or similar fuels used in aircraft engines, $LCV = 43$ MJ kg^{-1} is a good approximation. This magnitude is so large in relation to the specific heat of air (taken here to be $c_p = 1.005$ kJ kg^{-1}K^{-1}) that a small flow rate of fuel is sufficient to produce a substantial temperature rise in a much greater mass flow rate of air.

Important processes in a gas turbine which burns fuel can be represented by an equivalent closed-cycle gas turbine, and by doing this it is easy to represent the processes graphically. Figure 4.2

---

[1] To relate this to what is discussed in later chapters it is appropriate to mention that the temperatures and pressures used in connection with the gas turbine cycle in Chapter 4 are the *stagnation* values. Stagnation and static properties are explained in Chapter 6.

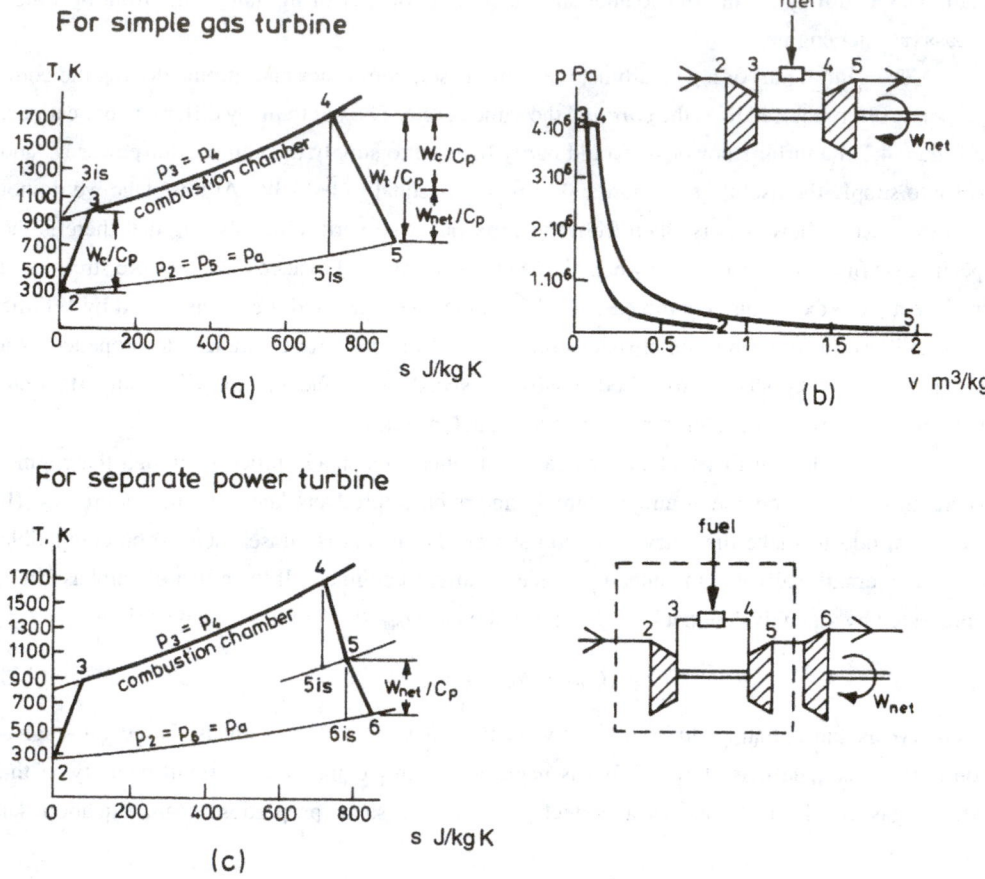

Figure 4.2. Scale diagrams of temperature–entropy and pressure–volume for gas turbine cycles. W is work per unit mass of air. Pressure ratio 40, $T_2 = 288$ K, $T_4 = 1700$ K, $\eta_c = \eta_t = 0.90$; $s = 0$ for $p = p_a$ and $T = 288$ K.

shows the temperature–entropy (T–s) and pressure–volume (p–v) diagrams for a closed-cycle gas turbine. At entry to the compressor the temperature is T_2 and the ambient pressure p_a. The upper pressure $p_3 = p_4$ is that at which the heat transfer (equivalent to the combustion) takes place; for the present simple example it is assumed that there is no pressure drop in the combustor. After the combustion process the temperature entering the turbine is T_4. The turbine expansion down to $p_a = p_2$ is between temperature T_4 and T_5. The heat exchanger between stations 5 and 2 in Figure 4.2a is to cool the low pressure gas leaving the turbine back to T_2 and is the alternative to the open-cycle arrangement where the turbine exhausts to atmosphere and the compressor draws in new air at ambient pressure and temperature. The work exchanges per unit mass of air are shown on the T–s diagrams in Figure 4.2 (though what is actually shown is work divided by specific heat, W/c_p).

In Figure 4.2 (and throughout Part 1: Design of Engines for a New Efficient Aircraft) the properties of the combustion products will be treated as pure air with the properties of a perfect gas: $c_p = 1.005$ kJ kg^{-1}K^{-1}, $\gamma = 1.40$, $R = 0.287$ kJ kg^{-1}K^{-1}. This is an approximation which can easily be removed (and is considered in more detail in section 4.4 below, and then in Chapter 11), but for the present purpose it is sufficiently accurate and is a substantial convenience.

4.2 ISENTROPIC EFFICIENCY AND THE EXCHANGE OF WORK

In the diagrams of Figure 4.2 process 2–3 is the compression, and 4–5 is the expansion through the turbine. In practice the compression and expansion processes occur virtually without heat transfer from the gas, that is to say they may be taken to be adiabatic. Also shown in Figure 4.2 are the hypothetical process 2–3is, which is an adiabatic and reversible (i.e. isentropic) compression, and the hypothetical process 4–5is, which is the isentropic expansion through the turbine. These isentropic processes are those which ideal compressors and turbines would perform. As can be seen, the actual compression process involves a greater temperature rise than that of the isentropic compressor for the same pressure rise,

$$T_3 - T_2 > T_{3\text{is}} - T_2;$$

in other words the work input to the actual compressor for each unit mass of air is greater than the work input for the ideal one. Similarly the actual turbine produces a smaller temperature drop than that in the ideal turbine, that is

$$T_4 - T_5 < T_4 - T_{5\text{is}},$$

and therefore for the same pressure ratio the actual turbine produces less work than the reversible adiabatic one.

For compressors and turbines it is normal to define efficiencies which relate actual work per unit mass flow to that of an ideal (i.e. loss-free) machine with an equivalent pressure change:

$$\eta_{\text{comp}} = \frac{\text{Ideal work}}{\text{Actual work}} \quad \text{and} \quad \eta_{\text{turb}} = \frac{\text{Actual work}}{\text{Ideal work}}. \tag{4.3}$$

Note that the efficiency definitions are different for a compressor or turbine so that their values are always less than unity. For an adiabatic machine the ideal reversible process is at constant entropy (i.e. isentropic) and the efficiencies of Equation 4.3 are referred to as *isentropic efficiencies*.

Treating the fluid as a perfect gas, for which $h = c_p T$, the isentropic efficiencies are

$$\eta_{\text{comp}} = \frac{T_{3\text{is}} - T_2}{T_3 - T_2} \quad \text{and} \quad \eta_{\text{turb}} = \frac{T_4 - T_5}{T_4 - T_{5\text{is}}}. \tag{4.4}$$

Nowadays the isentropic efficiencies in a high quality aircraft engine for use on a commercial aircraft are likely to be around 90% for compressors and turbines, and this round number will normally be used in this book when a numerical value is needed. For the simple gas turbine of Figure 4.2 the pressure rise across the compressor is equal to the pressure drop across the turbine and the corresponding pressure ratios are equal. For a jet engine, however, the pressure ratio across the turbine must be less than the pressure ratio across the compressor because some of the expansion is used to accelerate the jet. The pressure at exit from the turbine is denoted by p_5 and downstream of the propulsive nozzle the pressure is the atmospheric static pressure, p_a.

The isentropic temperature change can very easily be found once the pressure ratio is specified. It may be recalled that for an adiabatic and reversible process

$$p/T^{\gamma/(\gamma-1)} = \text{constant},$$

which, in the present case, means for the ideal compressor and turbine

$$T_{3is}/T_2 = (p_3/p_a)^{(\gamma-1)/\gamma} \quad \text{and} \quad T_4/T_{5is} = (p_4/p_a)^{(\gamma-1)/\gamma}.$$

Neglecting any pressure drop in the combustor gives $p_3 = p_4$ and writing $p_3/p_a = p_4/p_a = r$ gives

$$T_{3is}/T_2 = T_4/T_{5is} = r^{(\gamma-1)/\gamma}. \tag{4.5}$$

The power which must be supplied to the compressor is given by

$$\dot{W}_c = \dot{m}_{air} c_p \left(T_3 - T_2 \right) \tag{4.6}$$

and expressing this in terms of the isentropic temperature rise gives

$$\dot{W}_c = \frac{\dot{m}_{air} c_p \left(T_{3is} - T_2 \right)}{\eta_{comp}}$$

$$= \frac{\dot{m}_{air} c_p T_2 \left(T_{3is}/T_2 - 1 \right)}{\eta_{comp}} = \frac{\dot{m}_{air} c_p T_2 \left(r^{(\gamma-1)/\gamma} - 1 \right)}{\eta_{comp}}. \tag{4.7}$$

Similarly, the power available from the turbine, when the mass flow of fuel in the gas stream is neglected, is given by

$$\dot{W}_t = \dot{m}_{air} c_p \left(T_4 - T_5 \right)$$

$$\text{or} \quad \dot{W}_t = \eta_{turb} \, \dot{m}_{air} c_p \left(T_4 - T_{5is} \right) = \eta_{turb} \dot{m}_{air} c_p T_4 \left(1 - r^{-(\gamma-1)/\gamma} \right). \tag{4.8}$$

The turbine power must be greater than the power required to drive the compressor and the difference, W_{net}, which is available to drive the load or accelerate the jet, is from Equations 4.7 and 4.8

$$\dot{W}_{net} = \dot{m}_{air} \, c_p T_2 \left(\eta_{turb} \frac{T_4}{T_2} \left(1 - 1/r^{(\gamma-1)/\gamma} \right) - \frac{\left(r^{(\gamma-1)/\gamma} - 1 \right)}{\eta_{comp}} \right). \tag{4.9}$$

Certain features can be determined directly from the equation for the net power per unit mass flow rate. The pressure ratio is crucial and if this tends to unity the net power goes to zero. The ratio of turbine inlet temperature to compressor inlet temperature T_4/T_2 is important and, for a given pressure ratio, increasing the temperature ratio brings a rapid rise in net power. This is effectively how the engine is controlled, because increasing the fuel flow increases T_4 and thence the power. Note, however, that it is the *ratio* T_4/T_2 which is involved, so that at high altitude, when the inlet temperature T_2 is low, a high value of the ratio can be obtained with a *comparatively* low value of T_4. In fact the highest values of T_4/T_2 are normally achieved at top-of-climb, the condition when the aircraft is just climbing to its cruising altitude.

Exercises

4.1 Air enters a compressor at a temperature of 288 K and pressure of 1 bar. If the pressure ratio $r = 45$ (a plausible value for take-off in a new design engine) find the temperature at compressor outlet for isentropic efficiencies of 100% and 90%. What is the work input per unit mass flow for the irreversible compressor?

(**Ans:** 854.6 K; 917.5 K; $W_c = 632$ kJ/kg)

4.2 For the engine of Exercise 4.1 find the work per kg which could be extracted from a turbine for a pressure ratio of 45 when the turbine inlet temperature is 1750 K (a plausible value at take-off) and $\eta_{\text{turb}} = 0.90$. Compare the work per kg in compressing the air in Exercise 4.1. (**Ans:** $W_t = 1049$ kJ/kg)

4.3 THE GAS TURBINE THERMAL AND CYCLE EFFICIENCY

The efficiency of a closed-cycle gas turbine, such as are displayed in the T–s and p–v diagrams in Figure 4.2, can be written as the ratio of net power out to the heat transfer rate to the air in the process which replaces combustion:

$$\eta_{\text{cycle}} = \frac{\dot{W}_{net}}{\dot{m}_{air}\, c_p\, (T_4 - T_3)}$$

$$= \frac{\dot{W}_{net}}{\dot{m}_{air}\, c_p T_2\, (T_4/T_2 - T_3/T_2)}. \tag{4.10}$$

The temperature ratio across the irreversible compressor in Equation 4.10 is given by

$$T_3/T_2 = 1 + \frac{r^{(\gamma-1)/\gamma} - 1}{\eta_{\text{comp}}}.$$

The thermal efficiency of the open-cycle gas turbine is the ratio of the net power to the energy input by the combustion of fuel and is

$$\eta_{\text{th}} = \frac{\dot{W}_{net}}{\dot{m}_f LCV}. \tag{4.11}$$

Comparing Equations 4.10 and 4.11, and using Equation 4.2, $\dot{m}_f LCV = \dot{m}_{air}\, c_p (T_4 - T_3)$, it is clear that within the approximations we are adopting the thermal efficiency of the open-cycle gas turbine and the cycle efficiency of the closed-cycle gas turbine are equal,

$$\eta_{\text{th}} = \eta_{\text{cycle}}.$$

When the thermal efficiency was introduced in section 3.3 it was specific to the jet engine and all the power was transferred into the increase in kinetic energy of the flow through the engine, without considering the use of a turbine to extract mechanical power.

In our calculations it has been assumed that the mass flow of fuel is negligible compared to that of air. According to the approximations we have adopted the temperature rise in the combustion chamber is related to the fuel flow by Equation 4.2. Suppose air enters the compressor at $T_2 = 288$ K and the compressor produces a pressure ratio of 45 at an efficiency of 90%. The air then leaves the compressor at $T_3 = 917.5$ K. If the turbine inlet temperature $T_4 = 1750$ K the temperature rise in the combustor is 832.5 K, requiring an enthalpy rise of 837 kJ/kg. For a fuel with $LCV =$ 43 MJ/kg, the mass flow of fuel per unit mass flow of air is therefore 0.0194. This ratio uses the mass flow of air through the core; were the bypass ratio to be 10, for example, the ratio of fuel to total engine air would be down to 2 parts in 1000.

Exercises

4.3*a For the example of Exercises 4.1 and 4.2, which correspond to sea-level static operation on a standard day, calculate the thermal efficiency and the net work per kg of air flowing. Assume isentropic efficiencies of 90% for the compressor and turbine.   (**Ans:** $\eta_{th} = 0.498$; $W_{net} = 417$ kJ/kg)
      The calculation can be repeated for the following (a spreadsheet or programme is a good idea):

b $T_2 = 308$ K, $T_4 = 1750$ K (as for a take-off on a hot day), $r = 45$.
                              (**Ans:** $\eta_{th} = 0.483$; $W_{net} = 373$ kJ/kg)

c $T_2 = 245$ K, $T_4 = 1600$ K (representative of top-of-climb at 35,000 ft, $M = 0.78$), $r = 50$.
                              (**Ans:** $\eta_{th} = 0.514$; $W_{net} = 411$ kJ/kg)

d $T_2 = 245$ K, $T_4 = 1500$ K (representative of cruise conditions at 35,000 ft, $M = 0.78$), $r = 45$.
                              (**Ans:** $\eta_{th} = 0.500$; $W_{net} = 361$ kJ/kg)

e $T_2 = 245$ K, $T_4 = 1500$ K (representative of cruise conditions at 35,000 ft, $M = 0.78$), $r = 40$.
                              (**Ans:** $\eta_{th} = 0.497$; $W_{net} = 372$ kJ/kg)

f $T_2 = 245$ K, $T_4 = 1500$ K (representative of cruise conditions at 35,000 ft, $M = 0.78$), $r = 45$ for reduced compressor and turbine efficiency, $\eta_c = \eta_t = 0.85$.   (**Ans:** $\eta_{th} = 0.404$; $W_{net} = 280$ kJ/kg)

      Discuss these results of a–f, commenting on the effect of inlet temperature, turbine inlet temperature, pressure ratio and component isentropic efficiency.

Notes:
      1. The answers here show the highest work output for conditions at take-off on a standard day – note that this is actually work per unit mass flow, which is equal to the power per unit mass flow rate. In fact the power from any gas turbine falls rapidly with altitude because as the density of the air drops the mass flow rate of air through the engine falls.
      2. These thermal/cycle efficiencies are really quite high. For comparison, a high quality diesel engine in a large truck has an overall efficiency only just about 40%.
      3. The thermal efficiency increases with T_4/T_2 and net work output goes up even more rapidly.
      4. Reduction in component efficiency has a strong effect on thermal efficiency and work output, with the latter decreasing by about 22% for a 5% reduction in isentropic efficiency of compressor and turbine. The strong dependence is because the net work is the *difference* between turbine and compressor work, so proportional changes in W_{net} are much larger.
      5. Parts (a) and (b) show a 10% reduction in power per unit mass flow rate for a 20 K rise in air inlet temperature for the same turbine inlet temperature. In fact the drop in power output from the engine would be greater than this by about 3.5% because of the drop in density of the inlet air with increase in inlet temperature. The gas turbine is not alone in suffering a reduction in power output and efficiency as the inlet air temperature rises. In road vehicles we normally pay no attention to the maximum temperature in the engine (assuming that the engine has been developed to stand the worst conditions to which we are likely to expose it) but the peak temperature will rise with inlet temperature, assuming the same fuel input. Gasoline engines and diesel engines experience a marked reduction in power output as the inlet air density drops, which is most obvious at high altitudes; one of the Interstate freeways in the USA exceeds 12,000 feet and there the effect is pronounced.

The dependence of power per unit mass flow rate of air, $\dot{W}_{net}$, on T_4/T_2 and pressure ratio is shown by the curves in Figure 4.3a. As will be discussed in Chapter 5, the value of T_4/T_2 appropriate for cruise is approximately 6 for a modern engine. When the pressure ratio is small an increase in the

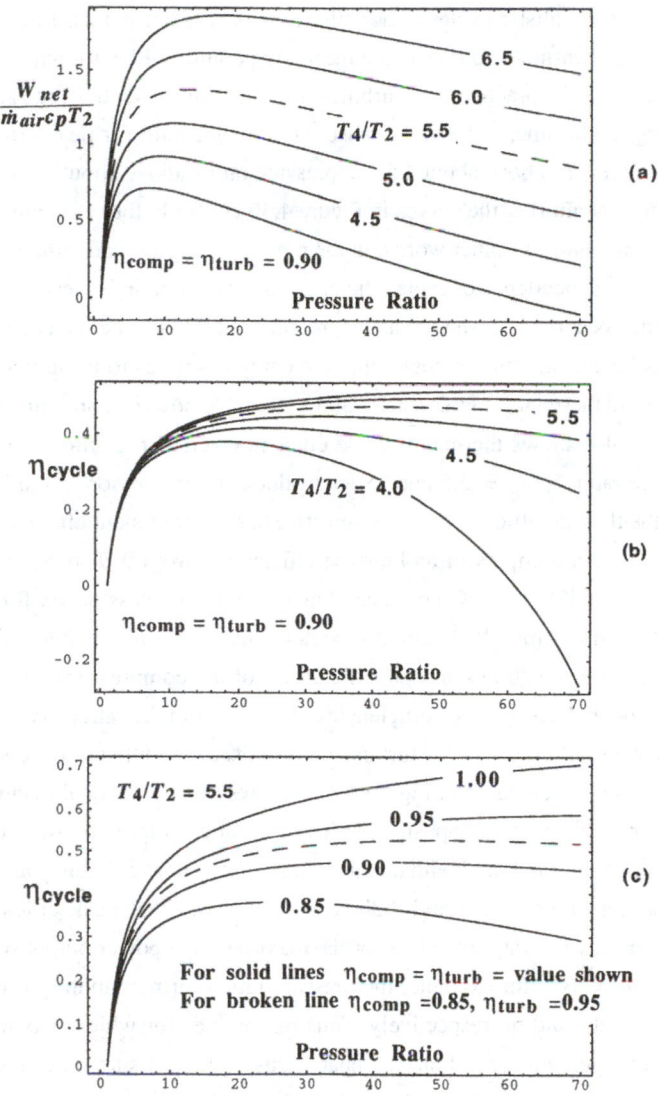

Figure 4.3. Non-dimensional power and cycle efficiency for an idealised gas turbine.

pressure ratio brings a rapid increase in $\dot{W}_{net}$, but the rate of increase falls as pressure ratio rises and there is a pressure ratio at which $\dot{W}_{net}$ peaks. The value of the pressure ratio for peak net power increases as T_4/T_2 rises but never exceeds about 20 for all practical values of temperature ratio.

Figure 4.3b shows the thermal or cycle efficiency for a gas turbine as a function of pressure ratio, with the curves denoting different values of the temperature ratio. In deriving these curves it is again assumed that $\eta_{comp} = \eta_{turb} = 0.90$. The importance of both the temperature ratio and pressure ratio in determining the efficiency of the engine are clear. (For the ideal gas turbine, in which $\eta_{comp} = \eta_{turb} = 1.00$, the trends are unrealistic and the cycle efficiency is independent of

T_4/T_2.) Put simply, for realistic cycles (where the components are not assumed to be isentropic) the thermal efficiency continues to rise as the mean temperature of heat input rises relative to the ambient temperature T_2. In practical gas turbines it is therefore essential for T_4, the temperature of the gas entering the turbine, to be high.[2] Also the pressure ratio for peak efficiency rises with T_4/T_2. For values of T_4/T_2 above about 4.5 and pressure ratios above about 20 the variation of η_{cy} with pressure ratio is slight (i.e. the curves in Figure 4.3b are fairly flat) to a considerable distance either side of the maximum. In other words, in the range of pressure and temperature ratios likely in aircraft practice the dependence of cycle efficiency on pressure ratio is weak. A pressure ratio for maximum efficiency occurs because increasing pressure ratio raises the mean temperature of heat addition, which is beneficial, but the power into the compressor has to be supplied by the turbine and irreversibilities in these take a larger share of the available power as pressure ratio is increased.

Figure 4.3c also shows thermal or cycle efficiency versus pressure ratio but this time for a fixed temperature ratio $T_4/T_2 = 5.5$ and various values of compressor and turbine efficiency. It can be seen that the thermal efficiency is very sensitive to the component efficiencies: at a pressure ratio of 40 a reduction in compressor and turbine efficiencies from 90% to 85% would lower the thermal efficiency from 47% to 37%, corresponding to about 21% less power for the same rate of energy input in the form of fuel. It should also be noted that the peak thermal efficiency occurs at ever higher pressure ratios as the isentropic efficiencies of the compressor and turbine are raised. Taking equal turbine and compressor efficiencies is an oversimplification. As the pressure ratio increases the isentropic efficiency of the turbine tends to rise, whilst that of the compressor falls by a similar amount. The broken curve in Figure 4.3c explores this, with equal magnitude changes of component isentropic efficiency of opposite sign in turbine and compressor. Although the alteration in η_{cy} is significant, the trends found with equal values of isentropic efficiency are not altered.

By comparing Figures 4.3a and 4.3b it may be noted that peak efficiency occurs at a substantially higher pressure ratio than that for the maximum net power output per unit mass flow rate of air: for $T_4/T_2 = 6.0$, for example, the pressure ratios for maximum power and maximum efficiency are around 15 and 50 respectively. Thus for engines for which maximum thrust is the goal (the normal requirement for military combat engines) the pressure ratio might be around 15 for $T_4/T_2 = 6.0$, whilst for engines for which efficiency is most important (most civil engines) the pressure ratio would be nearer 50. Maximum power occurs at lower pressure ratio than maximum efficiency for the following reason. At low pressure ratios the power output increases with pressure ratio because the efficiency rises so rapidly that more of the heat is converted to work. As the pressure ratio increases, the temperature at compressor outlet also increases and if there is an upper limit on turbine inlet temperature the permissible heat input (equivalent to the amount of fuel burned) reduces. At higher pressure ratios the heat input decreases more rapidly with pressure than

---

[2] Because here we are not dealing with a real closed cycle, but with an open system in which the heat input is replaced by a combustion process, the effect of temperature at the end of the combustion process should really be handled more carefully. Combustion produces flame temperatures typically in excess of 2300 K, which are higher than the temperature allowed into the turbine (typically no more than 1850 K for prolonged operation). The dilution of the combustion products to lower the temperature is associated with a rise in entropy and a loss in the capability of turning thermal energy into work.

the increase in efficiency (the conversion of heat into work) and where the two effects have equal magnitude defines the pressure ratio for maximum power output per unit mass flow rate of air.

The inlet temperature T_2 is determined by the altitude and the forward speed. At 35,000 feet and $M = 0.78$ this is $T_2 = 245.4$ K, close to the value used in cases (d) to (f) of Exercise 4.3. As discussed later in section 5.2, the turbine inlet temperature is fixed by metallurgical considerations (i.e. what temperature can the metal stand at a given level of stress), by cooling technology and by considerations of longevity – lower temperatures lead to longer life. With $T_4/T_2 = 6.11$, corresponding to $T_2 = 245.4$ K and $T_4 = 1500$ K, the peak thermal efficiency is about 0.501 and occurs at a pressure ratio of about 55; the variation in efficiency to either side of this pressure ratio is small and at a pressure ratio of 45 the efficiency is only reduced to 0.499. As the pressure ratio increases it becomes harder to design a satisfactory compressor and the isentropic efficiency tends to fall, so there is some advantage in designing to be on the lower pressure ratio side of the peak in η_{cy}. Furthermore, by putting the cruise at a pressure ratio slightly below the peak, the maximum climb condition can occur with the pressure ratio in an acceptable range.

There is an additional issue concerning pressure ratio, which will be discussed more fully in Chapter 6. For an aircraft flying at a Mach number of 0.78 the forward motion causes the pressure (more specifically the stagnation pressure) at inlet to the compressor to be raised by a factor of 1.495, compared to the surrounding atmosphere. At outlet from the engine (i.e. at nozzle outlet) the pressure is not raised in this way, but remains at the atmospheric pressure. The effect of forward speed is therefore to increase the compression ratio of the whole engine by 1.495; raising the effective pressure ratio from 45 to about 67 has, as explained above, only a small effect on the thermal efficiency.

4.4 THE EFFECT OF WORKING GAS PROPERTIES

The analysis up to now, and in most of what follows up to Chapter 10, treats the working fluid in the gas turbine as a perfect gas with the same properties as air at standard conditions. This is done to make the treatment as simple as possible, and it does yield the correct trends. In a serious design study, however, the gas would be treated as semi-perfect and the products of combustion in the stream through the turbine would be included. In the semi-perfect gas approximation c_p, R, and $\gamma = c_p/c_v$ are functions of temperature and composition but *not* of pressure. With this approximation, which is sufficiently accurate for all gas turbine applications, the useful relation between gas properties $p/\rho = RT$ is retained. As is discussed in more detail in Chapter 11, R is virtually independent of temperature and composition for the gases occurring throughout a gas turbine, and it may be taken to be constant at 0.287 kJ kg^{-1}K^{-1}. The specific heat capacity may be obtained from $c_p = \gamma R/(\gamma - 1)$.

Table 4.1 presents the results of Exercise 4.3 for the overall thermal efficiency and the net work produced per unit mass of air through the engine. These results under the heading 'γ constant' are based on air, with $\gamma = 1.40$. This value of γ, with $R = 287$ kJ kg^{-1}K^{-1} gives, $c_p = 1.005$ kJ kg^{-1}K^{-1}. Also shown in Table 4.1 (under the heading 'γ variable') are results for

Table 4.1 Comparison of thermal efficiency and net work based on constant gas properties and those related to local conditions

Exercise number	T_2 (K)	T_4 (K)	Overall pressure ratio, p_3/p_2	Component efficiencies $\eta_c = \eta_t$	γ constant		γ variable	
					η_{th}	W_{net} (kJ/kg)	η_{th}	W_{net} (kJ/kg)
4.3a	288	1750	45	0.90	0.498	417	0.483	545
4.3b	308	1750	45	0.90	0.482	373	0.472	502
4.3c	245	1600	50	0.90	0.514	411	0.502	518
4.3d	245	1500	45	0.90	0.500	361	0.490	451
4.3e	245	1500	40	0.90	0.497	372	0.485	459
4.3f	245	1500	45	0.85	0.404	279	0.412	366

the same engine pressure ratio and compressor and turbine entry temperatures but using accurate values for γ and c_p based on local temperature and composition. In these more accurate calculations (carried out by Professor J. B. Young in Cambridge) temperature and pressure are not related by $pT^{\gamma/\gamma-1} = $ constant, which is only correct when γ is independent of temperature, but by a process which is essentially exact. For these 'γ variable' results the mass flow rate of fuel is also added to the exhaust gas.

The table shows that treating the gas with constant properties throughout gives quite good estimates for the thermal efficiency, reflecting the variation with operating condition and, most significantly, with component efficiency. With the simple gas treatment, however, the work output is underestimated by around 20%, though the trends with pressure ratio and temperatures are similar to those calculated with the more realistic assumptions for the gas properties. It is possible to improve the constant-property gas model by having different values of c_p for the compressor and turbine and after Chapter 11 a better approximation for the flow in the turbine is used, which is to take $c_p = 1.244$ kJ kg^{-1}K^{-1} and $\gamma = 1.3$. For the calculations of the cycle to be more precise the mass flow through the turbine should be about 2% larger than through the compressor because of the fuel, but affecting the engine in the opposite sense is the effect of cooling air and combustor pressure loss. These complications are, however, unwarranted for the present purpose but are revisited for the New Efficient Aircraft engine in Chapter 19.

The underestimate of power predicted assuming a perfect gas with equal properties for the flow through the turbine and compressor is likely to lead to a core being designed which is about 20% bigger (in terms of the mass of air which passes through it) than that which would result if the correct gas properties were used. In fact the perfect gas design is not likely to give such an oversized core for a number of reasons. The model has neglected the cooling flows (which may take 20% of the air passing through the compressor). For an aircraft, air must be bled off the compressor to pressurise the cabin and to bring about de-icing of the wing and nacelle, whilst electrical and hydraulic power is taken from the engine. The drop in pressure in the combustor (perhaps 4% of the local pressure) is also neglected. Lastly the 90% isentropic efficiencies for compressor and turbine used in the design calculations are quite ambitious and might not be achieved; the net power has been shown to be strongly affected by compressor and turbine isentropic efficiency and this too means that the actual core might be larger than a design assuming $\eta_c = \eta_t = 0.90$.

4.5 THE GAS TURBINE AND THE JET ENGINE

As shown in section 4.4, high thermal efficiency depends on having a high temperature ratio and a pressure ratio appropriate to this. The net power generated by the core could all be used to accelerate the flow through it, in which case we would have produced a simple turbojet, as in Figure 5.1. The problem this creates is that with the high turbine inlet temperatures necessary to give high efficiency there is such a large amount of net work available that if all of this were put into the kinetic energy of the jet, the jet velocity would be very high. As a result the propulsive efficiency η_p would be low and therefore so too would be the overall efficiency. The way around this is to use the available energy of the flow out of the core to drive a turbine which is used to move a much larger mass flow of air, either by a propeller (to form a turboprop) or a fan in a duct as in the high bypass ratio engine used now on most large aircraft.

The noise power generated by jets is approximately proportional to the eighth power of jet velocity, so there is an incentive, in addition to propulsive efficiency, to reduce jet velocity.

Exercises
4.4 a A simple (no bypass) turbojet engine flies at 231 m/s, which is $M = 0.78$ at 35 kft. Find the jet velocity which would be produced with the temperatures given as appropriate for cruise in part (d) of Exercise 4.3. Assume that all the net work is used to increase the kinetic energy of the flow.

Calculate the propulsive efficiency η_p and the overall efficiency $\eta_o = \eta_p \eta_{th}$ at this flight speed. Explain why the propulsive efficiency and overall efficiency is low. Indicate ways in which η_o could be raised at this flight speed. (**Ans:** $V_j = 881$ m/s, $\eta_p = 0.416$, $\eta_o = 0.208$)

b Recalculate the efficiencies if the flight speed were 600 m/s ($M = 1.99$ at 31,000 feet).
(**Ans:** $V_j = 1041$ m/s, $\eta_p = 0.732$, $\eta_o = 0.366$)

Note: The turbojet is inefficient for propulsion at subsonic speeds even when the thermal efficiency is high. The same engine is comparatively efficient for propulsion at supersonic speeds.

SUMMARY OF CHAPTER 4

The gas turbine may be envisaged as a core with various loads fitted to it; one possible load is a nozzle to create a simple turbojet, another is a turbine driving a bypass fan. In a closed-cycle gas turbine a fixed amount of air is compressed, heated, expanded in a turbine and cooled; the efficiency of such a cycle is the *cycle* efficiency η_{cy}. Most gas turbines, and all jet engines, are open cycle, replacing the heating process by combustion of fuel and eliminating the cooling process by instead taking in fresh air to the compressor. For the open-cycle gas turbine the measure of performance is *thermal* efficiency η_{th}. Although describing fundamentally different engines, both η_{cy} and η_{th} vary in a similar way with pressure ratio and with the ratio of turbine inlet temperature to compressor inlet temperature, T_4/T_2. The terms thermal and cycle tend to be used as alternatives.

Isentropic efficiencies are the conventional method of relating actual compressor and turbine performance to the ideal; nowadays isentropic efficiencies of around 90% are expected. As pressure ratio increases there is a tendency for the compressor isentropic efficiency to drop a little and for the turbine isentropic efficiency to rise.

The gas turbine thermal efficiency, the ratio of the net power to the equivalent heat input rate, is strongly dependent on the isentropic efficiencies. It is also a strong function of T_4/T_2. For any given value of T_4/T_2 the pressure ratio to give maximum net power would be lower than the pressure ratio for maximum thermal efficiency. The pressure ratio for highest thermal efficiency increases at T_4/T_2 is raised.

To obtain high thermal efficiency it is essential to operate at high values of T_4/T_2. This invariably means that there is a high level of power output per unit mass flow through the core: were all this work to be used to accelerate only the core flow, it would lead to jet velocities unacceptably high for subsonic aircraft. The solution is to choose a combination of high temperature ratio and pressure ratio for the core, but to use the available power output via a turbine to drive a fan, supplying a relatively small increment in kinetic energy to a large bypass stream.

The elementary description of the gas turbine has made it possible to show what is needed to obtain large power output and high cycle efficiency and, equivalently, high thermal efficiency. Some important limitations of the present treatment are:

- no cooling flows (air taken from the compressor to cool or shield the turbine blades);
- a perfect gas with properties of ambient air ($\gamma = 1.40$) has been assumed throughout;
- the mass of fuel in calculating power from the turbine has been neglected;
- pressure drop in the combustor, which may be 4% of maximum pressure, is omitted;
- simple assumptions have been made for compressor and turbine efficiency;
- electrical power off-takes and air for cabin pressurisation have been neglected.

Despite these shortcomings the trends predicted are correct and the magnitudes for efficiency and available power are plausible. The manner in which this can be used effectively is explored in the next chapter.

APPENDIX

A brief summary of thermodynamics from an engineering perspective

This is not the place to give a detailed account of thermodynamic theory and principles, but a brief summary in the nature of revision may be helpful. For those who find the treatment here insufficient it is recommended that one of the very many texts be consulted. (There are so many books on engineering thermodynamics available that it is invidious to select only one, but some guidance does seem in order. The books by Moran et al. 2010 and Borgnakke and Sonntag 2008 can be recommended.) In the interest of brevity and simplicity details will be omitted wherever possible and no attempt is made to include every restriction and caveat. A feature of engineering thermodynamics is the common use of a closed control surface to enclose the process or device under consideration; when using a control surface attention is directed at what crosses the surface, both material flows of gas and liquid as well as work and heat being transferred across it.

The *first law* is a formal statement of the conservation of energy. For a *fixed* mass m of a substance this can be written in differential form as

$$\mathrm{d}Q - \mathrm{d}W = m\,\mathrm{d}e,$$

where Q is heat transferred to the substance, W is work extracted from the substance and e is the energy per unit mass. We will neglect changes in energy associated with chemical, electrical and magnetic effects. If potential and kinetic mechanical energy are also neglected for the moment, energy e is restricted to the internal energy denoted by u, the specific internal energy. The work here can be done by, for example, gas expanding and pushing back the atmosphere, or it could be work driving a shaft or electrical work.

The application we have for the first law in the gas turbine and jet engine is for steady flow of gas through a device (either a whole engine or a component like a compressor or turbine). To facilitate this we draw an imaginary control surface around the device or the process we are considering. The control surface is closed, so any matter entering or leaving the device must cross the control surface; properties entering are given subscript 1 whilst those leaving have subscript 2. We specify here that there is a steady mass flow $\dot{m}$ into and out of the control surface. We also have heat transfer rate *into* the control surface $\dot{Q}$ and work transfer rate *out from* the control surface $\dot{W}_s$. The first law is now written for the closed control surface as

$$\dot{Q} - \dot{W}_s = \dot{m}\left\{\left(h_2 + V_2^2/2\right) - \left(h_1 + V_1^2/2\right)\right\}. \tag{4.A1}$$

The terms on the right hand side are the energy transported across the control surface by the flow. The kinetic energy is recognisable as $V^2/2$ and h is the enthalpy. For a flow process it is appropriate to use the specific enthalpy $h = u + p/\rho$, where u is the internal energy per unit mass, p is the pressure and ρ is the density. Enthalpy allows for the displacement work done by flow entering and leaving the control surface. The work $\dot{W}_s$ in Equation 4.A1 is often referred to as shaft work (to distinguish it from the so-called displacement work done by flow entering and leaving the control surface) even though the work may be removed without a shaft, for example electrically.

For an ideal gas of fixed composition the enthalpy is a function of only the temperature. The formal expression is

$$\mathrm{d}h = c_p\,\mathrm{d}T,$$

where c_p is the specific heat capacity at constant pressure. If c_p is constant, as for a perfect gas, Equation 4.A1 can be written

$$\dot{Q} - \dot{W} = \dot{m}\{(c_p T_2 + V_2^2/2) - (c_p T_1 + V_1^2/2)\}. \tag{4.A2}$$

It is frequently useful to refer to the stagnation enthalpy h_0, defined by

$$h_0 = h + V^2/2$$

and the corresponding stagnation temperature

$$T_0 = T + V^2/2c_p$$

Hence
$$\dot{Q} - \dot{W} = h_{02} - h_{01}.$$

The *second law* leads to the property entropy. Entropy is useful because in an ideal process for changing energy from one form to another the net change of entropy of the system and the environment is zero. Since zero entropy increase gives the ideal limit, the magnitude of entropy rise gives a measure of how far short of ideal the real process is.

Entropy, as a property, can be expressed in terms of other properties, for example $s = s(T,p)$, where s denotes the specific entropy (entropy per unit mass). It is defined for an ideal process (normally referred to as a reversible process) by

$$\mathrm{d}s = \mathrm{d}Q/T, \tag{4.A3}$$

By using the first law in differential form the change in entropy in Equation 4.A3 can be re-written as

$$\mathrm{d}s = \mathrm{d}h/T - \mathrm{d}p/\rho T = c_p \mathrm{d}T/T - R\mathrm{d}p/p$$

and this can be integrated for a perfect gas to give

$$s - s_{ref} = c_p \ln\left(T/T_{ref}\right) - R \ln\left(p/p_{ref}\right), \tag{4.A4}$$

where s_{ref} is the entropy at the reference pressure p_{ref} and temperature T_{ref}.

As the energy of a gas is increased the temperature must also increase; if pressure is held constant Equation 4.A4 shows that entropy also rises. But if temperature is held constant and the pressure is reduced, entropy also rises. The implications of Equation 4.A4 can be understood a little better by considering the adiabatic (no heat transfer) flow of gas along a pipe or through a throttle; in both cases there is a drop in pressure but, because there is no work or heat transfer, no change in energy and therefore no change in temperature. The fall in pressure therefore leads to a rise in entropy. In fact the magnitude of the entropy rise is a measure of the pressure loss. An entropy rise in the absence of heat transfer is evidence of a dissipative process; the effect is irreversible, since reversing the process would not reduce the entropy. Ideal processes are therefore often referred to as reversible, whereas processes which lead to loss (that is, processes for which the entropy rises by more than follows from Equation 4.A3) are often referred to as irreversible.

Compressors and turbines have a small external area in relation to the mass flow passing through them. As a result the heat transfer from them is small and a good approximation is normally to treat both compressors and turbines as adiabatic. When ideal (hypothetical) machines are used as a standard to compare with real machines, it is normal to specify them to be adiabatic. The ideal compressor or turbine will have no dissipative processes – the flow is loss free and is normally described as reversible. A reversible adiabatic flow will experience no rise in entropy and such ideal processes are normally referred to as isentropic.

An ideal compressor or turbine, producing a reversible and adiabatic change in pressure, produces no change in the entropy of the flow passing through. With no change in entropy Equation 4.A4 leads immediately to

$$0 = \ln(T_2/T_1) - \frac{R}{c_p} \ln(p_2/p_1),$$

and since $c_p = \frac{\gamma R}{\gamma - 1}$,

$$T_2/T_1 = (p_2/p_1)^{(\gamma-1)/\gamma}.$$

CHAPTER 5   THE PRINCIPLE AND LAYOUT OF JET ENGINES

5.0 INTRODUCTION

This chapter looks at the layout of some jet engines, using cross-sectional drawings. This begins with relatively simple engines and leads to engines for a recent large aircraft, the Boeing 787 and an engine for the smaller Bombardier C-series. Two concepts are introduced in the chapter. One is the multi-shaft engine with separate low-pressure and high-pressure spools. The other is the bypass engine in which some, very often most, of the air compressed by the fan bypasses the combustor and turbines.

Any consideration of practical engines must address the temperature limitations on the turbine. The chapter ends with some discussion of cooling technology and of the concept of cooling effectiveness.

5.1 THE TURBOJET AND THE TURBOFAN

Figure 5.1 shows a cut-away drawing of a Rolls-Royce Viper engine. This is typical of the simplest form of turbojet engine, which was the norm in the 1950s when it entered service, with an axial compressor coupled to an axial turbine, all on the same shaft. (The shaft, the compressor on one end and turbine on the other are sometimes referred to together as a spool.) Even for this very simple engine, which was originally designed to be expendable as a power source for target drones, the drawing is complicated. For more advanced engines such drawings become unhelpful at this small scale and simplified cross-sections are therefore more satisfactory and will be shown. A simplified cross-section is also shown for the Viper in Figure 5.1, as well as a cartoon showing the major components.

More recent turbojet engines had two spools so that the compression and expansion were split into parts. For flight at sustained speeds well in excess of the speed of sound a turbojet engine remains an attractive option and a two-shaft example, the Rolls-Royce Olympus 593, is shown in Figure 5.2. Four of these engines were used to propel the Concorde at around twice the speed of sound. The low-pressure (LP) compressor and LP turbine are mounted on one shaft to form the LP spool. The LP shaft passes through the high-pressure (HP) shaft on which are mounted the HP compressor and the HP turbine. The compression process is split between two spools for reasons to do with operation at speeds below the design speed, including starting; this is discussed in some detail in Chapter 12.

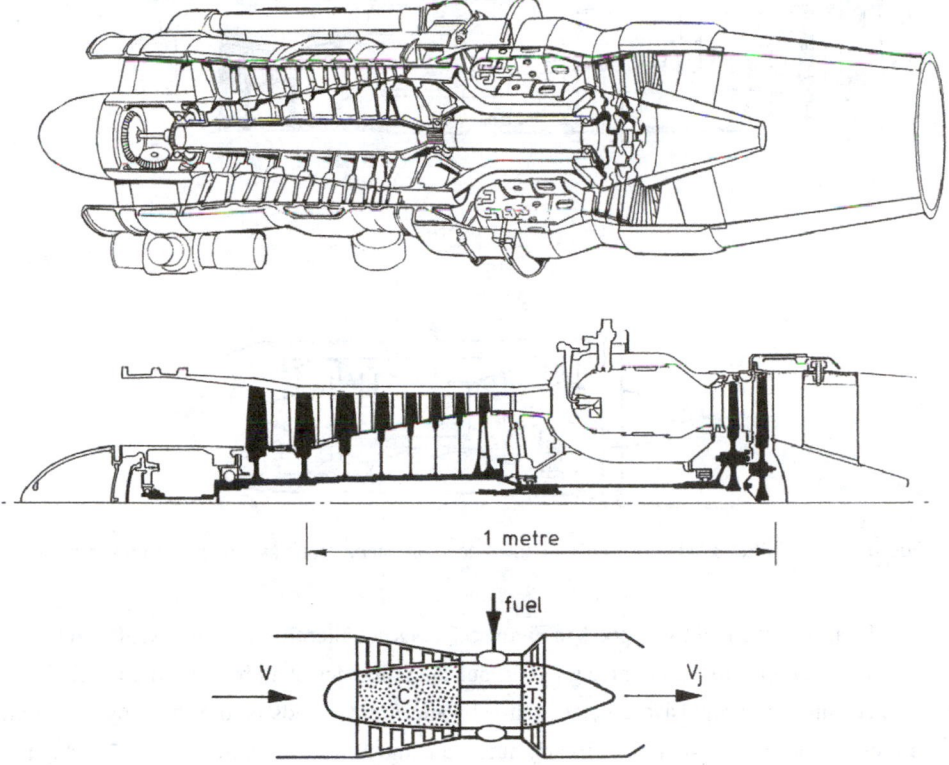

Figure 5.1. The Rolls-Royce Viper Mark 601 single-shaft turbojet shown as a cut-away, in simplified cross-section and as a schematic.

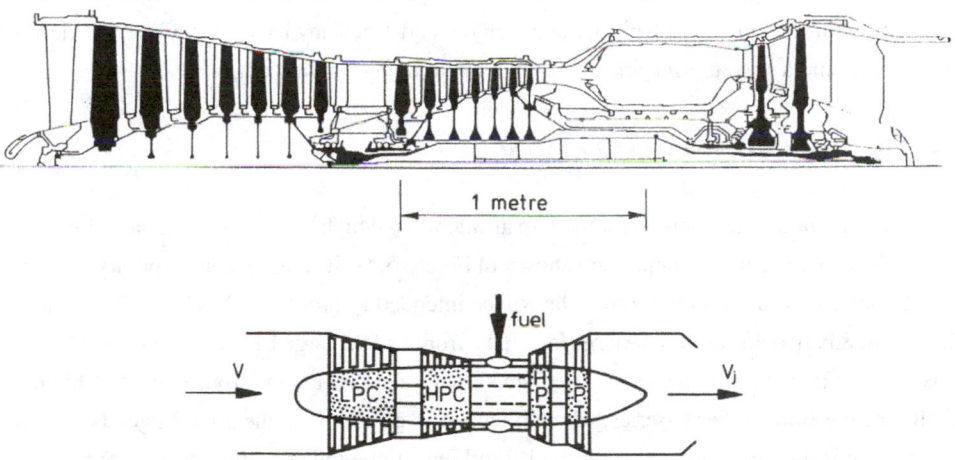

Figure 5.2. The Rolls-Royce Olympus 593 shown as a simplified cross-section and as a schematic.

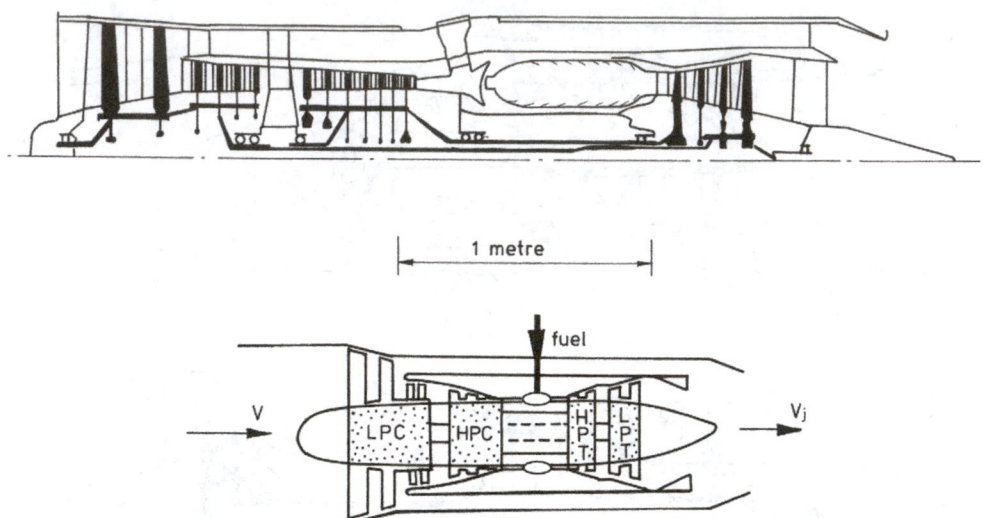

Figure 5.3. The Pratt & Whitney JT8D-1 shown as a simplified cross-section and as a schematic.

The arrangement of twin spools in Figure 5.2 does not alter the problem revealed in Exercise 4.4: the jet velocity is too high to give good propulsive efficiency at all but very high flight speeds. The way to raise the propulsive efficiency at subsonic flight speeds is to go to a bypass engine, sometimes known as a turbofan. An early turbofan, the Pratt & Whitney JT8D-1, which was manufactured in large numbers to propel the Boeing 727 and 737, is shown in Figure 5.3. In this engine some of the air compressed by the LP compressor is passed around the outside of the engine and does not go through the combustor, i.e. it is bypassed around the core. These early turbofan engines have a bypass ratio (the mass flow of air bypassed around the core divided by the mass of air going through the core) typically between 0.3 and 1.5. They have been widely used in older civil aircraft and a similar arrangement can be seen in many military engines.

5.2 THE HIGH BYPASS RATIO ENGINE

The arrangement which is normal for modern airliners is the high bypass ratio engine, with a bypass ratio of 5 or more. Three examples are shown in Figure 5.4, all roughly contemporary. An engine from General Electric and one from Rolls-Royce intended to propel the Boeing 787 are shown in Figure 5.4a&b. In both engines the large fan on the front and the large LP turbine at the back are very conspicuous; they are also heavy. Figure 5.4c shows a smaller engine from Pratt & Whitney, the 1524G for the Bombardier C-series, which breaks new ground with the use of a gearbox between the LP turbine and the fan. As a result the LP turbine can be much smaller, for reasons explained in Chapter 9. At the time of writing there are no large engines, like the Rolls-Royce Trent or the General Electric GEnx, which have a gearbox between the LP turbine and fan. For the gearbox the

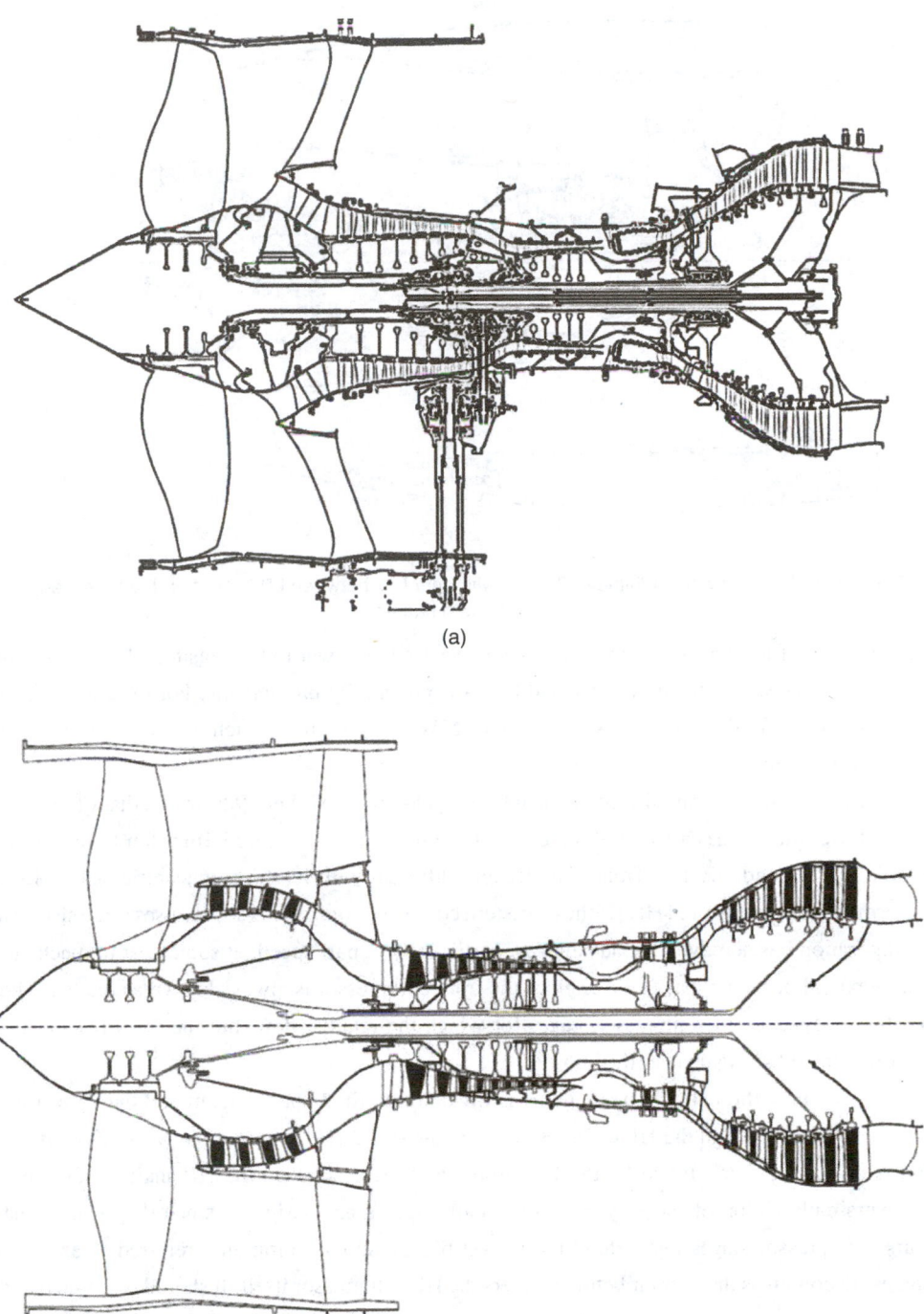

(a)

(b)

Figure 5.4. (a) The Rolls-Royce Trent 1000 (fan tip diameter 2.84 m). (b) The General Electric GEnx (fan tip diameter 2.82 m).

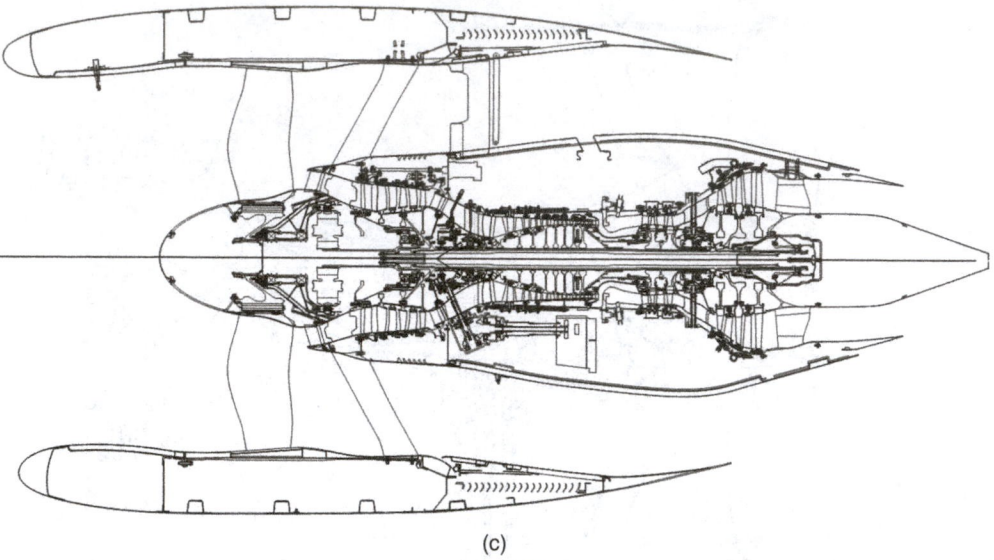

(c)

Figure 5.4. (c) The Pratt & Whitney GT 1524 with gear box between LP turbine and fan. This engine shown with nacelle.

transmission efficiency is of huge importance because the power to be transmitted is so large that removing the heat created by mechanical losses is potentially problematic. For the engine for the NEA the power to the fan at cruise is about 13 MW, so a gearbox which is as efficient as 99.5% would still require about 66 kW of cooling.

The Trent 1000 and the GEnx each have bypass ratio of about 9.5, whilst that of the P&W Geared Turbofan is has a value of about 12. The GEnx and the Geared Turbofan have only two spools, the LP and HP. The Trent is noticeably different with its three concentric shafts: LP, IP (intermediate pressure) and HP. Of these western companies only Rolls-Royce uses three shafts; the configuration has aerodynamic advantages, particularly at part speed, at some cost in mechanical complexity. For high bypass ratio engines the front compressor is always known as the fan. These are highly specialised compressor stages for which the relative flow into the rotor is supersonic near the tip and subsonic near the hub.

The flow through a bypass engine is divided into the bypass stream and the core stream, the latter going through the HP compressor, combustor and HP turbine. In a two-shaft engine the fan is on the LP shaft driven by the LP turbine and the core is on the HP shaft. In fact this is an oversimplification of the way most two-shaft engines are designed, since they usually have some compressor stages just behind the fan on the LP spool (commonly referred to as booster stages) to compress the core air before it enters the HP compressor itself. It should be noted that the terminology is a bit loose. The core or gas generator always includes the HP compressor, combustor and HP turbine but sometimes the term core is used to include the compression on the LP shaft (the fan root and the booster) and even the LP turbine. For the three-shaft engine the core includes IP and HP compressors and turbines.

Typically at cruise the fan and booster might have a pressure ratio of up to about 2.5 and the HP compressor a pressure ratio of up to 20 to give an overall pressure ratio of about 45. For the Trent the core air is compressed in three separate sections. At cruise the pressure ratio is about 1.5 in the fan (near the hub), around 7 in the IP compressor and a bit over 4 the HP compressor to give about 45 overall. In the two-spool engine considered here for the New Efficient Aircraft, and in later chapters, the overall pressure ratio is taken as 45, with the core pressure ratio equal to 18 and pressure ratio attributable to fan and booster equal to 2.5.

In the simplified treatment here it is assumed that all of the power from the HP turbine is used to drive the HP compressor. In fact a relatively small proportion is taken to drive fuel pumps, generate electricity and provide hydraulic power for the aircraft. Similarly it is assumed that all of the air compressed in the core passes through the turbine; in fact some is bled off to pressurise the cabin and de-ice (by warming) some surfaces of the wing and nacelle. Most of the power from the LP turbine is used in compressing the bypass flow and a small proportion is used to raise the pressure of the core in the root of the fan and in the booster.

At this stage of the design there is no need to consider whether to have a two-shaft engine with booster stages (the Pratt & Whitney and General Electric solution) or separate IP and HP compressors in a three-shaft engine (the Rolls-Royce solution). As noted above, the advantages of having separate IP and HP shafts appear mainly during off-design operation, including starting. It suffices for the present treatment to adopt the style which requires the fewest calculations and so to have two shafts with the fan and booster on the LP shaft and the remainder of the core-flow compression on a single HP shaft.

Exercises

5.1* At the start of cruise at $M = 0.78$ at 35,000 ft the temperature and pressure of the air entering the engine may be taken to be 245.4 K and 35.6 kPa. The pressure ratio across the fan in the bypass stream is 1.5 and the pressure ratio of core flow through the fan and LP compressor (booster) is 2.5. The pressure ratio of the HP compressor is 18. Assume isentropic efficiencies of 90% in each component.

a Find the temperature rise across the fan bypass stream, $T_{013} - T_{02}$, and the fan and booster for the core flow, $T_{023} - T_{02}$. Determine the temperature at entry to the HP compressor, T_{023}, and at exit from it, T_{03}, as well as the temperature rise in the HP compressor.

(**Ans:** $T_{013} - T_{02} = 33.5$ K, $T_{023} - T_{02} = 81.6$ k, $T_{023} = 327.0$ K, $T_{03} = 793.5$ K, $T_{03} - T_{023} = 466.5$ K)

b* The shaft power produced by the HP turbine must equal the power into the HP compressor. Since we are assuming a perfect gas with the properties of air and treating the combustion as equivalent to a heat transfer, the temperature drop in the HP turbine must equal the temperature rise in the HP compressor. If the temperature of the gas leaving the combustor (i.e. entering the HP turbine) is 1500 K, find the temperature at HP turbine outlet, T_{045}, and thence the pressure at outlet from the HP turbine, p_{045}. Assume a turbine efficiency of 90%. (Neglect any pressure drop in the combustor, i.e. $p_{04} = p_{03}$.)

(**Ans:** 1033.5 K, 363.5 kPa)

c The core exists to drive the LP turbine and the LPT is there to drive the fan. Most of the LPT power is used to pressurise the bypass stream but some is used to pressurise the core flow through the fan root and booster. The temperature rise of this core flow through the LP system is $T_{023} - T_{02}$, given in part (a). Find the temperature and pressure at a hypothetical station in the LP turbine when the power has been removed for the core flow and the remaining power is available for the fan to raise the pressure of the bypass flow. (**Ans:** 951.9 K, 263.6 kPa)

d Explain why the pressure ratio across the HP turbine is less than the pressure ratio across the corresponding HP compressor. Then show (but do not work out the numbers) how the difference will increase as either turbine entry temperature is increased or compressor entry temperature is reduced.

Note: The answers to Exercise 5.1c give the entry conditions into the part of the LP turbine for which power is notionally used to drive the bypass section of the fan. These will be used in Exercises 7.1 and 7.2.

5.3 TURBINE INLET TEMPERATURE

Turbine inlet temperature T_4 is important because increasing its value makes the pressure ratio across the core turbine smaller in relation to the pressure rise of the core compressor and thereby increases the power available from the LP turbine. Increasing T_4 also increases the thermal efficiency, provided that the pressure ratio increases by an appropriate amount. These trends were shown in the idealised treatment of the gas turbine cycle presented earlier in sections 4.2 and 4.3.

In Exercise 5.1 the temperature entering the turbine was given as 1500 K, a plausible value for cruise operation for long periods of time. For take-off, which is only expected to last for a short time, the temperature into the turbine is higher, and 1750 K is a representative value for the same engine. If a temperature as high as this were to continue for the long periods typical of cruise, additional amounts of turbine cooling would be needed. The temperature of the gas stream entering the turbine is significantly higher than the melting temperature of the material from which the blades are cast, which is typically about 1500 K.

It will be recalled from section 4.3 that it was the *ratio* of the turbine inlet temperature to the compressor inlet temperature T_4/T_2 which is crucial to engine performance. Reducing the T_2, as occurs in going to high altitude, has a similar effect to increasing T_4 on the ground. The temperature ratios at take-off, top-of-climb and cruise, for example, are less different than might be imagined, as the example in Table 5.1 from a representative high-bypass engine shows. It is apparent from this table that it is at top-of-climb when the temperature ratio is greatest; at this condition the engine is working hardest in a non-dimensional sense. This is an issue discussed further in Chapter 8. It can also be seen that the low engine inlet temperature at cruise means that T_4/T_2 is almost equal to that at the take-off on a standard day, even though the turbine inlet temperature is 150 K lower. The values in this table will be taken as appropriate for our design.

Table 5.1   Representative compressor and turbine inlet temperatures

	Engine inlet T_2	Turbine inlet T_4	T_4/T_2
Take-off (standard day, sea level)	288.15 K	1750 K	6.07
Top-of-climb (35,000 ft, $M = 0.78$)	245.4 K	1600 K	6.52
Start of cruise (35,000 ft, $M = 0.78$)	245.4 K	1500 K	6.11

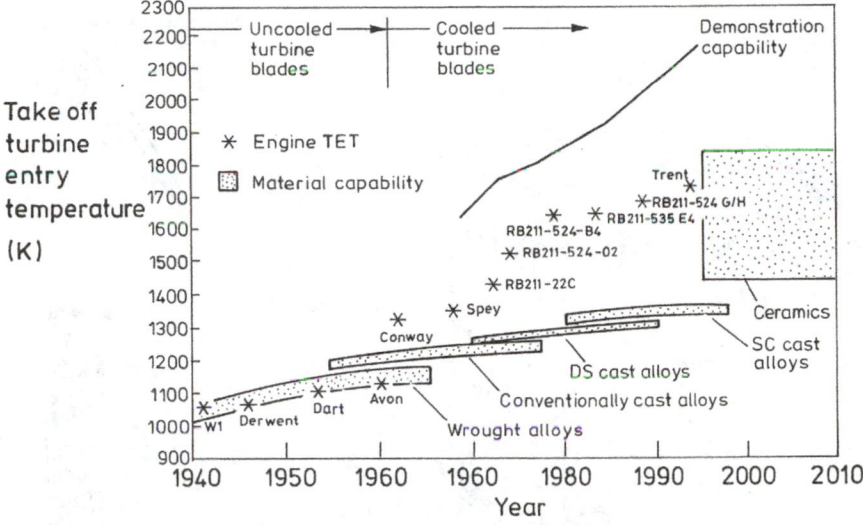

Figure 5.5. Turbine entry temperature[1] for Rolls-Royce engines since 1940: a figure drawn in about 1993.

The ability to operate at higher temperatures has been crucial in improving the performance of jet engines. The higher temperatures are possible in part because of better materials. More important, however, has been the use from the early 1960s of compressor air to cool the blades – it is the improvements in the way that cooling air is used that have brought the biggest gains.

Figure 5.5 shows how the turbine inlet temperature used by one company, Rolls-Royce, has increased over more than 50 years. Since the mid-1950s there has been an average increase of about 7 K rise per year in turbine entry temperature. The rise in gas temperature shown in Figure 5.5 is therefore all the more remarkable when it is appreciated that the turbines of newer engines are now operating in service for periods in excess of 20,000 hours. The life of hot blades is primarily limited by creep, by oxidation or by thermal fatigue. Creep is the continuing and gradual extension of materials under stress at high temperature. A rule of thumb for blades limited by creep is that turbine blade life is halved (at a given level of cooling technology and material) for each 10 K rise in temperature of the metal. Oxidation is also strongly affected by temperature. Thermal fatigue is not a function of the length of time the engine runs but how many operating cycles it goes through, in other words how many times the engine is started, accelerated and stopped with the attendant raising and lowering of the turbine temperature.

Originally the blades for high temperature use were forged, but better creep performance could be obtained when the blades were cast. (In addition the cooling passages can be cast inside the blade.) Then it was found that a better blade could be made by arranging for the crystals to form

---

[1] For cooled turbines the temperature needs to be defined carefully. What is shown here (and used throughout the book) is the stator outlet temperature (SOT). This is the temperature after a hypothetical complete mixing of the cooling air with the hot gas downstream of the first turbine stator row (the HP nozzle guide vane).

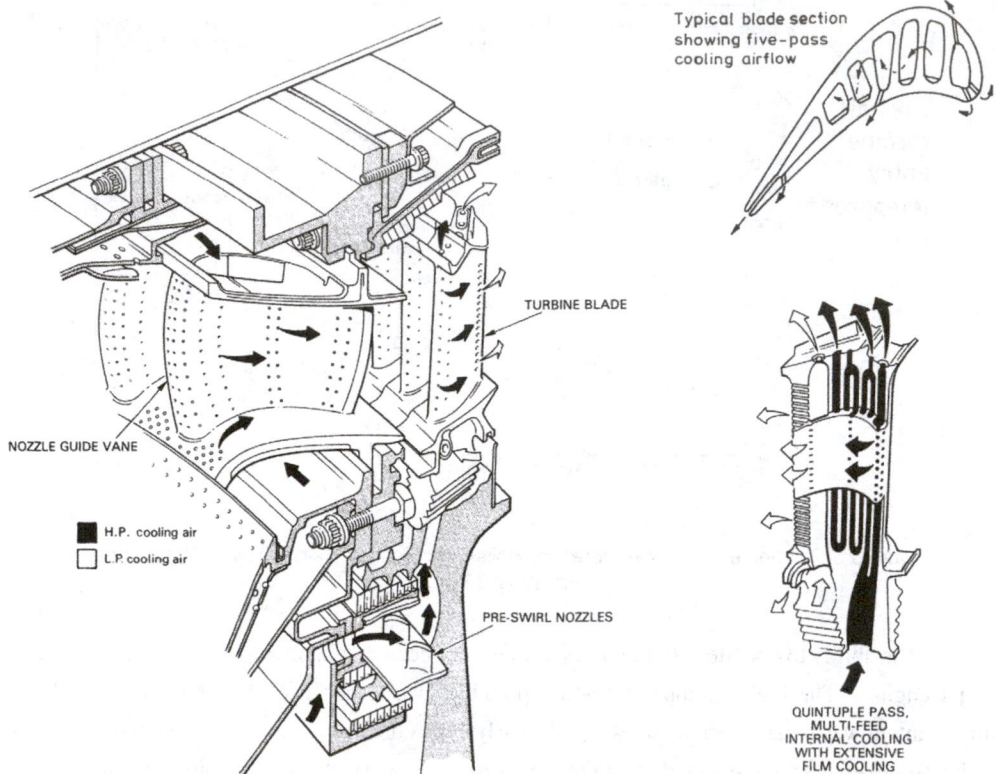

Figure 5.6. Arrangements for cooling an HP turbine and rotor blade. It should be noted that the rotor blade configuration is unique to Rolls-Royce, with a shroud on the tip. (From *The Jet Engine*, 1986)

elongated in the direction of the span, so-called directionally solidified blades. A still better blade was obtained by casting each blade as a single crystal, and this is now the norm for the HP blades.

As already noted, it is only possible for turbine blades to operate at such high temperatures because relatively cool air (at a temperature in excess of 900 K at take-off on a hot day) is taken from the compressor and fed to the inside of the turbine blades. Figure 5.5 shows the effect of cooling to raise gas temperature above allowable metal temperature and Figure 5.6 shows how this is achieved. Inside the blade there are complicated passages, with roughness elements to encourage heat transfer to the cooling air, which have the effect of lowering the metal temperature. More modern blades have even more complicated internal configurations to increase the heat transfer and to achieve more gradual variations in metal temperature.

The air which has passed inside the blade then emerges from many small holes in the surface. These are positioned so that a relatively cool film of air exists around the blade outer surface, thereby shielding the surface from the very hot gas. This process is known as film cooling. Between 15 and 25% of the core air may be taken from the compressor and used in this way; for the same level of technology, the amount of air used for cooling must rise as turbine inlet temperature is increased.

To some extent choice of turbine inlet temperature is a balance between engine performance and turbine life. For engine performance there is also a balance between turbine inlet temperature and cooling air requirements, since cooling air reduces the efficiency and thrust of an engine. Turbine cooling is an expensive technology to develop and, since raising the temperature can have such a big effect on engine performance (as demonstrated in Exercise 5.1), turbine cooling technology is one of the areas where competition and product differentiation are most intense.

One way of assessing the performance of blade cooling is to use the cooling effectiveness defined by

$$\varepsilon = \frac{T_g - T_m}{T_g - T_c}$$

where T_g is the temperature[2] of the hot gas stream, T_m is the temperature of the metal and T_c is the temperature of the cooling air. The level of effectiveness is determined by the sophistication of the cooling technology, but it is also increased when the amount of cooling air is increased. A value of between 0.6 and 0.7 is currently 'state of the art'. The expression for effectiveness also brings out the serious consequences of a rise in cooling air temperature, which is normally close to compressor delivery temperature. If the compressor efficiency turns out to be lower than expected the cooling air will be at a higher temperature. If the effectiveness is unchanged, the metal temperature will be raised, with potentially serious effects on the life of the turbine. In this case the effectiveness may have to be increased by use of more cooling air, but this has detrimental effects on the engine thermal efficiency, fuel consumption and thrust.

In Figure 5.5 a region is shown for ceramics. Ceramics appear very attractive for turbines, offering very high temperature performance as well as reduced density. Unfortunately the date of application to turbine blades tends to move to the right by about 12 months every year and Figure 5.5, which was produced about 1993, looks far too optimistic. Fundamentally the problem is that ceramics are vulnerable to defects and lack the safe ductile characteristics of metals. Some of the advantages of ceramics have been achieved by using thermal barrier coatings (TBC) on top of metal blades. The TBC has low conductivity and this reduces the temperature of the metal in internally cooled blades as well as providing a barrier against oxidation. For the same level of cooling flow and cooling technology the use of TBC has allowed an increase in turbine entry temperature of about 100 K. Although ceramics are still a long way off for use as blades or vanes, they are being used for seal segments around the annulus, allowing the cooling air to the endwalls to be reduced, and are expected to be used for vanes (i.e. not for rotating blades) in the mid-2020s.

In discussing the metals required for jet engines attention is normally focused on the vanes and blades for the HP turbine which are exposed to the hottest part of the combustion products. The highest stresses, however, are created in the rotating discs which carry the blades and these are an issue for both the turbine and compressor. The lowest temperature air which can be used to cool

---

[2] The temperature of the hot gas T_g and the temperature of the cooling air T_c are both *stagnation* temperatures, as are all the fluid temperatures used in this chapter. Stagnation temperature is discussed in Chapter 6.

HP turbine and compressor discs is approximately the compressor delivery temperature which rises with the engine overall pressure ratio. The *opr* is therefore set by the high-temperature capability of the metal from which the disks are made, though higher pressure ratios can be accepted if the efficiency of the compressor is raised. Until relatively recently, the 1990s, many compressor disks were made of titanium alloy and that limited compressor deliver temperature to around 870 K. More recently nickel based alloys have been used for compressor disks, or for the disks at the high-pressure end of the compressor, permitting air temperatures at compressor outlet up to about 950 K. This allows higher overall pressure ratios or take-off with full pressure ratio when ambient temperature is above the ISA value of 288 K. (If the compressor cannot tolerate the temperature for a hot-day take-off the engine must be de-rated to a lower value of thrust.) There is clearly a balance between allowable stress, weight and temperature, since nickel alloys are heavier than titanium; where the balance is struck must depend on the application and the relative desirability of low weight and of low fuel consumption.

Exercise

5.2 a Assuming a cooling effectiveness of 0.65 calculate the metal temperature of the HP turbine rotor when the temperature of the gas sensed by the rotor $T_g = 1650$ K and $T_c = T_{03}$ is compressor delivery temperature. The overall pressure ratio is 45 and isentropic efficiency of the booster and the HP compressor are both 0.90. The inlet temperature on the sea-level test bed is $T_{02} = 288$ K.   (**Ans:** $T_m = 1183$ K)

   b If the compressor efficiencies are reduced to 0.85 but the other quantities, including cooling effectiveness, are unchanged, find the turbine metal temperature. Estimate the reduction in creep life resulting from the reduced compressor efficiency if life is halved for each 10 K rise in temperature.
                              (**Ans:** $T_m = 1213$ K, creep life reduced by factor 8)

   c If the metal temperature were to be kept constant despite the reduction in compressor efficiency, what would the temperature of the gas T_g have to be reduced to?   (**Ans:** $T_g = 1564$ K)

Note: Assuming the effectiveness stays constant, 1 K change in cooling air temperature has a greater effect on metal temperature than 1 K change in temperature of the gas from the combustor into the turbine. A decrease in compressor efficiency would reduce net power from the engine core as well as raising the cooling air temperature. If the turbine entry temperature T_g were reduced to restore the turbine metal temperature there would be a further reduction in power.

SUMMARY OF CHAPTER 5

To make efficient use of high temperature ratios and pressure ratios in the core of the engine a bypass stream is normally used. Recent designs of subsonic civil aircraft engines normally have bypass ratios of 9 or more. The fans on the front normally have supersonic speeds near the tip, but subsonic flow near the hub. The fan is driven by the LP turbine, with a separate LP shaft passing through the core of the engine. The core may have one shaft, to give a two-shaft engine, or the core may have two shafts to give the three-shaft engine favoured by Rolls-Royce for its large engines. For simplicity, to reduce the number of calculations but with no assumption of superiority, this book is based on the two-shaft model. An eager student can easily redo the exercises with three shafts where the core pressure rise is in separate intermediate-pressure (IP) compressor and high-pressure (HP) compressors with the core expansion in separate HP and IP turbines. The two-shaft engine has a low-pressure compressor on the same shaft as the fan and this LP compressor is often referred to as the booster but the three-shaft engine does not need a booster.

The temperature of the gas entering the turbine is as high as the metal and the cooling arrangements will allow. At most operating conditions it is close to or above the melting temperature of the turbine blade material. During cruise the turbine entry temperature is typically about 250 K lower than at take-off; this is desirable to prolong the life of the turbine but it also keeps the non-dimensional turbine inlet temperature T_4/T_2 nearly constant. The highest temperature ratio is encountered at top-of-climb and at this condition the non-dimensional variables in the engine, such as pressure ratio and non-dimensional rotational speed, will be greatest.

The overall pressure ratios now used in engines are sufficiently high that the temperature of the gas leaving the compressor is near to the limit possible with current materials. For the design condition, with initial cruise at $M = 0.78$ and 35,000 ft, the turbine inlet temperature is chosen to be 1500 K with an overall pressure ratio of 45. For a two-shaft engine the overall pressure ratio may be plausibly divided into 2.5 for the core flow through the fan and LP compressor (booster), and 18 in the HP compressor. A pressure ratio of 45 for cruise can be expected to give about the same pressure ratio for take-off and a ratio of about 50 at maximum climb.

There are aspects of the engine which require some understanding of the way gases flow at high speed. The velocities inside the engine are typically near to the speed of sound and at such speeds the pressure *changes* are a substantial fraction of the *absolute* pressure. These pressure changes bring significant variations in density and the flow is said to be compressible. Compressible fluid flow is the subject of the next chapter.

CHAPTER 6 ELEMENTARY FLUID MECHANICS OF COMPRESSIBLE GASES

6.0 INTRODUCTION

In treating the gas turbine it is essential to make proper acknowledgement of the compressible nature of the air and combustion products. Compressible fluid mechanics is a large and highly developed subject, but here only that which is essential to appreciate the treatment and carry out the designs is given. There are also special approaches for handling the compressible, high-speed flow inside ducts which need to be introduced, and that is the purpose of this chapter. The most important book dealing with this topic is Shapiro (1953), but a more accessible account is given, for example, by Munson et al. (2009).

6.1 INCOMPRESSIBLE AND COMPRESSIBLE FLOW

For **liquids** the changes in density are normally negligible and it is possible to treat the flow as *incompressible*. Thus the equation for steady frictionless flow along a streamline,

$$V dV + dp/\rho = 0,$$

can be integrated directly, assuming the density is constant, to give Bernoulli's equation

$$\tfrac{1}{2}V^2 + p/\rho = p_0/\rho, \quad \text{a constant.}$$

p_0 is the *stagnation or total pressure* and corresponds to that pressure obtained when the flow is brought to rest in a frictionless or loss-free manner. The term $\tfrac{1}{2}\rho V^2$ is known as the *dynamic pressure* or *dynamic head*. A pitot tube records the stagnation pressure whereas a pressure tapping in a wall parallel to the flow records static pressure. The use of stagnation pressure is something like a book-keeping exercise – it indicates the pressure which would be achieved if the flow were decelerated to rest is a loss-free manner. The stagnation pressure also represents the pressure in a reservoir from which the fluid could be accelerated to velocity V_j and this is illustrated in Figure 6.1. The difference between stagnation pressure and static pressure is the dynamic pressure $\tfrac{1}{2}\rho V^2$. An analogy which is sometimes helpful can be drawn between the hydraulic system and a mechanical system: static pressure is analogous to potential energy and dynamic pressure is analogous to kinetic energy.

In dealing with gases at low speed (more precisely, at low Mach number) the density changes little and it is possible to use Bernoulli's equation as a reasonable approximation. The

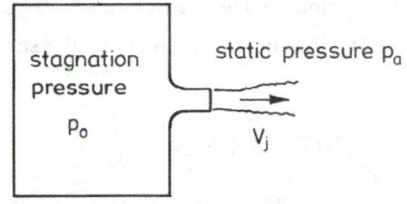

Equivalent to

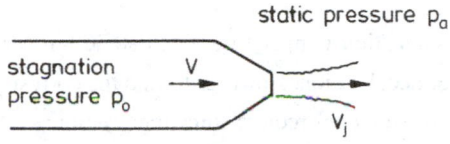

For INCOMPRESSIBLE FLOW $V_j{}^2 = (p_0 - p_a)/\frac{1}{2}\rho$

For COMPRESSIBLE FLOW $M_j{}^2 = \dfrac{2}{\gamma - 1}\,[(p_0/p_a)^{(\gamma-1)/\gamma} - 1]$

Figure 6.1. Schematic representation of stagnation conditions.

inaccuracy becomes significant when the dynamic pressure becomes a substantial fraction of the absolute pressure of the gas; this occurs when the Mach number exceeds about 0.3. For most of the processes inside jet engines the Mach number is nearer 1.0 than 0.3 and Bernoulli's equation is quite inappropriate and *must not be used*. (This is also true for the external aerodynamics of the aircraft itself.) To deal with compressible flow we have to start from a different approach.

6.2 STATIC AND STAGNATION CONDITIONS

The steady flow energy equation for the flow of gas, with no restriction to incompressibility, with no heat transfer and no external work transfer can be written

$$h_1 + V_1^2/2 = h_2 + V_2^2/2,$$

where h is the enthalpy. Each side of the equation can be rewritten more conveniently as

$$c_p T + V^2/2 = c_p T_0, \tag{6.1}$$

where T_0 is the *stagnation* temperature, being that temperature which the gas would attain if brought to rest without work and heat transfer. (The deceleration to get T_0 is adiabatic but not necessarily in an ideal or loss-free manner.) The specific heat can be rewritten $c_p = \gamma R/(\gamma - 1)$ and on rearranging and inserting this into the equation above the stagnation temperature may be written

$$T_0/T = 1 + \frac{\gamma - 1}{2}\frac{V^2}{\gamma RT}. \qquad (6.2)$$

The speed of sound is given by $a = \sqrt{\gamma RT}$ and the Mach number is given by $M = V/a$, so the equation for the stagnation temperature becomes

$$T_0/T = 1 + \frac{\gamma - 1}{2}M^2. \qquad (6.3)$$

Again, this equation relating stagnation temperature T_0 to static temperature T does not imply ideal or loss-free acceleration or deceleration. However to find the corresponding relation between stagnation pressure and static pressure does require some idealisation.

At this point we suppose that the acceleration or deceleration of the gas between the static state p and T and that at stagnation p_0 and T_0 is reversible and adiabatic (i.e. *isentropic*), for which we know that

$$p/\rho^\gamma = \text{constant} \quad \text{and} \quad p/T^{\gamma/(\gamma-1)} = \text{constant}. \qquad (6.4)$$

It then follows immediately that $p_0/p = (T_0/T)^{\gamma/(\gamma-1)}$

$$\text{or} \quad p_0/p = \left(1 + \frac{\gamma - 1}{2}M^2\right)^{\gamma/(\gamma-1)}. \qquad (6.5)$$

In incompressible flow it was often possible to use gauge pressures, but in all work with compressible flow it is important to remember to use the *absolute* pressures and temperatures.

It should be explained that until this section in the book a distinction between static and stagnation temperatures has not been drawn, or the description of conditions has been made in such a way that it was not necessary to distinguish between them. From now on we will be more careful and explicit. In calculating such quantities as the speed of sound or the density of the air through which the aircraft is travelling it is the *static* quantities which are relevant. In carrying out the cycle analyses, for example finding the work or the heat transfer in a gas turbine, it is the *stagnation* quantities which should be used. Returning to the book-keeping analogy, the use of stagnation properties like stagnation temperature or pressure gives a measure of the total amount of energy or capacity to accelerate the flow which is available. Stagnation properties are not the 'real' in the sense that one cannot measure them directly, but what one can measure are the static properties. In places where the velocity has been brought to zero, such as on the leading edge of a wing, in a pitot tube or in a large plenum, the static properties are equal to the stagnation ones and we colloquially say that we are measuring the stagnation quantities. The gas properties (density, speed of sound c_p), chemical reaction rates and thermal radiation all depend on the local *static* properties.

Stagnation pressure and temperature change with the frame of reference. For a stationary atmosphere (i.e. one with no wind, $V = 0$) the static and stagnation properties are equal, but an observer in an aircraft travelling through the stationary atmosphere would not see the static and stagnation properties as equal. The observer travelling at velocity V, Mach number M through an atmosphere with ambient (i.e. static) temperature T and pressure p would perceive the stagnation temperature given by

$$T_0 = T + V^2/2c_p$$
$$= T\left\{1 + \frac{\gamma - 1}{2}M^2\right\}, \tag{6.6}$$

and the stagnation pressure, calculable via the isentropic relation, as

$$p_0/p = (T_0/T)^{\gamma/(\gamma-1)}. \tag{6.7}$$

For flight at a Mach number of 0.78, for example, the engine inlet stagnation temperature T_{02} and pressure p_{02} are related to the ambient conditions by Equation 6.6 and 6.7, so that $T_{02}/T_a = 1.122$ and $p_{02}/p_a = 1.495$. The engine performance depends on these stagnation quantities. The flow at nozzle exit, however, responds to the ambient static pressure, p_a, and as a result of the forward motion the pressure ratio across the engine is increased by 1.495 for $M = 0.78$.

6.3 THE NOZZLE

For adiabatic flow the steady flow energy equation, Equation 6.1, leads to

$$V^2 = 2c_p (T_0 - T)$$
$$\text{or} \quad V = \sqrt{2c_p (T_0 - T)}, \tag{6.8}$$

where T is the local temperature of the air and T_0 denotes the stagnation temperature, which is uniform throughout the flow. Assuming no heat transfer, adiabatic flow, is a good approximation for most nozzles. Because the flow is accelerating it is also generally a good approximation to treat the flow as loss-free, that is reversible. The combination of adiabatic and reversible makes the flow isentropic so that

$$T/p^{(\gamma-1)}/\gamma = T_0/p_0^{(\gamma-1)/\gamma} = \text{constant}.$$

It is common with nozzles to have the inlet flow specified in terms of inlet stagnation pressure and temperature, p_0 and T_0, and the exit static pressure p_a. The exit static temperature is therefore given by

$$T = T_0(p_a/p_0)^{(\gamma-1)/\gamma}$$

and nozzle exit velocity can be written

$$V = \sqrt{2c_p T_0 \left(1 - (p_a/p_0)^{\frac{\gamma-1}{\gamma}}\right)}. \tag{6.9}$$

Exercises

6.1 Expand the expression for p_0/p in Equation 6.5 using the binomial expansion to express the relation in the form

$$p_0 - p = \tfrac{1}{2}\rho V^2 \{1 + a\,M^2 + b\,M^4 + \cdots\}.$$

What is the highest Mach number at which the *incompressible* expression for stagnation pressure (i.e. Bernoulli's equation) can be used if the error is not to exceed 1%? What is the error in calculating the stagnation pressure by Bernoulli's equation at $M = 0.3$?   (**Ans:** 0.20, 2.25%)

6.2* If the aircraft cruises at Mach number of 0.78 at an altitude of 35,000 ft find the stagnation temperature and pressure of the flow perceived by the aircraft. (These values were assumed in Exercise 5.1.)
                    (**Ans:** 245.4 K, 35.57 kPa)

6.3 a An aircraft flies at Mach 2 at 51,000 feet ($p_a = 11.0$ kPa, $T_a = 216.7$ K) propelled by a simple turbojet engine (i.e. no bypass). If the inlet is effectively isentropic find the stagnation temperature and pressure into the compressor. The engine compressor has a pressure ratio of 10 with an isentropic efficiency of 90%: find the stagnation temperature and pressure at compressor exit.   (**Ans:** 793.3 K; 0.861 MPa)

b In the combustor the velocities are low (so the stagnation and static pressures are equal) but the absolute stagnation pressure falls by 5%. At turbine entry the stagnation temperature is 1400 K and the turbine has an efficiency of 90%. Find the stagnation temperature and pressure downstream of the turbine.
                    (**Ans:** 996.7 K; 0.212 MPa)

c If the final propulsive nozzle is isentropic, find the velocity of the jet assuming that the expansion is to the static pressure after the nozzle, which is equal to that of the surrounding atmosphere. Calculate the gross and net thrust per unit mass flow, the propulsive efficiency and, from the temperature rise in the combustor, the overall efficiency.
          (**Ans:** $V_j = 1069$ m/s; $F_G = 1069$ N kg^{-1}s^{-1}; $F_N = 479$ N kg^{-1}s^{-1}; $\eta_p = 0.711$; $\eta_o = 0.464$)

6.4 THE CHOKED NOZZLE

A concept peculiar to compressible flow must be addressed at this stage, the maximum flow which can pass through a given cross-sectional area. Figure 6.1 shows flow from a reservoir or large volume entering a nozzle for which the minimum area occurs at the throat. We will assume here that the flow is one-dimensional (in other words, the flow is uniform in the direction perpendicular to the streamlines) and that it is adiabatic (no heat flux to or from the air). The mass flow rate per unit area of the throat is given by

$$\dot{m}/A = \rho V \qquad (6.10)$$

and in steady flow the maximum value of $\dot{m}/A$ evidently occurs at the throat. For adiabatic flow the steady flow energy equation, is given by Equation 6.8,

$$V = \sqrt{2c_p\,(T_0 - T)},$$

where T is the local temperature of the air and T_0 denotes the stagnation temperature, which is uniform throughout the flow. To consider mass flow we need density and for isentropic flow

$$T/\rho^{(\gamma-1)} = T_0/\rho_0^{(\gamma-1)} = \text{constant}$$

$$\text{or} \quad \rho = CT^{1/(\gamma-1)} \tag{6.11}$$

$$\text{where} \quad C = 1/\left(T_0/\rho_0^{(\gamma-1)}\right) = \text{constant.}$$

The mass flow rate per unit cross-sectional area of the throat can then be written

$$\dot{m}/A = CT^{1/(\gamma-1)} \sqrt{2c_p\left(T_0 - T\right)}$$
$$= C\sqrt{2c_p\left(T_0 T^{2/(\gamma-1)} - T^{(\gamma+1)/(\gamma-1)}\right)}. \tag{6.12}$$

The maximum mass flow rate per unit area can be found by differentiating $\dot{m}/A$ with respect to T, noting that the stagnation temperature T_0 is a constant. (It is actually easier, of course, to square $\dot{m}/A$ before differentiating – the turning point will be the same.) When this is done the maximum occurs when

$$T = T_0 \frac{2}{\gamma+1}. \tag{6.13}$$

Substituting this into Equation 6.8 yields

$$V = \sqrt{2c_p T \left(\frac{\gamma+1}{2} - 1\right)}$$

which simplifies to

$$V = \sqrt{c_p T \left(\gamma - 1\right)} = \sqrt{\gamma R T} = a, \tag{6.14}$$

the speed of sound.

In other words when the mass flow rate through an orifice or nozzle is at its maximum, the velocity in the throat is sonic. At this condition the nozzle or orifice is said to be choked. A maximum occurs in the mass flow rate per unit area because the rate of density decrease just matches the rate of velocity increase.

The ratio of inlet stagnation temperature to the static temperature at a choked throat is given in Equation 6.13. Because the flow is treated as isentropic the corresponding pressure ratio is

$$\frac{p}{p_0} = \left(\frac{2}{\gamma+1}\right)^{\frac{\gamma}{\gamma-1}}.$$

If the pressure ratio exceeds this value the nozzle will be choked. The value of the pressure ratio for choking is given by 1.893 for $\gamma = 1.4$ and 1.832 for $\gamma = 1.30$; these are also, of course, the pressure ratios which correspond to acceleration from stationary to sonic velocity.

If the pressure ratio across the nozzle is larger than that to produce $M = 1$ there is acceleration downstream of the throat. The rate of fall in density exceeds the rate of increase in velocity and it is necessary for the flow area to increase in the flow direction for there to be further acceleration. In other words, to accelerate a flow to supersonic velocity it is first necessary to reduce the area to a throat and then increase the area downstream. Such nozzles are familiar at the back of rockets used to launch satellites, missiles etc. and are referred to as a convergent–divergent (often abbreviated to con–di) nozzles or sometimes as Laval nozzles. If the nozzle is truncated at the throat, a normal convergent nozzle, acceleration to supersonic velocity can take place if the nozzle is choked. The area of the streamtube made up of the jet increases downstream of the nozzle. For a turbojet engine or low bypass ratio turbofan the final propulsive nozzles will be choked at most conditions of interest. For a modern high bypass engine the bypass nozzle will be choked at cruise but not at take-off, whilst the core nozzle will be un-choked at all conditions. Although the bypass nozzle of the high bypass engine is choked at cruise, where fuel efficiency is most important, it is not normally worthwhile fitting a con–di nozzle on subsonic civil aircraft. This is because at the pressure ratios involved the acceleration to supersonic is relatively efficient without a divergent section, as discussed in Chapter 11. For a high-speed military aircraft, however, the jet Mach number may be sufficiently high that a con–di nozzle is installed. Such con–di nozzles are made so they can be varied in shape and throat area to match the operating condition.

6.5 NORMALISED MASS FLOW PER UNIT AREA

Equation 6.12 for $\dot{m}/A$ can be expressed in terms of the local flow Mach number M and after some algebraic manipulation, the mass flow rate per unit area is given in non-dimensional form by the expression

$$\frac{\dot{m}\sqrt{c_p T_0}}{A p_0} = M \frac{\gamma}{\sqrt{\gamma-1}} \left(1 + \frac{\gamma-1}{2} M^2\right)^{-(\gamma+1)/2(\gamma-1)}. \tag{6.15}$$

This function is so widely used that it is convenient to denote it by the symbol $\overline{m}$ so that

$$\overline{m} = \frac{\dot{m}\sqrt{c_p T_0}}{A p_0}. \tag{6.16}$$

For a given gas (i.e. a given value of the ratio of specific heats) $\overline{m}$ can be expressed as a function of Mach number only. For a given Mach number mass flow per unit area, $\dot{m}/A$, is therefore proportional to stagnation pressure and the square root of stagnation temperature. The variation in $\overline{m}$ with Mach number for air ($\gamma = 1.40$) is shown in Figure 6.2. At the throat, when $M = 1$,

$$\overline{m}(M=1, \gamma) = \frac{\gamma}{\sqrt{\gamma-1}} \left(\frac{\gamma+1}{2}\right)^{-(\gamma+1)/2(\gamma-1)}. \tag{6.17}$$

Thus at a choked throat

$$\overline{m} = 1.281 \quad \text{for } \gamma = 1.4$$
$$\text{and} \quad \overline{m} = 1.389, \quad \text{for } \gamma = 1.3.$$

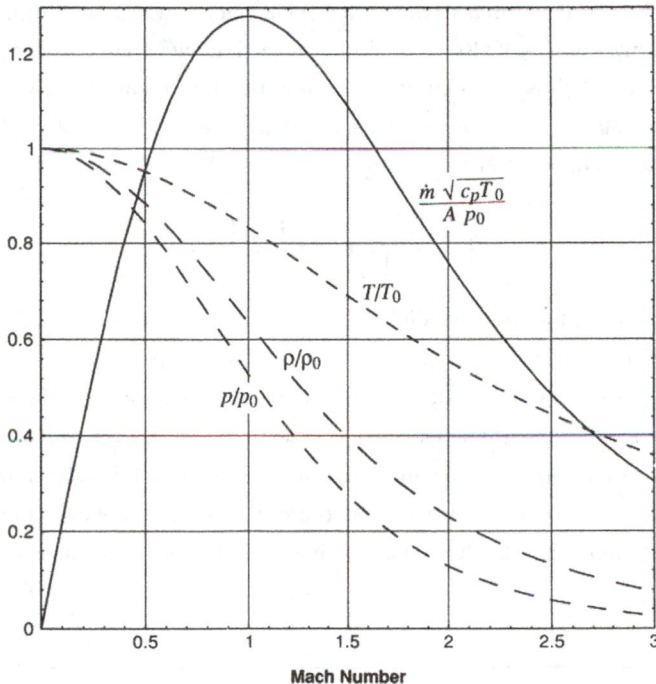

Figure 6.2. One-dimensional flow of a perfect gas, $\gamma = 1.4$.

Equation 6.17 shows very clearly that the mass flow through a choked nozzle of given area is proportional to the stagnation pressure of the flow and inversely proportional to the square root of the stagnation temperature. The *only* way to alter the mass flow through a choked nozzle of given size is to vary either the stagnation pressure or stagnation temperature. *Varying conditions downstream of the nozzle has no effect on mass flow rate or on conditions upstream of the throat.* This can be understood by realising that pressure disturbances travel at the speed of sound and, if the flow is supersonic, information from downstream in the form of pressure waves, cannot travel upstream past the throat to alter the incoming flow.

The non-dimensional mass flow per unit area $\overline{m}$ was expressed for fixed gas properties as a unique function of Mach number in Equation 6.15. This equation may be re-arranged (see Exercise 6.5) so that $\overline{m}$ is expressed in terms of the pressure ratio and ratio of specific heats if the process is effectively isentropic. If the stagnation pressure is p_0 and p is the static pressure somewhere along the nozzle (or turbine blade row) the local value of $\overline{m}$ can be written as

$$\frac{\dot{m}\sqrt{c_p T_0}}{A p_0} = \overline{m}\left(\frac{p}{p_{0in}}, \gamma\right) = \frac{\gamma}{\gamma - 1}\sqrt{2\left[\left(\frac{p}{p_{0in}}\right)^{\frac{2}{\gamma}} - \left(\frac{p}{p_{0in}}\right)^{\frac{(\gamma+1)}{\gamma}}\right]}. \tag{6.18}$$

If the Mach number were unity the static pressure could be that at the exit plane of a convergent nozzle, where the area is a minimum; for a convergent-divergent nozzle it would be at the throat;

or it could be at the minimum area section of a turbine blade row. The non-dimensional form of Equation 6.18 for mass flow will also be useful in connection with engine performance.

Figure 6.2 shows three other curves apart from that for $\bar{m}$. One curve is the ratio of static temperature to stagnation temperature as function of Mach number. This is for adiabatic flow, and the flow is not required to be isentropic,

$$T/T_0 = \left\{ 1 + \frac{\gamma - 1}{2} M^2 \right\}^{-1}.$$

The other two curves give the corresponding ratios of local static to stagnation pressure and density for isentropic flow along the nozzle, $p/p_0 = (T/T_0)^{\gamma/(\gamma-1)}$ and $\rho/\rho_0 = (T/T_0)^{1/(\gamma-1)}$ respectively.

For a final propulsive nozzle it is easy to appreciate that with a sufficient pressure ratio between upstream and the downstream atmosphere the nozzle will choke and, in particular, downstream conditions will not affect the incoming velocity or mass flow. It will be shown in Chapter 11 that for turbines it is also a good approximation to take $\bar{m}$ to be constant and equal to the choking value at the inlet to turbines. This has important implications for the off-design matching of gas turbines.

Exercises

6.4 Verify that $\dot{m}\sqrt{c_p T_0}/A p_0$ is non-dimensional. If for a choked con-di nozzle the area of the throat is equal to 1 m^2, find the areas of the nozzle at sections where the Mach numbers are 0.5, 0.9 and 2.0. Treat the flow as isentropic, i.e. $p_0 = $ constant, as well as adiabatic, i.e. $T_0 = $ constant.

(**Ans:** 1.340 m^2; 1.009 m^2; 1.688 m^2)

Note: The difference in area for $M = 1.00$ and $M = 0.90$ is very small indeed. Close to sonic velocity very small area changes produce large variations in Mach number and pressure.

6.5 Beginning with the form for $\bar{m} = \dot{m}\sqrt{c_p T_0}/A p_0$ given in Equation 6.15, derive Equation 6.18 which gives $\bar{m}$ in terms of the nozzle pressure ratio p_{0in}/p (inlet stagnation pressure divided by local static pressure).

6.6 For the simple turbojet engine of Exercise 6.3 (no bypass stream) find the area of the throat of the propulsive nozzle per unit mass flow through the engine. There is a divergent section downstream of the throat – find the final area if the expansion is isentropic to the static pressure of the surrounding atmosphere.

(**Ans:** 0.00369 m^2 kg^{-1}s; 0.010 m^2 kg^{-1}s)

6.6 THE FAN, THE BYPASS STREAM AND THE PROPULSIVE NOZZLE

In this section we slightly run ahead of ourselves because we look at some aspects of engine performance, particularly off-design performance. All we need are the ideas on thrust creation given in Chapter 3 and the elementary compressible fluid mechanics introduced in the present chapter. For modern high bypass engines around 90% of the net thrust comes from the bypass stream. Moreover the variation in thrust from the core stream with operating condition is in the same sense as the variation in thrust from the bypass stream: consequently proportional changes in thrust for the whole engine and for the bypass component are similar. As a result it is possible to get some insight into engine behaviour by considering only the bypass stream. The following exercises look at this and show how net thrust alters with flight conditions, including differences between take-off conditions and cruise and between cruise and climb. It is assumed that there are no losses in the engine intake, in the bypass duct or in the nozzle. For simplicity the fan adiabatic efficiency is taken to be constant at 90%, but modern fans would probably be somewhat better than this.

Take-off here is assumed to occur at $M = 0.25$, which for sea-level on an ISA day is 85 m/s. This condition is used to characterise the engine rather than the thrust when the aircraft is stationary because when stationary the intake does not behave as well as it does in forward motion. Thrust capability of an engine, however, is always quoted at sea-level static conditions (i.e. no forward motion) and this static value is obtained by a correction factor: the Boeing Equivalent Thrust is 1.255 times the thrust generated at sea-level flying at $M = 0.25$ and the Airbus value 1.25.

When the pressure ratio across the nozzle is larger than the choking value given by $p_0/p = 1.893$ the value of $\overline{m}$ is fixed and is 1.281 for air. For pressure ratios smaller than this the value of $\overline{m}$ is a function of the pressure ratio and may be determined by, for example, Figure 6.2 or Equation 6.18. Once $\overline{m}$ is known the mass flow rate per unit area, $\dot{m}/A$, can be found using p_0 and T_0 of the flow into the nozzle.

Exercises (The use of a spreadsheet or programme is recommended for 6.9 to 6.12)

6.7 A high bypass engine is 'throttled back' whilst the aircraft continues to fly at Mach 0.78 so the pressure ratio across the fan is reduced. Neglecting any losses in stagnation pressure upstream and downstream of the fan, calculate the stagnation pressure ratio across the fan at which the bypass nozzle will just un-choke.
(Ans: 1.267)

6.8 In Exercise 3.3 the effect of reducing cruise Mach number from 0.85 to 0.78 on *sfc* and overall efficiency was estimated. Considering only the bypass stream and assuming that the fan pressure ratio remains constant and large enough to choke the nozzle, find the reduction of mass flow of air attributable to the reduction in cruise speed.
(Ans: 0.941)

6.9 Consider an engine with a fan pressure ratio p_{013}/p_{02}, equal to 1.50 (such as might be on the NEA) and assume this pressure ratio is held constant. Take the isentropic efficiency of the fan to be 0.90. For cruise at $M = 0.78$ at 35,000 ft calculate the stagnation temperature and pressure into the bypass nozzle, T_{013} and p_{013}, neglecting losses in the bypass duct. Hence find the pressure ratio across the nozzle p_{013}/p_a, the jet velocity and the mass flow per unit area. Repeat this for sea-level on the test bed, $M = 0$ and for take-off when $M = 0.25$. When the nozzle is choked $\overline{m} = 1.281$ but when un-choked $\overline{m}$ can be found from Equation 6.18 or Figure 6.2.

(Ans:	M	p_{013} (kPa)	T_{013} (K)	p_{013}/p_a	V_j	$\dot{m}/A$ (kg s^{-1}m^{-2})
	0	151.9	327.7	1.500	268.3	325
	0.25	158.7	331.4	1.567	283.2	343
	0.78	53.4	278.9	2.242	339.8	129.1

6.10 Engines are extensively tested on static test beds. As noted above, the intakes do not always represent flight conditions consistently and the condition at which thrust is assessed is at $M = 0.25$. Use the results of Exercise 6.9 for a fan pressure ratios of 1.50 to find the ratio of thrust from the bypass stream at $M = 0$ and $M = 0.25$. Assuming the isentropic efficiency of the fan is constant.

(**Ans:** thrust ratio $= 1.283$)

Note: This ratio of the thrust is referred to as the thrust equivalence ratio. The value here for the bypass stream is close to the values used by Boeing and Airbus. The alteration in jet velocity because of the increase in inlet pressure with forward speed at take-off, is small. The big alteration in net thrust comes from the deduction of the ram drag, which is around a quarter of the gross thrust for $M = 0.25$. For a low pressure ratio fan the difference between gross and net thrust is larger than for a high pressure ratio fan. The Boeing and Airbus equivalent thrust ratios are based on a higher fan pressure ratio than 1.5.

6.11 Find the ratio of gross thrust and net thrust from the bypass stream for sea-level static to *net* thrust at $M = 0.78$ and 35,000 ft assuming fan pressure ratio is constant at 1.5.

(**Ans:** gross thrust ratio $= 1.99$; net thrust ratio $= 6.22$)

6.12 Logically the engine should be sized for cruise, where most fuel is burned, but climb to the cruise altitude requires additional net thrust. If the fan pressure ratio for cruise, at $M = 0.78$ and 35 kft, is 1.50, find the jet velocity for $fpr = 1.6$. Assume an isentropic efficiency of 90% for the fan and no losses downstream of the fan. Find the proportional increase in mass flow per unit area of bypass nozzle relative to the cruise condition and hence the proportional increase in net thrust per unit area of nozzle. What angle of climb and rate of climb would this produce?

(**Ans:** $V_j = 355.1 \text{ m s}^{-1}$; $\Delta \dot{m}/A = 5.6\%$; $\Delta F_N/A = 20.5\%$; climb angle 0.544°; climb rate 432 ft/min)

Summary of chapter 6

Bernoulli's equation is *not* in general valid for the flow of gases but if the Mach number is below about 0.3 the errors are often small enough to be accepted. In most cases in jet engines flow Mach numbers will be higher than this and stagnation pressure must therefore be determined using

$$p_0 = p \left\{ 1 + \frac{\gamma - 1}{2} M^2 \right\}^{\gamma/(\gamma-1)}.$$

Stagnation temperature and pressure provide a convenient way of accounting for the effects of velocity – they are the temperature and pressure of a gas were it to be brought to rest adiabatically and, in the case of pressure, reversibly. For the flow of a compressible gas through a converging duct or nozzle there is a maximum mass flow rate per unit area. This occurs when the velocity is sonic in the section of duct where the cross-sectional area is a minimum, normally called the throat. To accelerate a flow to supersonic velocity it is necessary to first converge the stream and then diverge it downstream of the minimum area where the velocity is sonic.

Mass flow per unit area may be made non-dimensional in a form which, for fixed gas properties, is a unique function of Mach number and ratio of specific heats,

$$\frac{\dot{m}\sqrt{c_p T_0}}{A p_0} = \bar{m}(M, \gamma) = M \frac{\gamma}{\sqrt{\gamma - 1}} \left(1 + \frac{\gamma - 1}{2} M^2 \right)^{-(\gamma+1)/2(\gamma-1)}. \tag{6.15}$$

If the Mach number M is the result of an adiabatic and reversible expansion from stagnation pressure p_0 to static pressure p, such as in a nozzle or turbine blade row, $\overline{m}$ may be expressed in terms of the pressure ratio and ratio of specific heats as

$$\frac{\dot{m}\sqrt{c_p T_0}}{A p_0} = \overline{m}\left(\frac{p}{p_0}, \gamma\right) = \frac{\gamma}{\gamma - 1}\sqrt{2\left[\left(\frac{p}{p_0}\right)^{\frac{2}{\gamma}} - \left(\frac{p}{p_0}\right)^{\frac{(\gamma+1)}{\gamma}}\right]} \qquad (6.18)$$

and this non-dimensional form for mass flow will also be useful in connection with engine performance.

The flow rate through a choked nozzle is determined by the upstream stagnation pressure and temperature and is independent of the conditions downstream of the throat. For jet engines the propulsive nozzles are normally choked; effectively so too are most turbines.

The understanding of the way in which a choked orifice works is indispensable to the treatment of jet engines, as will become apparent in the later chapters. It will also be found that the non-dimensional expressions above for the mass flow, Equations 6.15 and 6.18, will be useful in expressing the mass flow through the engine.

The design pressure ratio of the fan alters the way in which the engine performance changes with forward speed. A fan designed for a low pressure ratio has a greater rise in net thrust as flight speed is reduced than a fan designed for a higher pressure ratio. The effect is to make it comparatively easy to provide adequate thrust for take-off and the difficult or stretch condition becomes the top-of-climb.

CHAPTER 7   SELECTION OF FAN PRESSURE RATIO, SPECIFIC THRUST AND BYPASS RATIO

7.0 INTRODUCTION

In this chapter we set about selecting the fan pressure ratio, *fpr*, which is one of the most important parameters in the design of the engine. Fan pressure ratio determines the jet velocity and thus, for a given thrust, the diameter of the engine. Traditionally in courses and in books it is bypass ratio which has been selected and from this fan pressure ratio has been derived. Indeed this was the approach adopted in editions 1 and 2 of *Jet Propulsion*. For reasons which will be explained in this chapter, it is far better to use *fpr* as the independent design variable.

The relationship between overall efficiency, η_0, thermal efficiency, η_{th}, and propulsive efficiency, η_p, can be formalised for a turbojet, as in Equation 3.9, as

$$\eta_0 = \eta_p \times \eta_{th}. \tag{7.1a}$$

In Chapter 4 we showed that to obtain a high thermal efficiency, $\eta_{th,}$ we must operate with a high overall pressure ratio and high turbine inlet temperature. The overall pressure ratio and the temperature ratio T_{04}/T_{02} have been selected in Chapter 5. This combination of temperature and pressure would lead to a high jet velocity if all the available energy at turbine exit were used to accelerate the core flow, as in the turbojet in Exercise 7.1 below. In Chapter 3, however, we have shown that high jet velocities give low propulsive efficiency for subsonic aircraft and that an efficient engine will generate its thrust by accelerating a large mass flow of air by only a small amount. The combination of constraints on η_{th} and η_p give rise to the high bypass engine for subsonic propulsion. Essentially, the high-energy core flow is expanded through the LP turbine and the power extracted is used to drive the fan which raises the pressure of the large flow of bypass air. The pressure rise of the bypass stream is small compared with the overall pressure rise in the engine. This small pressure accelerates the air through the bypass nozzle to produce a bypass jet of modest velocity. Some of the LP turbine power is used to compress flow which passes through the core, first near the root of the fan and then in the LP compressor. The LP compressor is usually known as the booster.

In Chapter 3 the thermal efficiency was defined as the kinetic energy increase in the exhaust divided by the energy release from the burning of fuel. For the bypass engine it is helpful to redefine the thermal efficiency η_{th} of the engine so that the numerator becomes the kinetic energy that *would* be produced if the core flow were to be expanded to ambient pressure before the work is extracted to drive the bypass section fan. In computing this kinetic energy a hypothetical station is implied in the LP turbine when the gas has powered the fan root and booster but has not yet powered the bypass stream of the fan. This allows us to introduce the transfer efficiency, η_{tr}, to account for the losses in extracting power from the gas in the LP turbine and converting it in the fan into the pressure rise of the bypass stream. (We will ignore here the additional pressure loss of the flow in the bypass duct). The expression for the overall efficiency[1] of the bypass engine is therefore given by

$$\eta_0 = \eta_p \times \eta_{th} \times \eta_{tr}. \tag{7.1b}$$

The overall efficiency for a bypass engine, as given by Equation 7.1b would exceed that for a turbojet engine with the same core thermal efficiency because the improvement in the propulsive efficiency η_p far outweighs the losses associated with energy transfer η_{tr} from the core flow to the bypass stream, as will be shown in Exercises 7.1 and 7.2.

The design choice of fan pressure ratio can be thought of as a choice of bypass jet velocity. The choice of core jet velocity is a decision about what fraction of the available power is to be extracted in the LP turbine and how much energy is to be left in the core flow to accelerate the core jet. The core and bypass jet velocity magnitudes are always chosen to be similar to minimise noise and to maximise propulsive efficiency.

The core power per unit mass flow depends on the overall pressure ratio and turbine entry temperature. Once we have selected the fan pressure ratio and the ratio of bypass to core jet velocity, the bypass ratio follows from the power per unit mass flow rate through the core.

7.1 FAN PRESSURE RATIO AND BYPASS RATIO

The bypass ratio is still widely used as a descriptor of the engine type. The bypass ratio greatly affects the appearance, size and weight of the engine: the length of a military engine with low bypass ratio is typically four or five times its diameter, whereas high bypass ratio engines typically have lengths between about 1.5 and two times the diameter. The first generation of big commercial engines (Pratt & Whitney JT9D, General Electric CF6 and Rolls-Royce RB211 generation) had bypass ratios of around 5, but the more recent generation from GE (the GE90 and GEnx) and from Rolls-Royce (the Trent 1000 and Trent XWB) have bypass ratios near to 10. We may assume that a bypass ratio at least as great as this will be employed in a new engine for the New Efficient Aircraft. For practical reasons related to the minimum acceptable blade speed of the LP turbine, bypass ratios in excess of about 10 are not attractive at the present time with direct-drive between LP turbine and fan. To go significantly above 10 probably requires a gearbox between the LP turbine and the fan, to allow the LP turbine to be smaller and to turn faster. It is noteworthy that the Pratt & Whitney 1524G shown in Figure 5.4c, which has a gearbox between the LP turbine and the fan, has a bypass ratio of about 12.

---

[1] This is not the only way that overall efficiency can be separated out. A viable alternative is to have a core efficiency where the numerator is the LP turbine power to the fan bypass stream plus the core jet kinetic energy and the denominator is the fuel flow times calorific value. The transfer efficiency is then just the fan efficiency and propulsive efficiency is unchanged.

To bring home how confusing the use of bypass ratio can be, it is worth considering the history of so-called high bypass engines from the first generation (JT9D, CF6 and RB211) with a bypass ratio of about 5 to the latest ones with a bypass ratio of about 10. The fan pressure ratio at cruise for the first generation of high bypass engines was about 1.5 and it is remarkable that this fan pressure ratio for cruise is approximately the same as that of the latest generation of large civil engines, for example the Trent 1000 and the GEnx. In other words, the jet velocity for the engines separated in age by about 40 years will be almost the same and the propulsive efficiency must be nearly equal. In the intervening period the design fan pressure ratio did not, however, remain constant. As the thrust from the older generation of engines was increased, while keeping the fan diameter nearly constant, the pressure ratio was increased and so too the jet velocity. For example, the RB211, using the same fan diameter, had take-off thrust increased from about 42,000 lb in 1972 to 60,600 lb in 1990, whilst at the same time cruise *sfc* was reduced by about 11%. Since the increase in thrust must have been due to an increase in jet velocity, the propulsive efficiency can only have decreased over this time. The improvement in *sfc* was therefore largely due to an improvement in core thermal efficiency, achieved by higher overall pressure ratio, higher turbine entry temperature and higher compressor and turbine isentropic efficiency. This evolutionary process would have been easier to understand if the fan pressure ratio and not bypass ratio had been the descriptor chosen.

Although fan pressure ratio is a good descriptor of engine type, an alternative is to use the specific thrust. This is defined as the net thrust divided by overall mass flow,

$$specific\ thrust = F_N/\dot{m}_{air} = V_j - V.$$

This has found wider application in military engines, where the variation between designs is larger. The primary variable determining V_j and specific thrust is the fan pressure ratio.

7.2 ENGINE LAYOUT AND STATION NUMBERING

It is necessary to introduce a system for designating the various stations through the engine and Figure 7.1 shows this for a simple turbojet and for a simplified high bypass engine; for the latter only some of the stations are labelled. The system of numbering stations may not seem the most desirable or most simple, but it is used here because it is close to that internationally accepted. For the two-shaft turbofan configuration the core flow passes through a low-pressure compressor (usually known as the booster) downstream of the fan and driven by the LP shaft. Station 23 is in the core flow downstream of the booster stages while station 13 is downstream of the fan in the bypass duct.

7.3 ENGINE OVERALL SPECIFICATION

The processes are illustrated by the temperature–entropy diagrams in Figure 7.2 for a high bypass engine. The upper diagram shows the core flow and the lower diagram shows the bypass stream

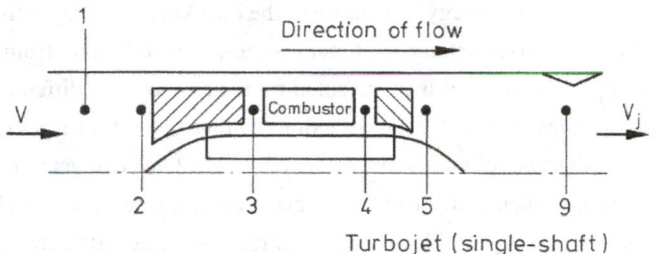

Turbojet (single-shaft)

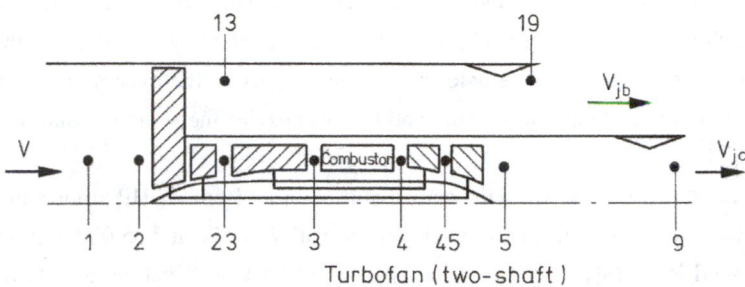

Turbofan (two-shaft)

Figure 7.1. Standard numbering schemes for stations in jet engines.

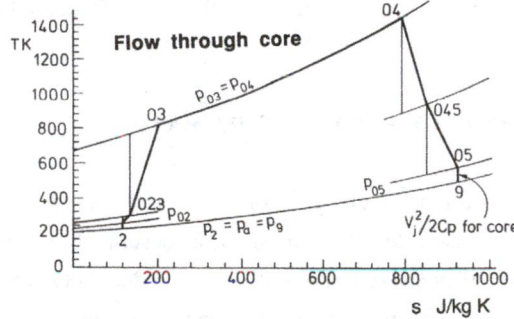

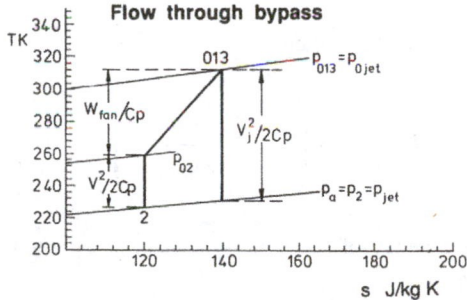

Figure 7.2. Temperature–entropy diagrams for the high bypass ratio engine. Core and bypass jet velocities set equal. Cruise at $M = 0.85$ at 31,000 $fpr \approx 1.8$, core pressure ratio $= 32$, turbine inlet temperature $= 1407$ K, fan, compressor and turbine efficiencies $= 90\%$.

with $fpr \approx 1.8$. The bypass jet kinetic energy is made up of the inlet kinetic energy from the aircraft forward speed, which is assumed to be retained without loss, and the work input from the fan. For the core the kinetic energy of the core jet is represented by the temperature difference $T_{05} - T_9$, where T_{05} is the stagnation temperature at LP turbine exit (note that $T_{05} = T_{09}$ since there is almost not energy loss in the core exhaust duct.) As can be seen in Figure 7.2, the core jet kinetic energy is small compared with the energy changes in the turbines or core compressors. There is a large energy input in the HP compressor $T_{03} - T_{023}$ and large energy extractions in the HP turbine, $T_{04} - T_{045}$, and LP turbine, $T_{045} - T_{05}$. Largest of all is the temperature rise in the combustor, $T_{04} - T_{03}$. The LP turbine power must be made equal to the power to the fan and booster. The kinetic energy, per unit mass flow, of the core jet is chosen to be similar to the kinetic energy of the bypass jet to obtain near optimum propulsive efficiency and the velocity of the bypass jet is determined by fan pressure ratio. Downstream of the core and bypass nozzles the static pressure is the ambient value, p_a.

The conditions at outlet from the core, that is at outlet from the HP turbine and inlet to the LP turbine, station 45 here, were determined for cruise at $M = 0.78$ and 35,000 ft in Exercise 5.1b. They were found to be $T_{045} = 1033.5$ K and $p_{045} = 363.5$ kPa. What we now have to do is to find the pressure ratio in the LP turbine p_{045}/p_{05} and thence stagnation pressure and temperature downstream of it, p_{05} and T_{05}, such that the core jet velocity is appropriate compared to the bypass jet velocity. The LP turbine pressure ratio determines the power from the LP turbine and therefore the power to the fan.

7.4 THE SPECIFICATION OF FAN PRESSURE RATIO

The design process begins with the specification of the ratio of stagnation pressures across the fan, p_{013}/p_{02}. The stagnation pressure into the fan p_{02} is, as described in Chapter 6, higher than the ambient static pressure p_a. If pressure losses in the bypass duct between the fan and the bypass nozzle may be neglected, the pressure across the nozzle is given by

$$\frac{p_{013}}{p_a} = \frac{p_{013}}{p_{02}} \frac{p_{02}}{p_a} = fpr \frac{p_{02}}{p_a}.$$

The temperature into the nozzle can be written

$$T_{013} = T_{02} \left(1 + \frac{fpr^{\frac{\gamma-1}{\gamma}} - 1}{\eta_f} \right), \tag{7.2}$$

where η_f is the isentropic efficiency of the fan, $T_{02} = T_a(1 + \frac{1}{2}(\gamma - 1)M^2)$ and for isentropic deceleration $p_{02}/p_a = (T_{02}/T_a)^{\gamma/\gamma-1}$.

The bypass flow is assumed to accelerate isentropically through the nozzle to ambient pressure downstream of it. As will be discussed in Chapter 11, for the nozzle pressure ratios here, less than three, the irreversibilities will be small. The bypass jet velocity is therefore given by

$$V_{jb} = \sqrt{2c_p T_{013} - T_9}$$

$$= \sqrt{2c_p T_{013} \left(1 - (p_a/p_{013})^{\frac{\gamma-1}{\gamma}}\right)} \tag{7.3}$$

using pressure ratio and temperature from Equation 7.2. For fixed ambient conditions and flight Mach number the velocity in the bypass jet stream is entirely specified by the fan pressure ratio, assuming constant fan isentropic efficiency.

The jet velocity of the core stream V_{jc} depends on stagnation temperature and pressure into the core nozzle, T_{09} and p_{09}. It is plausibly assumed that there is no loss in pressure or heat transfer between the LP turbine exit and the entry to the nozzle, so $p_{09} = p_{05}$ and $T_{09} = T_{05}$. The nozzle exit pressure is well approximated by $p_9 = p_a$. The core jet velocity is therefore given by

$$V_{jc} = \sqrt{2c_p T_{05} \left(1 - (p_9/p_{05})^{\frac{\gamma-1}{\gamma}}\right)}. \tag{7.4}$$

The bypass jet velocity is fixed by Equation 7.3 after specifying the fan pressure ratio. The core jet velocity can now be chosen in relation to the bypass jet velocity by varying the pressure ratio across the LP turbine, p_{045}/p_{05}. The core and bypass jet velocities are always similar in magnitude but here we make them equal

$$V_{jc} = V_{jb},$$

which is simple and corresponds to maximum propulsive efficiency. The appropriateness of this choice is assessed later. At this stage the pressure ratios and jet velocities are determined but so far not the bypass ratio.

The power of the LP turbine is $\dot{m}_c c_p (T_{045} - T_{05})$, where $\dot{m}_c$ is the mass flow of air passing through the core, is equal to the power taken by the fan and booster. The resulting equation is

$$\dot{m}_c c_p (T_{045} - T_{05}) = \dot{m}_c c_p (T_{023} - T_{02}) + bpr\, \dot{m}_c c_p (T_{013} - T_{02}), \tag{7.5}$$

where T_{023} and T_{013} are the temperatures leaving the booster and the fan in the bypass stream respectively. The temperature drop in the LP turbine is determined by the inlet temperature to it T_{045} and the LP turbine pressure ratio p_{045}/p_{05} using the isentropic efficiency. The only remaining unknown in Equation 7.5 is the bypass ratio.

The procedure for calculation can be summarised, assuming core exit conditions T_{045} and p_{045} are given, as:

    i. Choose the fan pressure ratio, which determines bypass jet velocity.
    ii. Choose the ratio of core jet velocity to bypass jet velocity (here made equal).
    iii. Guess a value for LP turbine pressure ratio, p_{045}/p_{05}.

iv. From p_{045}/p_{05} find T_{05} and p_{05}/p_a.

v. Compute core jet velocity V_{jc} and compare it to the bypass jet velocity V_{jb}. If V_{jc} is too large increase p_{045}/p_{05} (i.e. lower p_{05}) and return to (iv).

vi. From p_{045}/p_{05} and T_{045} find LP shaft power and then compute bypass ratio.

vii. Compute gross thrust and net thrust per unit mass flow through the core.

Throughout this chapter, including the exercises which follow, the aircraft is taken to be flying at $M = 0.78$ at 35,000 ft. We assume that the pressure ratio for the core flow through the fan and booster is $p_{023}/p_{02} = 2.5$ and Exercise 5.1 showed for this pressure ratio the temperature rise from fan inlet to booster exit to be $T_{023} - T_{02} = 81.6$ K. The design pressure ratio across the HP compressor is chosen to be 18. The conditions delivered to the LP turbine are $T_{045} = 1033.5$ K and $p_{045} = 363.5$ K.

Exercises

7.1 Consider a pure turbojet engine (no bypass stream) with overall pressure ratio of 45 and turbine entry temperature of 1500 K. This produces entry conditions into the propulsive nozzle given by exit conditions from the gas generator in Exercise 5.1c; $T_0 = 951.9$ K and $p_0 = 263.6$ kPa. Flight is at a Mach number of 0.78 at 35,000 ft. Assume the nozzle expands the flow isentropically to ambient pressure. (Note that the inlet temperature used in Exercise 5.1 already took the effect of flight speed into account, i.e. the values of inlet pressure and temperature given were to the stagnation values.)

Calculate the following:

a jet velocity (where expansion is from the propulsive nozzle inlet to ambient *static* pressure;

b propulsive efficiency;

c gross thrust F_G and net thrust F_N per unit mass flow, $\dot{m}_c$;

d overall efficiency, given by (net thrust) × (flight speed) ÷ $c_p(T_{04} - T_{03})$;

e specific fuel consumptions, *sfc* (see section 3.3) taking for the fuel $LCV = 43$ MJ/kg;

$$\text{(Ans: } V_j = 974.9 \text{ m/s; } \eta_p = 0.383; \, F_G/\dot{m}_c = 975 \text{ N kg}^{-1}\text{s;}$$
$$F_N/\dot{m}_c = 744 \text{ N kg}^{-1}\text{s; } \eta_0 = 0.242; \, sfc = 22.2 \text{ g s}^{-1}\text{kN}^{-1}\text{)}$$

Note: The thermal efficiency of this engine $\eta_{th} = \eta_0/\eta_p = 0.63$ is high because pressure ratio and turbine entry temperature are high, but the propulsive efficiency is low because of the high jet velocity.

7.2* Consider a high bypass ratio engine with overall pressure ratio of 45 and turbine entry temperature of 1500 K. This produces entry conditions into the LP turbine which are the exit conditions from the gas generator in Exercise 5.1b; $T_0 = 1033.5$ K and $p_0 = 363.5$ kPa. Flight is at $M = 0.78$ and 35,000 ft.

Assume that the inlet is loss free and the nozzles expand the flow isentropically to ambient pressure. Take the isentropic efficiencies for the fan and for the LP turbine to be each 0.90. Assume equal velocity for the core and bypass jets.

Calculate the following for fan pressure ratios of 1.4, 1.5 and 1.6 for the flow entering the bypass duct:

a bypass jet pressure, temperature and velocity V_{jb} and thence the propulsive efficiency;

b pressure and temperature ratios for the LPT (use above iterative solution to $V_{jc} = V_{jb}$);

c bypass ratio (assuming a constant pressure ratio of 2.5 through the fan and booster for the flow that enters the core);

d gross thrust F_G and net thrust F_N from the whole engine (core plus bypass) per unit mass flow *through the core* (i.e. comparing engines for same size core), $\dot{m}_c$;

e specific thrust, $F_N/\dot{m}_a$;

f overall efficiency, given by (net thrust) × (flight speed) ÷ $c_p(T_{04} - T_{03})$;

g *sfc* (see Section 3.3) taking for the fuel $LCV = 43$ MJ/kg.

The use of a spreadsheet or programme is recommended.

(Ans:	V_j (m/s)	η_p	p_{045}/p_{05}	T_{045}/T_{05}	bpr	$F_G/\dot{m}_c$ (kN kg^{-1}s)	$F_N/\dot{m}_c$	$F_N/\dot{m}_a$ (m/s)	η_0	sfc (g s^{-1}kN^{-1})
fpr = 1.4;	323	0.834	10.94	1.804	13.8	4.77	1.355	91.8	0.441	12.18
fpr = 1.5;	340	0.810	10.57	1.789	11.2	4.14	1.322	108.6	0.431	12.49
fpr = 1.6;	355	0.789	10.23	1.776	9.4	3.71	1.293	123.9	0.421	12.77

7.3 In Exercise 2.2 the net thrust required from each engine was shown to be $F_N = 38.9$ kN for initial cruise at 35,000 ft and $M = 0.78$. Use this with the results of Exercise 7.2 to find the mass flow of air at start of cruise for the bare engine (i.e. ignore drag of the nacelle). Ignore mass of fuel flow.

(**Ans:** fpr = 1.4, $\dot{m}_{air} = 424$ kg/s; fpr = 1.5, $\dot{m}_{air} = 358$ kg/s; fpr = 1.6, $\dot{m}_{air} = 314$ kg/s)

7.5 THE IMPACT OF FAN PRESSURE RATIO

The approach adopted in Exercise 7.2 has been used to generate Figures 7.3, 7.4 and 7.5 for a range of fan pressure ratios. For each fan pressure ratio the LP turbine pressure ratio is varied until the bypass and core jet velocities are equal to a prescribed ratio (taken to be one, as in Exercise 7.2). All results are for constant LP turbine entry conditions, T_{045} and p_{045}, and for flight at $M = 0.78$ at 35,000 ft.

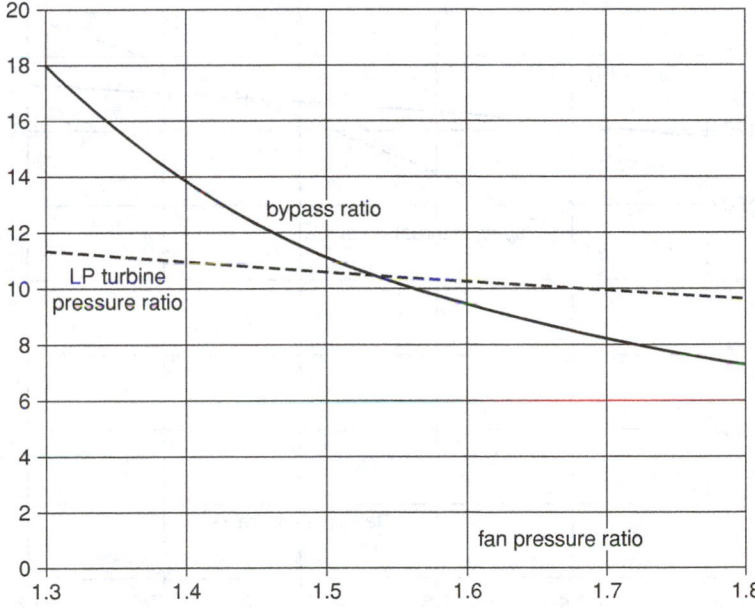

Figure 7.3. Predicted variation in LP turbine pressure ratio and bypass ratio with fan pressure ratio for a constant core, opr = 45, $T_{04} = 1500$ K. $M = 0.78$ at 35,000 ft. Bare engine. Core and bypass jet velocities equal.

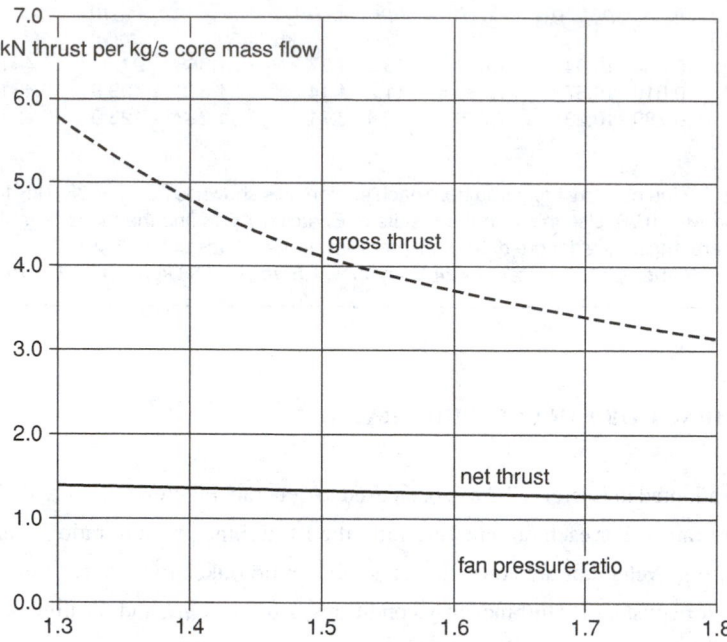

Figure 7.4. Predicted variation in gross and net thrust with fan pressure ratio for a constant core, $opr = 45$, $T_{04} = 1500$ K. $M = 0.78$ at 35,000 ft. Bare engine. Core and bypass jet velocities equal.

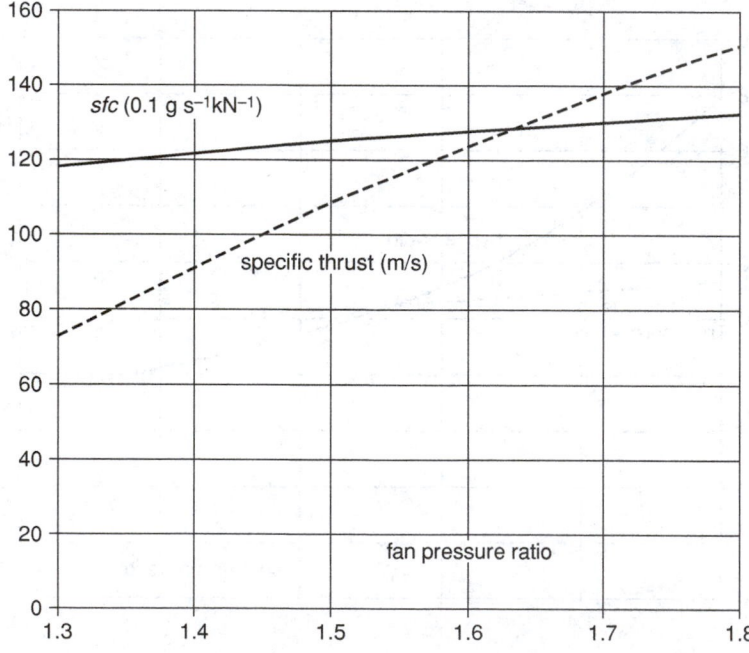

Figure 7.5. Predicted variation in specific thrust and sfc with fan pressure ratio for a constant core, $opr = 45$, $T_{04} = 1500$ K. $M = 0.78$ at 35,000 ft. Bare engine. Core and bypass jet velocities equal.

The dependence of bypass ratio on fan pressure ratio is shown in Figure 7.3, with the anticipated rise in this as the fan pressure ratio falls; reducing the fan pressure ratio from 1.8 to 1.3 increases bypass ratio from about 7 to about 18. As the bypass jet velocity is decreased with reduction in fan pressure ratio the core jet velocity reduces in step with it according to the chosen velocity ratio. As a result the stagnation pressure upstream of the core nozzle is reduced as fan pressure ratio is reduced, and for the same engine overall pressure ratio, this requires that the pressure drop across the LP turbine increase. Reducing fan pressure ratio from 1.8 to 1.3 gives an increase in LP turbine pressure ratio from about 9.6 to 11.4.

Gross and net thrust for unit flow through the core are shown in Figure 7.4 as a function of fan pressure ratio. Gross thrust is a strong function of fan pressure ratio, nearly doubling as *fpr* is reduced from 1.8 to 1.3 mainly because *bpr* has increased so much. The net thrust, however, varies much less, only about 14% over the same range of *fpr*. At cruise conditions it is net thrust which is most relevant, but gross thrust is also shown because during take-off, when the ram drag is small, net thrust is nearly equal to the gross thrust. For the same core and for the same thrust at cruise, reducing fan pressure ratio gives a much greater thrust at take-off or on the test bed, a topic addressed in Chapters 8 and 12.

The variation in net thrust in Figure 7.4 is for a constant core, which means that the mass flow rate of fuel is the same for all fan pressure ratios. In consequence the variation in net thrust translates directly into the variation in specific fuel consumption *sfc* shown in Figure 7.5. The corresponding reduction in *sfc* for a reduction in *fpr* from 1.8 to 1.3 is about 14%. (The reduction in *sfc* for the bare engine with fan pressure ratio reduced from 1.8, a value typical of the previous generation of engines, to 1.5 for the current engine, is a bit under 6%). Also shown in Figure 7.5 is the variation in specific thrust with fan pressure ratio, which more than halves over the range of *fpr* from 1.8 down to 1.3. For a given required net thrust this would require a doubling of mass flow rate through the engine.

The magnitude of *sfc* in Figure 7.5 and the values predicted in Exercise 7.2 are similar to those for engines recently entered into service. The absolute levels are, as already noted, sensitive to the isentropic efficiencies of the compressor and turbine. We have also so far ignored the effect of cooling air and the drop in pressure in the combustor. Moreover we have used constant c_p and γ for the gas, which is a gross simplification. The simplifications give a higher bypass ratio for the same fan pressure ratio than that which would come from a more complete analysis. Happily though, the levels predicted with a more complete method are sufficiently similar to those derived in our simplified approach that we may have confidence that the *trends* observed here are correct.

7.6 THE BARE ENGINE AND THE EFFECT OF THE NACELLE

Consideration so far has been for the so-called *bare* engine, which corresponds to considering the momentum of the incoming streamtube and the momentum of the jets downstream of the nozzles.

Engines are surrounded by a nacelle and the combined engine and nacelle is sometimes referred to as the *powerplant*. There is an aerodynamic drag on the outside of the nacelle as well as in the intake and bypass duct and these need to be included in any procedure to find the optimum fan pressure ratio. As noted earlier, the mass flow must increase as the fan pressure ratio decreases to give the same net thrust and, as a result, the engine becomes larger and heavier.

For the present purpose we will lump together as D_{nac} the drag from the nacelle and the losses in the inlet and bypass duct. It is plausible to assume that D_{nac} will be proportional to the 'wetted' surface area, A_w of the nacelle so, for a fixed style of nacelle and for fixed flight Mach number,

$$D_{nac} \propto A_w \rho V^2,$$

where ρ is the ambient air density and V is the flight velocity. For a given style of nacelle the wetted area will be proportional to the fan inlet area $\pi d^2/4$. For a fixed style of fan, fixed flight speed and constant altitude, the fan area is proportional to the mass flow of air through the engine, i.e. $\pi d^2/4 = \dot{m}/\rho V$, and therefore the wetted area is $A_w \propto \dot{m}/\rho V$. The mass flow rate through the engine is given by $\dot{m} = F_N/X$, where F_N is the net thrust and X is the engine specific thrust. The engine specific thrust, $X = V_j - V$, is determined by fan pressure ratio. The nacelle drag can therefore be written as

$$D_{nac} = kV(F_N/X),$$

where k is an empirical constant depending on the design of fan and nacelle. The effective net thrust from the engine, after deducting the nacelle drag, is therefore

$$F_{Neffective} = F_{Nbare}(1 - kV/X).$$

The empirical constant needs to be found from experiment or detailed computation, but choosing $k = 0.04$ gives plausible estimates consistent with tests. For the present New Efficient Aircraft cruising at $M = 0.78$ at 35,000 ft, $V = 231$ m/s and $kV = 9.25$, so the effective net thrust is

$$F_{Neffective} = F_{Nbare}(1 - 9.25/X). \tag{7.6}$$

The reduction in thrust resulting from intake drag means that the mass flow through the engine must be increased, in other words, the engine needs to be larger, in the ratio of Equation 7.6. Furthermore, the specific fuel consumption is increased in the same ratio and the effect of this is shown in the plot of *sfc* against fan pressure ratio in Figure 7.6. The effective *sfc* is increased on the order of 10% relative to the bare engine for a *fpr* of about 1.5. For the bare engine the *sfc* is still reducing quite steeply with *fpr* at the lower end of the range, whereas for the powerplant (the engine with a nacelle) the decrease with *fpr* is smaller. It is clear that the effective *sfc* is highly susceptible to the empirical values and assumptions used. Nevertheless it is only when powerplant weight is considered, as in the next section, that a clear minimum in *sfc* emerges.

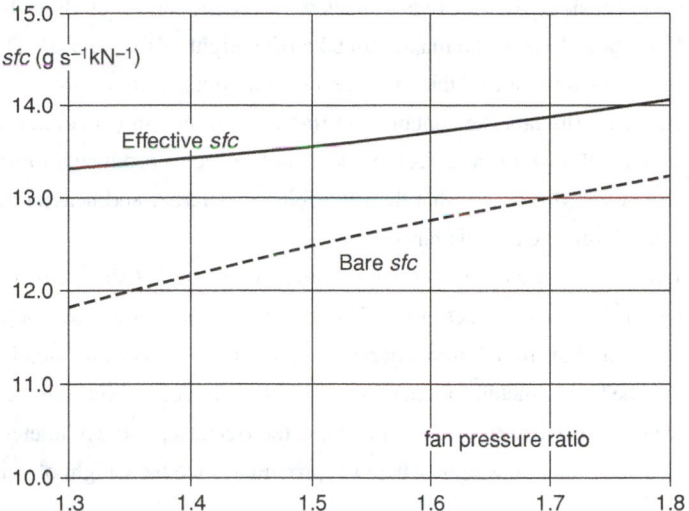

Figure 7.6. Predicted variation in *sfc* with fan pressure ratio for a constant core, $opr = 45$, $T_{04} = 1500$ K.
$M = 0.78$ at 35,000 ft. Core and bypass jet velocities equal.
Assumed: $sfc_{effective} = sfc_{bare}/\{1.00 - 9.25/X\}$ where $X = $ specific thrust.

Exercises

7.4* In Exercise 2.4 the required net thrust to propel the NEA at $M = 0.78$ at 35,000 ft was found to be $F_N = 38.9$ kN. Exercises 7.2 and 7.3 were for a 'bare' engine, which neglected the drag and losses associated with the nacelle. In consequence the air mass flow through the engine (effectively determining the size of the engine) was underestimated. Assuming the loss and drag associated with the intake, bypass duct and nacelle to be given by Equation 7.6, find the mass flow needed to create the effective net thrust, $F_{Neffective} = 38.9$ kN.

(**Ans:** $fpr = 1.4$, $\dot{m}_{air} = 472$ kg/s; $fpr = 1.5$, $\dot{m}_{air} = 392$ kg/s; $fpr = 1.6$, $\dot{m}_{air} = 339$ kg/s)

7.5* The specific fuel consumptions calculated in Exercise 7.2 was for the bare engine, neglecting intake, bypass duct and nacelle drag. Use Equation 7.6 to estimate the powerplant *sfc* when this drag is accounted for.   (**Ans:** $fpr = 1.4$, 13.55 g s^{-1}kN^{-1}; $fpr = 1.5$, 13.65 g s^{-1}kN^{-1}; $fpr = 1.6$, 13.8 g s^{-1}kN^{-1})

7.6* The mass flow of air determines the diameter of the fan. It is plausible to assume that the axial Mach number of the flow entering the fan rotor is 0.6. Use Equation 6.15 to find the non-dimensional mass flow $\overline{m}$ for this Mach number. Assuming that the ratio of fan hub diameter to fan tip diameter is 0.3, find the diameter of the fan based on the mass flows in Exercise 7.4.

(**Ans:** $\overline{m} = 1.078$; $fpr = 1.4$, $D_{tip} = 2.92$ m; $fpr = 1.5$, $D_{tip} = 2.66$ m; $fpr = 1.6$, $D_{tip} = 2.48$)

7.7 EFFECT OF ENGINE WEIGHT AND SELECTION OF FAN PRESSURE RATIO

It is possible to forget that the object of the engine maker is *not* to make the engine with the lowest fuel consumption but the one which gives the largest yield to the airline. One feature that leads to higher yield is a reduced fuel burn and the weight of the engine affects fuel burn since the engine weight has to be balanced by lift from the wings and this lift brings a consequent drag. The weight of the powerplant is considerable; for two engines with nacelle, and including the weight of the pylon, the weight is about 10% of the maximum take-off weight of the aircraft. The weight of the bare engine is approximately half of this. The largest components of weight are associated with the low-pressure system: the fan, the containment ring around the fan (to contain debris in case a fan blade detaches) and the LP turbine. (As the fan is optimised to produce a lower pressure ratio is rotates more slowly and this means that the LP turbine gets larger and heavier unless a gearbox is used between the LP turbine and the fan.)

A consequence of reducing fan pressure ratio, as Exercise 7.4 shows, is that a larger mass flow of air is required for the same net thrust. This means that the engine will be bigger and, as a good approximation, the fan frontal area will be proportional to mass flow rate. The weight tends to be dominated by the fan diameter, for the reasons outlined above, and if things scaled in a simple manner weight would increase as the cube of fan diameter. Because, as the diameter increases more effort and cost goes into reducing weight, a better approximation to the weight of existing engines is

$$W_{engine} \propto d^{2.4}.$$

The effect of the powerplant weight is assessed here by simply taking the contribution to overall aircraft drag attributable directly to aircraft weight. (An increase in powerplant weight leads to a bigger and heavier aircraft, with higher thrust and bigger engines – this additional effect is not considered here so the total impact of variation in weight is likely to be greater than the present method predicts.) The corrected thrust, after deducting the drag attributable to engine weight W_{engine} is therefore given by

$$F_{Ncorrected} = F_{Neffective} - W_{engine}/(L/D).$$

For modern engines a plausible weight, with nacelle and pylon, for a fan diameter of 3 m is 12 tonne and using the $d^{2.4}$ scaling the impact of changes in diameter can be assessed. The lift-drag ratio is given for the NEA as 21.6 and this has then been used with the effective sfc to produce Figure 7.7. Although the empirical estimates for engine weight and its dependence on fan diameter are merely approximate, Figure 7.7 shows how radically inclusion of weight affects the trend with fpr, so an optimum fan pressure approximately equal to 1.6 ratio does emerge for minimum corrected fuel burn.

As the engine gets larger, that is for lower fpr, it becomes more difficult to integrate the nacelle under the wing without a large reduction in the lift-drag ratio of the wing. If the engine were to be very large, the design of the aircraft itself could begin to be affected: for example, the wings would have to be higher off the ground and this would require a longer undercarriage, which is heavy and may make the aircraft, including the fuselage, too high to be conveniently accommodated in existing airports. Engines of very large diameter might be too large to be air-freighted without dismantling. To be set against these dis-incentives, reducing engine noise (discussed in the Appendix to the book) gives an incentive to lower fan pressure ratio. For these reasons it is not possible to select the correct fan pressure ratio just on the basis of a calculation similar to that in Exercises

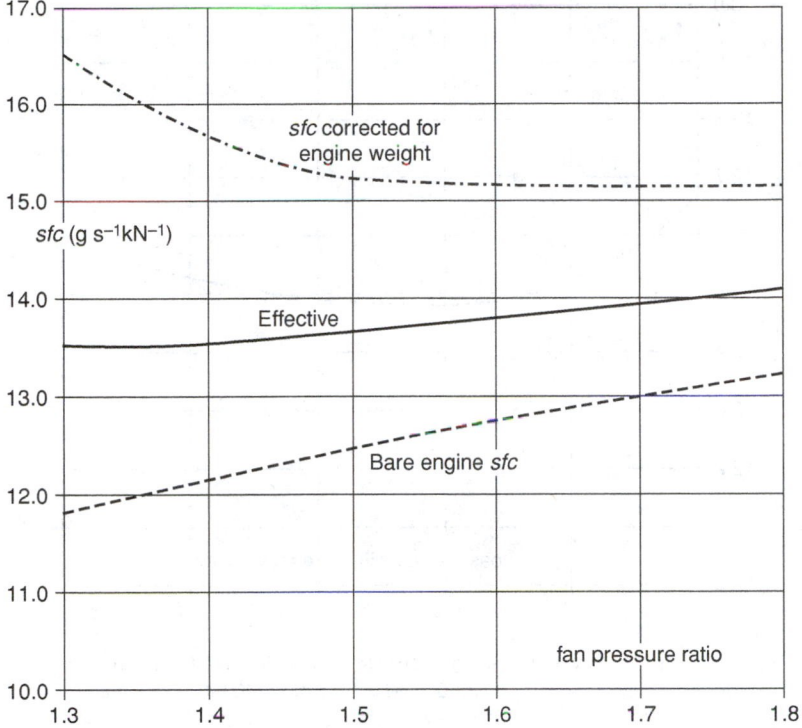

Figure 7.7. Predicted variation in *sfc* with fan pressure ratio with allowance for drag due to engine weight. Engine weight assumed to be given by 12 $(d/3)^{2.4}$ tonne, where d is fan diameter. For a constant core, $opr = 45$, $T_{04} = 1500$ K. $M = 0.78$ at 35,000 ft. Core and bypass jet velocities equal.

7.2 to 7.6, even if the calculation were more refined. Detailed empirical knowledge is needed of the nacelle and aircraft drag as well as knowledge of the engine and aircraft weights: for example, when noise must be considered the performance of the aircraft on take-off and approach is needed. Therefore, for the remainder of Part 1 of the present book, a value for fan pressure ratio which is in line with current practice for new engines seems appropriate, because this is based on the thorough optimisations carried out in industry. The engine in the New Efficient Aircraft will therefore be based on *fpr* = 1.5.

Exercises

7.7 a Assuming a fan pressure ratio of 1.5, cruise at $M = 0.78$ and an altitude of 35,000ft calculate the bypass nozzle area. The mass flow including the core mass flow may be obtained from Exercise 7.4. The nozzle is choked so $\overline{m} = \dot{m}\sqrt{c_p T_0}/Ap_0 = 1.281$. (**Ans:** bypass nozzle area = 2.78 m²)

b If the stagnation pressure and temperature downstream of the LP turbine are 34.4 kPa and 578 K show that the stagnation to static pressure ratio across the core propulsive nozzle is $p_{0in}/p_a = 1.44$. Hence find the core propulsive nozzle area. Note that this is not choked so Equation 6.18 should be used. (**Ans:** core nozzle area = 0.593 m²)

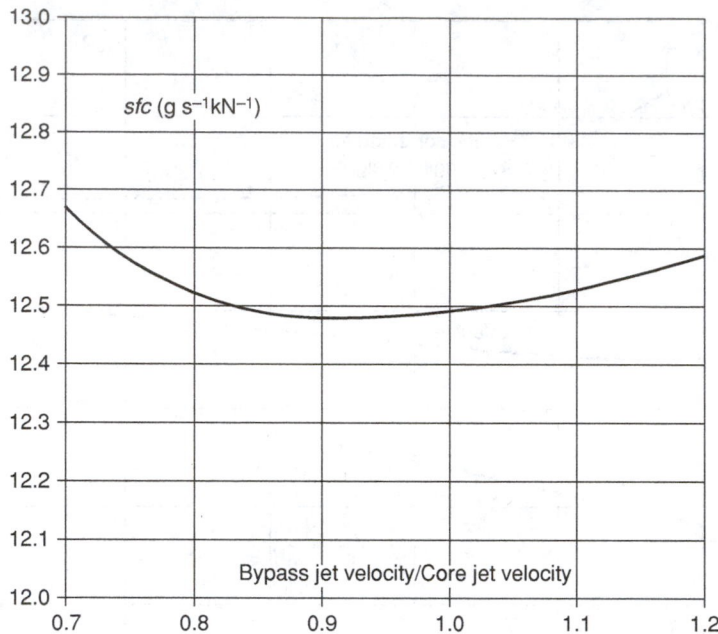

Figure 7.8. Predicted variation in bare *sfc* with jet velocity ratio for a constant core: $fpr = 1.5$, $opr = 45$, $T_{04} = 1500$ K. $M = 0.78$ at 35,000 ft. Bare engine.

7.8 THE EFFECT OF UNEQUAL BYPASS AND CORE JET VELOCITY

In Exercises 7.2. to 7.6, as well as in Figures 7.3 to 7.7, it has been assumed that the jet velocities of the core and bypass jets are equal. Figure 7.8 shows the variation in *sfc* for the bare engine when V_{jb}/V_{jc} is varied for the same core used earlier in the chapter and maintaining fan pressure ratio at 1.5.

The lowest sfc and highest net thrust occur[2] when $V_{jb}/V_{jc} \approx 0.92$ but, compared to the values obtained earlier with the simple assumption that $V_{jb}/V_{jc} = 1.0$, the improvement is less than 0.1%. Any serious design would, of course, exploit the advantage of the difference in jet velocity, but for the purpose of the exercises here, the simple assumption is entirely satisfactory.

Although the differences in specific fuel consumption shown in Figure 7.8 are small, there are significant alterations to the engine parameters when the velocity ratio is altered. This is shown in Table 7.1 in which quantities are compared for a bare engine with equal core and bypass jet velocities and one in which the optimum ratio, $V_{jb}/V_{jc} \approx 0.92$ is selected. The table is for a constant fan pressure ratio of 1.5. The inlet conditions to the LP turbine and to the fan are those used in Exercises 7.2 to 7.6 and Figures 7.3 to 7.7. Although the difference in net thrust and *sfc* are small, there are marked differences in both the bypass stream and in the core downstream of the LP turbine.

---

[2] It is often supposed that the optimum velocity ratio is equal to the product of LP turbine efficiency and fan efficiency, which in this example would equal 0.81. This product of efficiencies has some relevance to the optimum transfer of work to the two gas streams, to give the highest kinetic energy for the constraints imposed. What is wanted for the jet engine is the highest momentum from the two jet streams given the same constraints, and for this the velocity ratio is found to be higher than the product of efficiencies.

Table 7.1 Comparison of engine parameters with unequal and equal jet velocities for constant net thrust. Engine net thrust after allowance for nacelle drag 38.9 kN

Engine mass flow rate		$V_{jb} = 0.92V_{jc}$ 382 kg/s	$V_{jb} = V_{jc}$ 392 kg/s
Core stream	V_{jc}	369 m/s	340 m/s
	p_{05}	36.7 kPa	34.4 kPa
	T_{05}	586 K	578 K
	$T_{045} - T_{05}$	447 K	456 K
	Nozzle area	0.530 m^2	0.593 m^2
Bypass stream	V_{jb}	340 m/s	340 m/s
	p_{013}/p_{02}	1.50	1.50
	T_{013}	278.9 K	278.9 K
	Nozzle area	2.71 m^2	2.78 m^2
	Bypass ratio	10.92	11.2
Fan	Tip diameter	2.63 m	2.66 m

Note: fpr = 1.50, other parameters as in Exercises 7.2 and 7.4.

Of particular note is the change in the fan diameter, showing the fan with unequal velocities to be smaller by 30 mm. This would have a noticeable effect on both the weight (about 3%) and drag of the nacelle (about 2%); in other words the advantage of non-equal jet velocity is because it gives a slightly smaller and lighter engine rather than one of reduced *sfc* of the bare engine. The reasoning can be taken further and in order to reduce weight and drag the ratio V_{jb}/V_{jc} can be selected to be below that for the lowest *sfc* of the bare engine. Although significant, this is a subtlety that it is safe to neglect in this book.

SUMMARY OF CHAPTER 7

The principle behind the design is as follows. The choice of overall engine pressure ratio and the turbine inlet temperature essentially fixes the pressure and temperature into the LP turbine. (Here this was done in Chapter 5.) Since the core jet velocity will always be similar in magnitude to the bypass velocity (here chosen to be equal) power per unit core mass flow from the LP turbine is effectively fixed. This power goes mostly to the fan bypass stream and is proportional to the bypass mass flow times the bypass temperature rise. Hence, once the bypass pressure ratio of the fan, *fpr*, is selected, the bypass mass flow and the bypass ratio are determined.

Decisions have to be made regarding the division of power between the core and the bypass streams. The optimum ratio of bypass to core jet velocity for minimum *sfc* of the bare engine is computed to be about 0.92. It is convenient here to assume equal velocities for the bypass and core and the error associated with this is only about 0.1% of net thrust or *sfc*.

As the fan pressure ratio is reduced the propulsive efficiency rises and the *sfc* of the bare engine falls, but the *sfc* of the bare engine does not show a minimum with *fpr*. The aerodynamic drag associated with the intake and nacelle increases as the fan pressure ratio is reduced and the mass flow through the engine must increase to achieve the same net thrust. For the values used here the powerplant (the engine plus nacelle) does not show a *fpr* for minimum *sfc*. When the effect of *fpr* on engine weight is included a *fpr* for minimum fuel burn in cruise is found.

The precise optimum *fpr* depends on knowledge and data which will be owned and managed by engine and airframe companies. For the purposes of this book, $fpr = p_{013}/p_{02} = 1.5$ is assumed for cruise, similar to recent optimised new engines. The pressure ratio adopted for the core stream through the fan and booster, p_{023}/p_{02} is kept fixed at 2.5.

Minimising the noise has become crucial and has come to exercise a major effect on the engines for large aircraft. The noise is reduced if the jet velocity is low, requiring low *fpr*. This could lead to aircraft which are not optimised for fuel burn or range and the ultimate limit on the size of the fan may be for logistic reasons (e.g. will the engine fit in a cargo aircraft?) not associated with performance. Also, as the engine gets bigger it becomes harder to integrate it under the wing or to install it without raising the height of the wing when on the ground.

It is still common to use the bypass ratio as the way to characterise engines. Certainly this gives a good impression of what the engine will look like, since the size of the core and the size of the fan are roughly proportional to the mass flows through them. However, in terms of propulsive efficiency, *sfc* or noise generation the bypass ratio is flawed as a descriptor. This is because, as the technology has advanced, the overall pressure ratio and turbine entry temperature of new engines have both increased and, as a result, the core is producing more power for the same mass flow through it.

A better descriptor of engine performance (for propulsive efficiency, *sfc* and noise) is the specific thrust. This is the net thrust per unit mass flow through the engine, which is equal to the difference between the average jet velocity and the velocity of flight. In consistent SI units the specific thrust is expressed in m/s.

The engine can now be regarded as defined in terms of its size, pressure ratio, bypass ratio and specific thrust. We can obtain considerable insight into the way engines behave by using dimensional analysis, the subject of the next chapter.

Key Engine Parameters at Design Point (Cruise $M = 0.78$ at 35,000 ft)

Fan pressure ratio (bypass)	1.5
Overall pressure ratio	45
Booster pressure ratio	2.5
Turbine inlet temperature	1500 K
Mass flow	392 kg/s
Fan tip diameter	2.66 m

CHAPTER 8 — DYNAMIC SCALING AND DIMENSIONAL ANALYSIS

8.0 INTRODUCTION

It would be possible to calculate the performance of an engine in the manner of Exercises 7.2 to 7.7 for every conceivable operating condition, e.g. for each altitude, forward speed and rotational speed of the components. This is not an attractive way of considering variations and it does not bring out the trends as clearly as it might. An alternative is to predict the variations by using the appropriate dynamic scaling – apart from its usefulness in the context of engine prediction, the application of what is conventionally called dimensional analysis is illuminating. The creation of groups which are actually non-dimensional is less important than obtaining groups with the correct quantities in them. The reasoning behind these ideas is discussed in Chapter 1 of Cumpsty (2004). For compatibility with the usual terminology, however, the phrases 'dimensional analysis' and 'non-dimensional operating conditions' will be retained here.

Using the ideas developed on the basis of dynamic scaling it is possible to estimate the engine performance at different altitudes and flight Mach numbers when the engine is operating at the same non-dimensional condition. From this it is possible to assess the consequences of losing thrust from an engine at cruise and, in some cases, to estimate how the engine will behave on a static test bed.

8.1 ENGINE VARIABLES AND DEPENDENCE

Figure 8.1 shows a schematic engine installed under a wing. For simplicity this is a mixed-flow engine, where the core and bypass streams are mixed before entering a single propulsive nozzle. Later the treatment can be generalized by considering the two streams separately. Alternatively, with bypass ratios to 10 or more, the errors may not be large if the analysis concentrates on the bypass stream and assumes the ratio of core jet velocity to bypass jet velocity stays approximately constant.

The only effects of the pylon are assumed here to be the transmission of a force between the engine and the wing and the passage of fuel to the engine. The primary control to the engine is the variation in the fuel mass flow rate, $\dot{m}_f$, and until relatively recently the pilot controlled only the fuel flow. Based on the fuel flow, together with the inlet air temperature and pressure, the settings of any variable bleeds or variable vanes in the engine were determined by a mechanical controller attached to the engine. Now the whole process can be controlled by an electronic controller that determines the fuel flow and the settings of any variables. The electronic controller does not, however, alter the fundamental dependence of, for example, rotational speeds of the engine on the magnitude of fuel flow, the atmospheric conditions and the speed of flight. We can therefore say that any variable in the engine can be expressed as a function of $\dot{m}_f$, p_a, T_a, and V (fuel mass flow rate, ambient static pressure and temperature and flight speed). The static pressure and temperature T_a characterise the atmosphere from which the engine must draw its working fluid and to which it must discharge the propulsive jet. However it is preferable to work in terms of variables more directly associated with the engine. So, at the inlet the stagnation conditions p_{02} and T_{02} are used. From Equations 6.3 and 6.5 it can be seen that these depends on p_a, T_a and forward speed so the use of stagnation variables at inlet includes the effect of forward speed.

At the rear of the engine we must be concerned with the static pressure at exit from the propulsive nozzle, that is with ambient static pressure, p_a, and *not* the stagnation pressure. (Although the ambient temperature has a large effect on the engine through the inlet flow it has no additional effect on the flow leaving the nozzle.)

If the propulsive nozzle is not choked the pressure at the exit plane of the nozzle may be assumed to be equal to ambient pressure, p_a, and the ambient pressure can have an upstream influence on the turbine. As a result, the pressures and temperatures within the engine are dependent on the inlet stagnation properties, p_{02}, and T_{02}, the fuel mass flow rate, $\dot{m}_f$, and the ambient pressure, p_a. Put another way, including stagnation pressure p_{02} and static pressure p_a is including the effect of forward speed (or Mach number); equivalently one of the pressures could be given together with the flight speed or the Mach number.

If the propulsive nozzle is choked, the static pressure downstream of the nozzle exit plane can have no effect upstream, which means that the flow out of the turbine is unaffected by the ambient static pressure. Hence with a choked propulsive nozzle, ambient conditions only affect the engine inlet and we can say that the performance of the engine depends only on $\dot{m}_f$, p_{02}, and T_{02}; in other words the separate influence of static pressure and of forward speed disappears.

For most turbojets, and for turbofans with relatively low bypass ratio, the pressure ratio across the propulsive nozzle is sufficiently large that for most important operating conditions the nozzle is choked. (The bypass ratio is low enough in current combat turbofans and in early civil turbofans, such as the Pratt and Whitney JT8D.) In such cases we can adopt the simplification that comes with assuming a choked nozzle. For modern bypass engines, with separate core and bypass propulsive nozzles, and with low fan pressure ratio, for example *fpr* $\approx$ 1.5, the final propulsive nozzles will not always be choked. The bypass nozzle will be choked at cruise but not for static conditions and not even at the take-off condition when the flight Mach number is about 0.25. The core propulsive nozzle for the modern civil engine will probably not be choked at any condition. This is because the velocity of the core jet is designed to be similar in magnitude to the bypass jet velocity and since the core stream is significantly hotter than the bypass stream, the pressure ratio across the core nozzle must be significantly lower than that of the bypass. Because the core nozzle

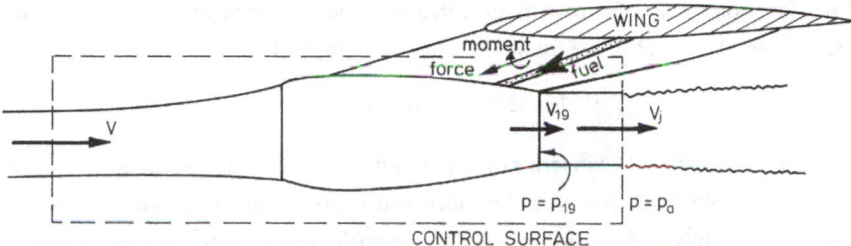

Figure 8.1. Engine mounted on wing pylon showing control surface for calculation of thrust. (Engine shown has mixed core and bypass streams through the final nozzle.)

is therefore unchoked the ambient static pressure is able to affect the pressure ratio across the LP turbine.

8.2 Non-dimensional Variables of the Engine

In addition to the external variables that *control* the performance of the engine there are a host of internal variables that *describe* its performance: pressures, temperatures, mass flow, shaft speeds and fluid properties. As discussed already it is common practice to use dimensional analysis to assemble internal and external variables into a set of carefully chosen dimensionless groups that characterise the performance of the engine and reveal important trends.

Producing non-dimensional groups is easy; it is finding the appropriate variables and dependencies from which the non-dimensional groups are created which is more challenging. It is particularly interesting when the dependence changes, as it does here, with the choking of the propulsive nozzle; when it is unchoked the engine depends on two non-dimensional groups but when it is choked only one non-dimensional group is sufficient to fix the performance.

To illustrate dependence on appropriate variables, consider the mass flow of air through the engine. First consider the general case where the propulsive nozzles are not choked so the engine is affected by ambient static conditions. The mass flow can be written as a function of the controlling variables identified above, where fuel mass flow rate is treated as being principal amongst them:

$$\dot{m}_{air} = f_a \left(\dot{m}_f, p_{02}, T_{02}, p_a \right). \tag{8.1a}$$

The same dependence is true of other variables such as the turbine inlet temperature,

$$T_{04} = f_b \left(\dot{m}_f, p_{02}, T_{02}, p_a \right). \tag{8.1b}$$

It is worth clarifying one aspect of notation used here. When the function f is written in Equations 8.1 it indicates a dependence on the variables in the bracket but the form the function implied in Equations 8.1a and 8.1b, (and 8.1c below) will be different. We do not need to know the precise form of the function f_a, f_b or f_c. If we assume that f_b could be rearranged to isolate $\dot{m}_f$ on the left hand side, we could then substitute for $\dot{m}_f$ in Equation 8.1a to obtain

$$\dot{m}_{air} = f_c \left(T_{04}, p_{02}, T_{02}, p_a \right). \tag{8.1c}$$

To rewrite Equation 8.1c in terms of dimensionless groups we could conduct a dimensional analysis of the variables that make up the functional relationship. However, we can skip to the answer by judiciously selecting the appropriate dimensionless groups: two are simple ratios of temperature and pressure and the group for mass flow will be recognised as the non-dimensional mass flow from Chapter 6. Area is denoted by D_2 where D is a diameter.

$$\frac{\dot{m}\sqrt{c_p T_{02}}}{D^2 p_{02}} = F_c \left\{ T_{04}/T_{02}, p_a/p_{02} \right\}. \tag{8.2a}$$

It is worth looking briefly at what Equation 8.2 means. In general the non-dimensional mass flow of air depends on the ratio of turbine inlet to compressor inlet temperature and the ratio of ambient static and inlet stagnation pressure. If the flight Mach number is constant the ratio p_a/p_{02} is constant. In this case the only variable left in the function is the ratio of turbine inlet temperature to compressor inlet temperature and the non-dimensional engine mass flow rate of air depends solely on this one variable.

As Equations 8.1 shows, the dependence of mass flow of air could be written as a function of fuel mass flow rate and in that case we obtain

$$\frac{\dot{m}\sqrt{c_p T_{02}}}{D^2 p_{02}} = F_a \left\{ \frac{\dot{m}_f LCV}{\sqrt{c_p T_{02}} D^2 p_{02}}, p_a/p_{02} \right\}. \tag{8.2b}$$

This time, for constant flight Mach number, the non-dimensional mass flow of air is a function only of the non-dimensional mass flow of fuel. Note that the form of the non-dimensional fuel mass flow is quite different from that of air. This is because the fuel group relates energy flux; the numerator is the energy released in burning the fuel whilst the denominator has the form of power derived from a velocity (related to speed of sound) times an area and a pressure. This demonstrates very clearly how crucial physical understanding is to creation of appropriate groups.

By reasoning similar to that employed in Equations 8.1 and 8.2 *every* relevant variable may be related in terms of dimensionless groups: as well as air and fuel mass flow rates, every pressure, temperature and shaft speed can put in suitable non-dimensional form. A selection of the variables customarily used is provided in Table 8.1. Here N refers to a rotational speed of a shaft and D to a characteristic length, such as a diameter. $\dot{m}_{air}$ and $\dot{m}_f$ are the mass flow rates of air and fuel respectively. The specific heat at constant pressure c_p is used with stagnation temperature to give a measure of energy per unit mass. From this $\sqrt{c_p T_0}$ is a measure of velocity and so too is $\sqrt{R T_0}$, both being proportional to the speed of sound at the reference stagnation condition. Note that shaft rotational speed is non-dimensionalised by relating tip speed to a speed of sound at stagnation conditions, $ND/\sqrt{\gamma R T_0}$. In general the engine operating points are characterised by functions of two dimensionless groups, these could, for example, be T_{04}/T_{02} and p_a/p_{02} as in Equation 8.2,

Table 8.1 Dimensionless numbers used to describe engine performance

$\bar{m} = \dfrac{\dot{m}_{air}\sqrt{c_p T_{02}}}{D^2 p_{02}}$	Non-dimensional mass flow rate where D is a characteristic dimension of the engine. With $D^2 = A$ this is Equation 6.17.
$\dfrac{p_a}{p_{02}}$ or M	The ratio of atmospheric to inlet stagnation pressure, which can also be represented by the Mach number given the relationship identified in Equation 6.5. This term is only required if the propulsive nozzle is unchoked.
$\dfrac{p_{03}}{p_{02}}, \dfrac{T_{04}}{T_{02}}, \dfrac{p_{04}}{p_{05}}$	A range of temperature and pressure ratios defining turbomachinery performance and critical temperatures within the engine. Note in particular that $opr = p_{03}/p_{02}$ and $fpr = p_{013}/p_{02}$.
$\dfrac{ND}{\sqrt{\gamma R T_{02}}}$	Non-dimensional shaft speed where N is the shaft speed in rad/s. It is proportional to the ratio of rotor tip speed to a speed of sound based on the inlet stagnation temperature.
$\dfrac{\dot{m}_f\, LCV}{\sqrt{c_p T_{02}} D^2 p_{02}}$	Non-dimensional fuel flow rate. The numerator represents the rate of energy supply via the fuel. The denominator represents a velocity multiplied by a force and so represents a rate of doing work, i.e. power.
$\dfrac{\dot{m}_{air} V_{19} + p_{19} A_N}{D^2 p_{02}}$	A dimensionless gross thrust where V_{19} and p_{19} are the velocity and pressure at the exit plane of a choked nozzle as discussed in Section 8.3.

but could be $ND/\sqrt{(\gamma R T_0)}$, a group based on $\dot{m}_f$ or on any other temperature or pressure ratio. If these are held constant all the other non-dimensional numbers are constant too, meaning that every pressure ratio, temperature ratio and shaft speed ratio in the engine, opr and fpr for example, is fixed.

If the propulsive nozzle is choked, the pressure ratio p_a/p_{02} has no effect on the engine and this term in Equations 8.2 can be eliminated. In other words it is no longer necessary for the flight Mach number to be constant for the value of T_{04}/T_{02} or non-dimensional fuel flow rate to fully specify engine performance. To be more concrete, suppose the value of T_{04}/T_{02} were specified as the design value at cruise with the nozzle unchoked. With the same temperature ratio on a sea-level test bed the engine would *not* be at the design condition since p_a/p_{02} would not be the same. If, however, the propulsive nozzle were choked then the dependence on p_a/p_{02} would be removed and T_{04}/T_{02} alone would suffice to define the engine at cruise and on a static test bed.

8.3 Non-dimensional Treatment of Thrust for Choked Nozzles

A choked propulsive nozzle was common for older commercial engines and is still normal for combat engines today. The bypass nozzle is choked at cruise, even for the newest commercial engines, and since the thrust from the bypass stream is an order of magnitude greater than that of the core, some benefit can still be gained from an approach based on choked nozzles. For simplicity here we consider a mixed engine with a common nozzle for the core and bypass. The velocity, pressure and area used at nozzle outlet are those for the bypass with subscript 19. With a choked convergent nozzle (i.e. not a convergent–divergent nozzle) some of the expansion of the stream takes place downstream of the nozzle exit plane. The exit plane forms the throat and, if the pressure ratio is larger than that for choking, the velocity is sonic at the exit plane. The jet is accelerated downstream of the exit plane to become supersonic as the static pressure tends to the ambient pressure. The supersonic jet velocity can be calculated by assuming an isentropic expansion down to the atmospheric pressure p_a, as was done in earlier chapters, because the pressure ratios for commercial engines are small enough that the irreversibilities are negligible. One of the principal variables we would like to scale between different conditions is the thrust of the engine. We need to distinguish between the gross thrust F_G and the net thrust F_N. The net thrust, it will be recalled from Chapter 3, is related to the gross thrust by

$$F_N = F_G - \dot{m}_{air}V,$$

where V is the flight velocity and $\dot{m}_{air}V$ is often referred to as the *ram drag* or *inlet-momentum drag*.

Retaining the assumptions that the mass flow of fuel is very small compared with the mass flow of air, the gross thrust is defined by

$$F_G = \dot{m}_{air}V_j.$$

The net thrust cannot be a function of the engine operating condition alone because the flight speed is involved. It is therefore more convenient to work with the gross thrust in obtaining non-dimensional groups and convert this to net thrust when needed. With choked propulsive nozzles the implicit dependence of engine non-dimensional variables on Mach number (or p_a/p_{02}) may be removed. The dynamic scaling of the engine gross thrust is therefore done in terms of p_{02} but the dependence of fully-expanded jet velocity V_j on p_a complicates this.

It is possible to apply conservation of momentum flux out of the nozzle and the pressure force acting on the nozzle, as in Figure 8.1, to obtain the gross thrust

$$F_G = \dot{m}_{air}V_j = \dot{m}_{air}V_{19} + (p_{19} - p_a)A_N, \tag{8.3}$$

where A_N is the nozzle area, V_{19} is the velocity at the nozzle exit plane and $(p_{19} - p_a)$ is the pressure difference between the nozzle exit plane and the surrounding static pressure of the atmosphere. Note that when the nozzle is choked V_{19} is the sonic velocity, which is less than fully-expanded value V_j achieved in the jet a short distance downstream because the jet expands to be supersonic with a drop in pressure from p_{19} to atmospheric pressure. For a choked nozzle the pressure at the convergent nozzle exit plane, p_{19}, is the static pressure corresponding to sonic flow, so for $c_p = 1.4$, $p_{19}/p_{019} = 1/1.893 = 0.528$. The area used in the Equation 8.3 is the nozzle outlet area because, as the jet expands in area downstream of the nozzle, the same ambient pressure will act over the sides of the jet as over the cross-section where the flow has expanded to atmospheric pressure; in other words the projected area of the sides of the jet exactly equal its increase in cross-sectional area.

Rearranging Equation 8.3 for gross thrust gives

$$F_G + p_a A_N = \dot{m}_{air} V_{19} + p_{19} A_N. \tag{8.4}$$

The significance of using gross thrust plus $p_a A_N$ now emerges, for all the terms on the right hand side, $\dot{m}_{air}$, V_{19}, and p_{19}, are wholly determined by the conditions *inside* the engine if the nozzle is choked. In consequence, once the operating point of the engine is fixed they depend only on inlet stagnation pressure and temperature. The non-dimensional form of the right hand side of Equation 8.4 can easily be shown to be

$$\frac{\dot{m}_{air} V_{19} + p_{19} A_N}{D^2 p_{02}} \tag{8.5}$$

or equivalently, in terms of the gross thrust,

$$\frac{F_G + p_a A_N}{D^2 p_{02}}. \tag{8.6}$$

It is clear that in the above expressions the term D^2 merely denotes a characteristic area, which can conveniently be replaced by the nozzle area A_N.

This treatment of thrust exemplifies a feature of dimensional analysis which is normally ignored: making the non-dimensional groups is comparatively easy, the problem is deciding what are the relevant variables. The relevant variable in the present case is $F_G + p_a A_N$ and *not* thrust alone. Analogous with equations of the form 8.2 we can say that

$$\frac{F_G + p_a A_N}{D^2 p_{02}} = F\left(T_{04}/T_{02}, p_a/p_{02}\right) \tag{8.7}$$

and with the propulsive nozzle choked the dependence on p_a/p_{02} is removed. Once the gross thrust has been found, the mass flow of air and the flight speed allows the net thrust to be obtained.

When the nozzle is unchoked it is a good approximation to take pressure at the nozzle exit plane to be equal to ambient static pressure

$$p_{19} = p_a \tag{8.8}$$

and in this case $\dot{m}_{air} V_{19} = F_G$.

To illustrate some of the functional dependence, Figure 8.2 shows results plotted in two different ways for a mixed-flow bypass engine (i.e. an engine in which the core and bypass flows are mixed before the propulsive nozzle) which entered service in the early 1980s. The single propulsive nozzle will be choked over operating conditions of interest. In each case specific fuel consumption, *sfc*, is shown versus thrust, and curves are drawn for constant flight Mach number. The remaining curves are for constant turbine inlet temperature in one case, and for constant speed of the HP shaft in the other. It can be seen that the lines of constant turbine temperature and constant speed are exactly parallel to one another, confirming that once one variable is chosen the other variables of the engine are also fixed. It may be noted from Figure 8.2 that the *sfc* rises with flight Mach number

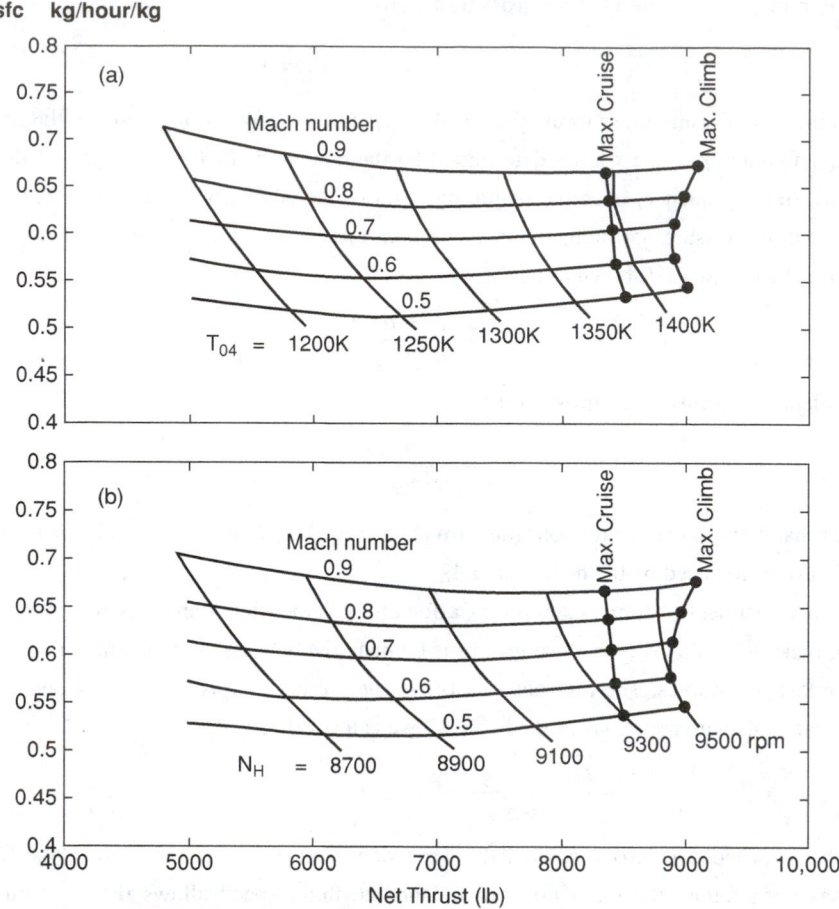

Figure 8.2. Performance maps for an existing mixed-flow engine, showing *sfc* versus thrust for constant altitude flight at various flight Mach numbers. Bypass ratio about 4.5. Case (a) shows the variation in terms of turbine inlet temperature, case (b) shows the same range in terms of HP rotational speed.

for constant engine speed and T_{04}. This is because the ram drag is increased and the net thrust reduced as Mach number increases. It may also be noted that for the same Mach number the *sfc* is slightly higher at maximum thrust than at somewhat lower thrusts. There are two factors which explain this. One is that most of the time during cruise the engine is not called upon to produce its maximum allowable thrust for the altitude, and it is therefore sensible to set the component maximum efficiency to occur at speeds and pressure ratios for typical cruise, which are lower than those for maximum thrust which would be encountered only for short times during climb. The other explanation is that reducing thrust requires a reduction in jet velocity which has the effect of increasing propulsive efficiency.

Exercises

8.1 Verify that for constant flight Mach number the ratio of inlet stagnation temperature to ambient static temperature, T_{02}/T_a, and inlet stagnation pressure to ambient static pressure, p_{02}/p_a, are constant. Use the constancy of p_{02}/p_a to show that the gross thrust will scale as $F_G/D^2 p_{02}$ for a constant engine non-dimensional operating point and Mach number, and thence that $F_G/D^2 p_a$ is constant too. (In other words gross thrust is proportional to ambient pressure for constant Mach number and engine condition.)

The net thrust, F_N, is equal to the gross thrust minus the ram drag, $m_{air}V$. Show that for a constant flight Mach number and engine condition $\dot{m}_{air}\sqrt{c_p T_a}/D^2 p_a = $ constant and hence that $\dot{m}_{air}V/D^2 p_a = $ constant. Use this to show that for a constant flight Mach number and engine non-dimensional operating point the net thrust, F_N, is also proportional to ambient pressure.

8.2 a A mixed-flow turbofan has a stagnation pressure upstream of the propulsive nozzle equal to 2.5 times inlet stagnation pressure when cruising at $M = 0.85$ through an atmosphere where the ambient temperature is 219 K and the pressure is 24 kPa. The mass flow of air entering the engine is 150 kg/s and the *sfc* is 0.800 kg hr^{-1}kgf^{-1}. The turbine inlet temperature is 1300 K and the stagnation temperature of the jet is 600 K. The temperature and pressure of the flow through the nozzle may be assumed uniform. Find the gross thrust and net thrust. Also find the nozzle area.

(**Ans:** $F_G = 94.3$ kN, $F_N = 56.4$ kN, $A_N = 0.945$ m^2)

b The engine is operated on a sea-level static test bed at the same non-dimensional condition (i.e. with the same pressure ratios and temperature ratios) as in part (a). Find the turbine inlet temperature. What is the gross thrust on the test bed and the ratio of this to net thrust at cruise. Find the air mass flow rate and fuel flow rate.

(**Ans:** $T_{04}=1495$ K, $F_G=212$ kN, thrust ratio $= 3.76$, $\dot{m}_{air}=368$ kg/s, $\dot{m}_f=3.61$ kg/s, $sfc=0.601$ kg s^{-1}kg^{-1})

8.4 PRACTICAL SCALING PARAMETERS

Engineers tend to resent writing more than is needed. In the non-dimensional group for mass flow of air it is clear that c_p is constant and D^2 will be constant for a given engine. Therefore the group

$$\frac{\dot{m}_{air}\sqrt{T_{02}}}{p_{02}} \tag{8.9}$$

is often used to describe mass flow, even though it has units. Similarly for the fuel flow the calorific value will normally be constant and an abbreviated form of the group is used,

$$\frac{\dot{m}_f}{\sqrt{T_{02}}p_{02}}, \tag{8.10}$$

again with dimensions. Notice, however, that the term involving the inlet stagnation temperature is in the numerator for the air flow but in the denominator for the fuel flow. This difference only emerges from a physical understanding of the processes involved *prior* to setting up the non-dimensional groups – recognising that the contribution of the fuel flow really contributes only as an energy input. A blind process following rules for dimensional analysis would treat the mass flow rate of fuel and the mass flow rate of air as equivalent and would not produce this difference.

The same practice of omitting constant terms is used for other variables so, for example, the non-dimensional speed is abbreviated to $N/\sqrt{T_{02}}$. This leads to the corrected speed

$$N/\sqrt{\theta}, \tag{8.11}$$

where $\theta = T_{02}/T_{02ref}$, and T_{02ref}, is a reference inlet temperature, typically 288 K. The magnitude of θ is usually fairly close to unity so corrected speed retains the advantage of being close to the actual speed, and therefore the intuitive sense of magnitude is retained. The same connection with the actual magnitude can be claimed for the corrected mass flow (units kg/s) derived from the non-dimensional expression

$$\frac{\dot{m}_{air}\sqrt{\theta}}{\delta}, \tag{8.12}$$

where $\theta = T_{02}/T_{02ref}$ and $\delta = p_{02}/p_{02ref}$.

Specific fuel consumption, $sfc = \dot{m}_f/F_N$ is clearly a parameter which has dimensions and is widely used. For comparing values at constant non-dimensional operating point the mass flow of fuel can be derived from 8.10 and the net thrust from the approach in Section 8.3.

8.5 LOSS OF THRUST FROM AN ENGINE

Any aircraft in civil operation has to be able to fly safely even if one of the engines fails completely, the engine-out condition. A particularly critical requirement is that it must be able to complete a take-off, climb and then land safely even if the engine fails at the worst possible moment, just as the plane is about to leave the ground. It must be able to do this without the pilot altering the engine throttles, so normal take-off is always with at least the minimum thrust to achieve the take-off safely if one engine were to fail; for a twin this means that with both engines operating the thrust is at least twice the minimum necessary. As already noted, at take-off both the core and bypass nozzles for a modern engine are likely to be unchoked and this means that a simple treatment with dimensional analysis is not really possible since performance depends on forward speed. This calls for the off-design treatment addressed in Chapter 12 and again in Chapter 19.

The other principal concern is the loss of thrust from an engine during cruise where things can be learned about the engine using dimensional analysis, as discussed below. Because the bypass nozzle is choked during flight at altitude, the flow through the engine fan and the bypass duct is unaffected directly by the ratio of inlet stagnation pressure to atmospheric pressure, p_a/p_{02}. The effect of this pressure ratio on the core stream is also proportionately small because of the large engine overall pressure ratio. The thrusts from both the bypass and the core are strongly affected by the inlet stagnation pressure p_{02}.

The aircraft must have adequate range to fly to an alternative airfield and it must be able to maintain adequate altitude to get over mountains on route. As discussed in Chapter 2, the aircraft will normally cruise at an altitude such that the lift-drag ratio is near its maximum. At this condition the engines should be sized such that when they are at their most efficient they produce just enough thrust to equal the drag. If the aircraft is a twin, like the New Efficient Aircraft, the loss of one engine would require a doubling of net thrust from the remaining engine to maintain cruise altitude. If it were a four-engine plane, like the Boeing 747 or Airbus A380, loss of one engine would require thrust to be increased by 4/3. To meet the loss of thrust from one of the engines the other(s) can operate at higher thrust; in practice this normally means allowing a higher turbine inlet temperature. Increasing T_{04} on its own is not normally enough if severe shortening of engine life is to be avoided. In a complete analysis of engine-out operation the optimum altitude and speed would be calculated to achieve lowest aircraft drag and highest engine thrust from the remaining engine (or engines if there are more than two) with modest increase in T_{04}. Nevertheless something can be learned by examining engine-out behaviour with the remaining engines maintained at the same non-dimensional condition as prior to the loss of another engine.

The normal procedure on losing thrust from an engine is for the aircraft to descend to a lower altitude where the higher density of the air allows the aircraft to fly more slowly and the remaining engine or engines to increase their thrust, with only a modest increase in T_{04}. To illustrate this we consider the case where the aircraft and the engines remain at cruise non-dimensional conditions as the aircraft descends to lower altitude. For the core to remain near its non-dimensional design condition T_{04}/T_{02} would be constant at its cruise value.

Cruise normally takes place at a lift coefficient, $C_L = L/(^1\!/_2 \rho V^2)$, such that the lift-drag ratio (more correctly VL/D) is close to its maximum. When the aircraft descends to a lower altitude the density increases so the speed is reduced to keep the lift coefficient close to the value for cruise which keeps the lift-drag ratio near its maximum as Figure 2.2 shows. If the aircraft weight does not change significantly as altitude is reduced and the lift coefficient is held constant it can be shown that $p_a M^2$ is constant; in other words the flight Mach number is inversely proportional to the square root of ambient pressure. Moving to lower altitude therefore reduces flight speed and this reduces the ram drag, giving an increase in net thrust relative to gross thrust. The variation in flight speed from an initial cruise altitude of 35,000 ft is shown by a curve in Figure 8.3.

Figure 2.2 shows some increase in L/D as the Mach number is reduced, but this would be offset by the drag of the 'dead' engine and drag due to the rudder force needed to balance the moment of the asymmetric engine thrust on the aircraft. It is therefore assumed here that L/D is constant as altitude is reduced. It is also assumed that after losing thrust in one engine the aircraft descends rapidly to the lower altitude required, so the aircraft weight does not change significantly. Then, if L/D remains constant, the total net thrust required for steady flight from the remaining engines at the new altitude is exactly equal to that at the cruise condition prior to the engine failure: this is the assumption adopted here. For a twin engine aircraft the thrust required from the remaining engine is therefore twice that required at cruise prior to loss of thrust in one engine.

The specific case considered here is when an engine is lost during cruise, which is taken here to be at an altitude of 35,000 feet. Because the core nozzle of a modern engine is not choked

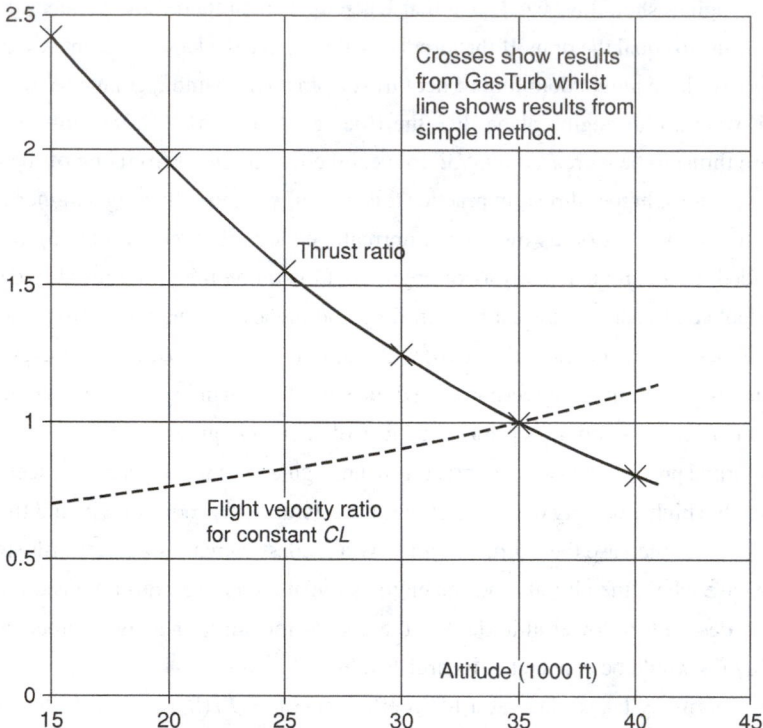

Figure 8.3. Predicted variation of flight speed and net thrust as function of altitude for the NEA after loss of thrust from one engine. Prior to shut-down of an engine the aircraft is cruising at 35,000 ft (10.67 km). Solid line shows bypass thrust with $fpr = 1.5$ held constant. Engine as in Exercise 7.2.

this means that the pressure ratio across the LP turbine is altered as flight Mach number is altered because this changes the ratio between the inlet stagnation pressure, p_{02}, and the ambient pressure acting on the outlet of the nozzle, p_a. We therefore restrict attention to the bypass flow, noting that the bypass nozzle is choked at the Mach numbers corresponding to high-altitude flight when p_{02} is substantially larger than p_a. We assume that the fan pressure ratio remains constant at its design value, $fpr = 1.5$, for the new flight altitude and consider *only* the thrust from the bypass. Since the bypass and core jet velocities are of similar magnitude and the bypass ratio is of order 10, this is quite a good approximation.

The variation in engine thrust as speed and altitude are changed is shown by a curve of F_N/F_{Nd} in Figure 8.3. This is the ratio of net thrust from the bypass stream when a constant fan pressure ratio is maintained compared to that at design. For a two-engine aircraft which loses thrust in one engine, the required value of F_N/F_{Nd} for steady flight is equal to 2. From Figure 8.3 it can be seen that the aircraft which was at 35,000 ft altitude must descend to about 19,000 ft, where the flight speed would be reduced to about 173 m/s from 231 m/s at cruise at 35,000 ft. This altitude would not allow the aircraft to fly over many mountain ranges and twin aircraft need to be careful. Were it a four-engine aircraft the increase in thrust to compensate for the loss of an engine must only be 4/3 and the new altitude with one engine out of action would be 28,170 ft. (Everest is 29,029 ft and a small increase in turbine entry temperature would take the four-engine aircraft over even that.)

To give some further support to the use of only the bypass stream, and of keeping fan pressure ratio constant, in assessing the effect of loss of engine thrust, some calculations have been performed using the program GasTurb which makes allowance for the variation in the core as flight condition changes. At five altitudes the flight Mach number has been calculated using $p_a M^2 = $ constant and the turbine entry temperature has been adjusted to keep T_{04}/T_{02} constant. The results of these calculations for F_N/F_{Nd} are shown as the crosses in Figure 8.3. The agreement is clearly satisfactory. Part of the explanation for this good agreement is that the pressure ratios computed by GasTurb, which allows for the non-choked propulsive nozzle, stay nearly constant, even though the core nozzle is not choked and flight Mach number is not constant. Thus, from GasTurb at 19,000 ft with $T_{04}/T_{02} = 6.11$ the fan pressure ratio is calculated to be 1.50, the same as the design value. The engine overall pressure ratio is calculated to have risen to 45.9 at the reduced altitude, fairly close to the design value of 45. In other words the engine has not deviated far from a constant non-dimensional condition, even though the core nozzle is not choked.

Useful information may evidently be obtained by assuming all non-dimensional groups to be constant if one is held constant. It is particularly plausible that the core of the engine, the HP compressor, combustor and HP turbine, will be operating near a constant non-dimensional condition if T_{04}/T_{02} is constant. A non-dimensional expression including fuel mass flow is shown in Table 8.1 and using this the ratio of mass flow rate of fuel for flight at 35 kft and 19 kft with T_{04}/T_{02} held constant is

$$\frac{\dot{m}_{f19}}{\dot{m}_{f35}} = \frac{p_{0219}}{p_{0235}} \sqrt{\frac{T_{0219}}{T_{0235}}},$$

where stagnation temperatures and pressures correspond to engine inlet at the two altitudes. The fuel flow is then fixed by the altitudes and the flight Mach number. From the change in fuel flow, and knowing the thrust, the variation in *sfc* can be found. Although the fuel flow is greater at the lower altitude, the thrust increase is greater still and as a result *sfc* is reduced. However, as discussed in Chapter 2, the aircraft range is proportional to V/sfc and the drop in V is larger than the drop in *sfc*.

Exercises

8.3 In Exercise 7.4 the *net* thrust required by the NEA from each engine at the start of cruise at an altitude of 35,000 ft and a flight Mach number of 0.78 was taken to be 38.9 kN. After allowing for the effect of the nacelle, the mass flow of air at this condition for an engine with *fpr* = 1.5 is found to be 392 kg/s. For the same flight Mach number, and with the engine at the same non-dimensional operating condition, calculate the mass flow of air and the gross and net thrust from each engine at an altitude of 41,000 ft ($p_a = $ 17.9 kPa, $T_a = 216.7$ K). Find also the ratio of fuel flow to that at 35,000 ft.

(**Ans:** $\dot{m} = 296.3$ kg/s, $F_G = 97.44$ kN, $F_N = 29.26$ kN, fuel flow ratio = 0.748)

If the lift-drag ratio of the aircraft is unaltered at 21.6, what is the maximum aircraft mass which can be propelled in steady flight at $M = 0.78$ and 41,000 ft with both engines operating at the same non-dimensional conditions?   **(Ans:** maximum mass $= 128.9$ tonne)

8.4 a In Exercise 7.7 the area of the bypass nozzle is computed to be 2.78 m². This is with a fan pressure ratio of 1.5 and cruise Mach number of 0.78 so that the bypass nozzle is choked. The bypass nozzle is convergent (no divergent section) so the Mach number is unity at the exit plane. The engine is to be tested on a sea-level static test bed and the fan performance should be the same as at cruise requiring that the fan pressure ratio and the non-dimensional mass flow at fan inlet and exit should be the same as at cruise. Find the bypass nozzle area ratio required for the static tests Use $\overline{m} = 1.281$ for the choked case at cruise and Equation 6.18 for the static case.   **(Ans:** 1.045)

    b At the design condition, 35,000 ft and $M = 0.78$, using the method of Chapter 7, the pressure ratio across the HP turbine is found to be 4.41 and across the LP turbine 10.58. The overall pressure ratio in the engine, $p_{04}/p_{02} = 45$. Confirm that a static test of the engine can *never* give the non-dimensional core behaviour seen in flight.

Note: Exercise 8.4 highlights a problem with the analysis of engines having low fan pressure ratio, such as that forming the basis of this chapter. Because the propulsive nozzles are not choked the simple treatment with non-dimensional variables cannot be used for conditions like take-off because the engine non-dimensional operating point alters. The analysis of Chapter 12, "Engine Matching Off-Design," does allow this.

8.5 a Suppose the NEA is cruising at $M = 0.78$ and 39,000 ft ($p_a = 19.7$ kPa, $T_a = 216.7$ K) with the optimum value of C_L. An engine is shut down and the plane descends to a lower altitude where it flies with the lift-drag ratio of the aircraft unchanged. If the new altitude is 23,925 ft, find the ambient pressure and temperature and the flight Mach number. If T_{04}/T_{02} remains at the design value of 6.11 find the turbine entry temperature in the operating engine.   **(Ans:** 39.4 kPa, 240.8 K, $M = 0.552$, $T_{04} = 1561$ K)

    b Find the ratio of fuel mass flow rate at 39,000 ft and 23,925 ft. Use this to determine the ratio of specific fuel consumption and then the ratio of range from position at which the engine failure occurs, assuming that the aircraft L/D is unaltered.
       (Ans: fuel flow ratio $= 1.687$, *sfc* ratio $= 0.843$, range ratio $= 0.884$)

Note: GasTurb calculates the specific fuel consumption, whilst holding $T_{04}/T_{02} = 6.11$, to be 12.8 g s⁻¹kN⁻¹ at 35,000 ft and 11.3 g s⁻¹kN⁻¹ at 19,000 ft, a ratio of 0.883.

SUMMARY OF CHAPTER 8

Using non-dimensional groups, or dimensioned groups derived from them, it is sometimes possible to estimate performance at speeds and altitudes different from those at which engines were designed or tests were performed.

    The crucial step is choosing or arranging the relevant variables. Once these are known it is straightforward to make them into non-dimensional groups. The significance of choosing the correct combination of variables is demonstrated when a non-dimensional group involving thrust is involved; the significance is also apparent when fuel mass flow is being considered. The non-dimensional expression for the fuel mass flow is very different from that for the air mass flow, reflecting the different effects of fuel and air in the engine. Simply forming non-dimensional groups without regard for the physical process will not always give the correct dependence.

In general any non-dimensional group for variables in the engine, such as mass flow of air or temperature ratio, will depend on two other non-dimensional groups. In the special case when the propulsive nozzle(s) is choked the dependence on ambient static pressure (or flight Mach number) disappears and one independent group suffices. Many engines operate with choked nozzles and the engine behaviour is then determined entirely by, for example, a group made up of the fuel flow, or the inlet stagnation temperature to the turbine and the inlet stagnation temperature to the compressor. Engine behaviour here refers to operating point and could include temperature ratios, pressure ratios in the engine, non-dimensional blade speeds (blade Mach numbers), as well as the mass flow of air.

If the propulsive nozzle is not choked the second independent variable can be the flight Mach number; if the flight Mach number is held constant then the engine behaviour is again effectively specified by one non-dimensional group such as temperature ratio.

To determine the gross thrust with a choked nozzle the ambient static pressure must also be included appropriately in the relevant variable, $F_G + p_a A_N$. To get the net thrust from gross thrust requires, in addition, the forward speed or flight Mach number. At a fixed Mach number the gross and net thrust from a given engine is proportional to the ambient static pressure.

With choked propulsive nozzles the non-dimensional treatment of engines is comparatively straightforward, with one non-dimensional variable sufficient. Modern engines, with low fan pressure ratio and correspondingly low pressure ratio across the core nozzle, are more difficult and two non-dimensional groups are normally required; the bypass nozzle is choked for cruise but not for take-off, whilst the core is not choked even at cruise. There therefore remains an influence of flight Mach number (or equivalently p_{02}/p_a) even when T_{04}/T_{02} is held constant. Because most of the thrust comes from the bypass stream adequate estimates can be made for engine performance at lower altitude after the loss of an engine at cruise by considering only the bypass flow. The core, with its large pressure ratio, is relatively little affected by the associated change in Mach number. For modern engines with low fan pressure ratio the assumption of constant non-dimensional operation is *not* adequate to relate conditions at cruise to take-off or to operation on a static test bed.

With loss of an engine during cruise the conditions for steady flight can be found. Losing thrust from one engine of a two-engine aircraft leads to a reduction in altitude of about 16,000 feet for steady flight at the same value of T_{04}/T_{02} and a loss of range of about 17%.

The engine has now been designed in outline and the next step in Chapter 9 is to design a compressor and a turbine to meet the requirements which were assumed in the cycle analyses.

CHAPTER 9   TURBOMACHINERY: COMPRESSORS AND TURBINES

9.0 INTRODUCTION

The compressor, which raises the pressure of the air before combustion, and the turbine, which extracts work from the hot high-pressure combustion products, are at the very heart of the engine. Up to now we have assumed that it is possible to construct a suitable compressor and turbine without giving any attention to how this might be done. In this chapter an elementary treatment is given with the emphasis being to find the overall features of the compressors and turbines, including the number of stages, the suitable rotational speeds, the diameters and some indication of the flowpath. The details of blade shape will not be addressed. Further information is obtainable in a recent book by Dixon and Hall (2013) and at a more specialised level for compressors in Cumpsty (2004).

The description of turbomachinery is based on the fan, compressor and turbine for the engines of the NEA. To avoid unnecessary duplication, the design is restricted to the case with a fan pressure ratio of 1.5 at cruise. At this condition the stagnation pressure and temperature entering the engine are $p_{02} = 35.6$ kPa and $T_{02} = 245.4$ K. From Exercise 7.4 the mass flow through the engine to give the required thrust is 392 kg/s and of this 32.1 kg/s goes through the core ($bpr = 11.2$). The turbine inlet temperature is 1500 K, the overall pressure ratio is 45 and the HP compressor pressure ratio is 18. From Exercise 7.6 the fan diameter is 2.66 m.

For the large engine that we are considering the most suitable compressor and turbine will be of the *axial* type. These are machines for which the flow is predominantly in the axial and tangential directions, and stand in contrast to radial machines for which the flow is radial at inlet or outlet.

Because the pressure rises in the direction of flow for the compressor there is always a great risk of the boundary layers separating, and when this happens the performance of the compressor drops precipitously and it is said to stall. To obtain a large pressure rise (or, as it is more commonly expressed, pressure ratio) the compression is spread over a large number of *stages*. A stage consists of a row of rotating blades (the *rotor*) and a row of stationary blades (the *stator*). In a modern engine compressor there may be between 10 and 20 stages between the fan outlet and the combustor inlet. Each rotor or stator row will consist of many blades, typically anywhere between 30 and 100.

In the turbine the pressure falls in the flow direction and it is possible to have a much greater pressure ratio across a turbine stage than a compressor stage; quite commonly a single turbine stage can drive six or seven compressor stages on the same shaft. A turbine required to produce a larger drop in pressure than is appropriate will work less well; that is the efficiency will be lower than a turbine with a more moderate pressure drop. Nevertheless a turbine always produces a power output and the flow will be in the intended direction. In the case of a compressor, however, an attempt to get more pressure ratio than is appropriate may result in rotating stall or surge, either of which is quite unacceptable. Because the operation of a turbine is normally easier to understand than a compressor, the principles of the turbine will be described before the compressor. First, however, it is appropriate to consider how the blades work.

9.1 THE BLADES FOR AXIAL COMPRESSORS AND TURBINES

The purpose of compressor and turbine blades is to turn the flow. By turning or deflecting the flow a force is exerted on the gas by the blade. In addition the turning of the flow alters the cross-sectional area of the streamtube going through the blade row; in the case of compressors the area is increased whilst for turbines the area is reduced. Reducing the area leads to an increase in gas speed, whilst increasing the area reduces the speed. The deflection and the acceleration of the flow are complementary and will be considered in this chapter.

Figure 9.1 shows the shape of sample compressor and turbine blades, together with some measurements of their performance. We need to distinguish between the directions in which the metal of the blades point, denoted by β here, and the direction in which the gas flows, denoted by α. In both cases the angles are measured from the *axial* direction. Relative to any blade row, compressor or turbine, we define the angles by

$$\text{gas angle into blade } \alpha_1 \qquad \text{blade metal angle in } \beta_1$$
$$\text{gas angle out of blade } \alpha_2 \qquad \text{blade metal angle out } \beta_2$$

The turning or deflection produced by the blades is the difference in flow direction between inlet and outlet, $\alpha_1 - \alpha_2$. The difference in blade angle between inlet and outlet, $\beta_1 - \beta_2$, is referred to as the camber.

The angle between the inlet gas and blade angles is defined as the incidence,

$$i = \alpha_1 - \beta_1,$$

which, when a blade row is being considered, is normally treated as an independent variable. The corresponding angle between the flow of gas leaving the blades and the blade outlet angle is defined as the deviation, given by

$$\delta = \alpha_2 - \beta_2$$

and this is normally treated as a principal dependent variable. The deviation is almost invariably defined to be positive with the indication that the flow does not turn quite as much as the blade camber suggests it should. The deviation depends on the design of the blade row and on the incidence and the inlet Mach number.

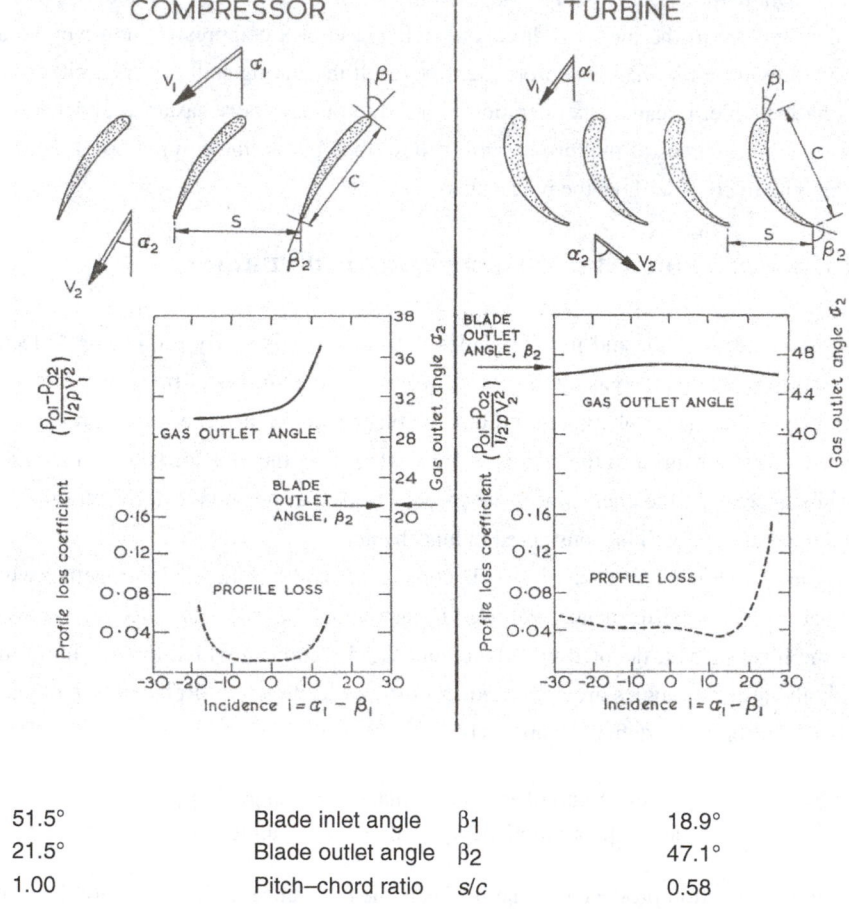

51.5°	Blade inlet angle	β_1	18.9°
21.5°	Blade outlet angle	β_2	47.1°
1.00	Pitch–chord ratio	s/c	0.58

Figure 9.1. Blading for axial compressors and turbines.

In Figure 9.1 outlet flow direction and stagnation pressure loss coefficient are shown versus incidence for compressor and turbine blades. The stagnation pressure loss is made non-dimensional by the flow dynamic pressure,[1] by convention inlet dynamic pressure for the compressor blades and exit dynamic pressure for the turbine blades. Loss coefficient is almost independent of incidence for the turbine blades, but for the compressor the incidence rises rapidly at positive incidence which corresponds to thickening of suction-surface boundary layers and ultimately flow separation, sometimes called blade stall.

Figure 9.1 also shows that the deviation is much larger for compressor blades than turbine blades, but in both cases the outlet flow direction (deviation) changes relatively little as the inlet flow direction (incidence) is altered, at least until the flow in compressor blade row gets close

---

[1] The results in Figure 9.1 were obtained in a low-speed cascade for which the flow was effectively incompressible. More generally the loss should be non-dimensionalised by $p_0 - p$ which is applicable to high-speed flow.

to separation. The compressor deviation in the compressor blade row of Figure 9.1 is actually about one third of the camber of the blades even for small incidence. This is not harmful, so long as the magnitude is estimated accurately, but is much larger than would be produced by more modern blading. (The old style of compressor blades used for Figure 9.1 had curvature right up to the trailing edge, but modern blades straighten out in the aft portion of the blades. The deviation from modern compressor blades is consequently much smaller, typically no more than $1°$ or $2°$.)

The large tolerance of the blades to incidence in Figure 9.1 is largely because the flow in the tests was almost incompressible; for the higher flow Mach numbers likely to be used in an engine the incidence would be maintained much smaller, particularly for the compressor blades. Throughout this chapter it suffices to consider and to draw the blades as if the incidence and the deviation were zero. Choosing the incidence to be zero is a design choice, at least at the design point, but deviation is a consequence of blade shape and of incidence.

It has been mentioned already that the rising pressure in the flow direction for the compressor blade makes the flow more difficult to control and limits the amount of turning and deceleration that can be achieved. It will be seen in Figure 9.1 that the compressor blade row at zero incidence turn the flow by about $20°$ whereas the turbine blade row turns the flow by about $63°$. (More modern turbine blades often turn the flow more than $90°$ with low loss and at Mach numbers around unity.) Another indication of the problems of compressors is the much narrower range of incidence for which the loss is small. When the loss starts to rise rapidly it is evidence of massive boundary layer separation; in the case of the compressor blades this coincides with a steep reduction in the amount of flow turning produced which can be expressed as a rapid increase in deviation.

The blades in Figure 9.1 show another important aspect of their function. For compressor blades the flow is turned so that it is more nearly axial at outlet than at inlet so the streamtube area is increased. For turbine blades it is the other way around, and the flow is more nearly tangential at outlet giving a reduction in streamtube area. It is a common design choice to make the axial velocity V_x nearly constant through a blade row (and indeed through most of the compressor or turbine). It is usually possible to neglect the radial velocity in jet engines unless there is a radial (centrifugal) compressor; but radial compressors are only used on small engines.

In the case of a row of compressor or turbine blades the velocity is given by

$$V_1 = V_x / \cos \alpha_1 \text{ at inlet and by } V_2 = V_x / \cos \alpha_2 \text{ at outlet}$$

If there is no variation in axial velocity,

$$V_1 / V_2 = \cos \alpha_2 / \cos \alpha_1.$$

For a compressor, given that $|\alpha_2| < |\alpha_1|$, it also follows that $V_2 < V_1$. In other words the flow is decelerated in a compressor blade row. In contrast in the turbine the flow is turned away from the axial direction, $|\alpha_2| > |\alpha_1|$, so that the flow is accelerated in a turbine blade row and because its velocity is increased its static pressure and static temperature fall.

9.2 FRAMES OF REFERENCE

It has been pointed out that blade rows accelerate the flow, positively in turbine blade rows and negatively in compressor rows. For the turbine there is a limit to how much acceleration is desirable in one blade row whereas for a compressor there is a practical limit on how much deceleration can occur without separation occurring. The essential feature of the turbomachine is that the rotor blades are moving relative to the stator blades and the effect can be best understood by adopting a frame of reference fixed to each blade row: the stationary frame of reference fixed to the stators and the moving or relative frame of reference fixed relative to the rotors. It appears to be conceptually easier to think of the turbine than the compressor, so we begin by considering it. Figure 9.2 shows one and a half stages of a turbine: a stator row, downstream of this a rotor row and then a second stator. The view is looking radially inward along the span of the blades and in practical turbines the rows are close together, the gap is perhaps 20% of the blade chord.

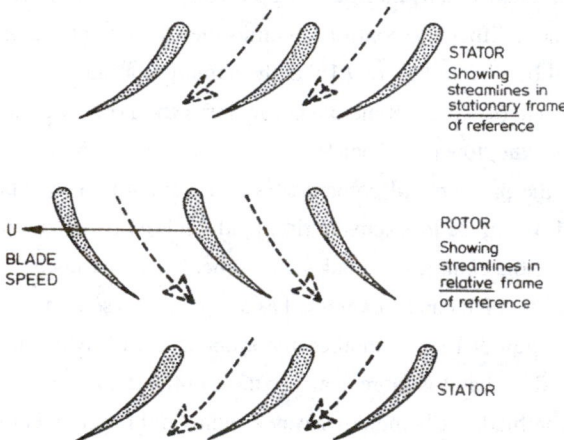

Figure 9.2. A schematic representation of one and a half stages of an axial turbine indicating the *steady relative* streamlines in each row.

The velocity of the rotor in the tangential direction U is normally not very different from the local speed of sound whilst the axial velocity (in the direction down the page in Figure 9.2) is smaller than this and normally approximately uniform from inlet to outlet. It can be seen that in both the rotor and the stator rows the flow is more nearly tangential at outlet than inlet, so for each turbine row $V_2 > V_1$. Because the rotor is moving in the same direction as the tangential velocity leaving the stator, the *relative* velocity into the rotor is less than the *absolute* velocity out of the stator. This makes it possible to accelerate the *relative* flow in the rotor row without getting excessive velocities. The *absolute* velocity leaving the rotor is less than the *relative* velocity out of the rotor and it is this lower *absolute* velocity which enters the second stator.

The flow into the rotor is unsteady and so too is the flow in the second stator – it is a flow of great complexity. Fortunately it was discovered a long time ago that an approximation with sufficient accuracy for engineering purposes is to treat the flow as steady in a frame of reference fixed to the blade row being examined. This approach is invariably used in the gas and steam turbine industries and is applicable to compressors as well as turbines.

For a stator row we use the *stationary* frame of reference and the velocities observed in this frame are conventionally described as *absolute*. For the rotor we use a frame of reference which moves with the rotor at speed U, and in this frame the velocity components are referred to as *relative*. The nomenclature adopted here is convenient for use by those still unfamiliar with turbomachinery, so absolute velocity out of the first stator would be written V_2 whereas the relative velocity will be written V_2^{rel}. The same nomenclature is adopted for relative flow angles α_2^{rel} and tangential components of relative velocity $V_{\theta 2}^{rel}$.

The axial velocity and radial velocity (though radial velocity is normally very small) are the same in both absolute and relative frames of reference – it is the tangential and resultant velocities which are altered. Static temperature and static pressure are also invariant with respect to frame of reference, though stagnation properties do change.

9.3 THE EULER WORK EQUATION

For both compressors and turbines the work exchange is described by the Euler equation, which we derive here. Figure 9.3 shows a hypothetical rotor of very general shape rotating with an angular velocity Ω. The flow enters at radius r_1 with a velocity in the tangential direction $V_{\theta 1}$ and leaves at radius r_2 with tangential velocity $V_{\theta 2}$. Consider an imaginary packet of fluid of mass $\delta m = \dot{m}\,\delta t$ that enters the rotor. This packet has a moment of momentum about the axis of rotation given by $\delta m r_1 V_{\theta 1}$. The corresponding moment of momentum for the same packet leaving at radius r_2 with velocity $V_{\theta 2}$ is $\delta m r_2 V_{\theta 2}$.

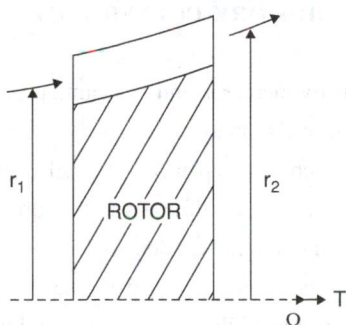

Figure 9.3. A hypothetical rotor for which flow enters at radius r_1 and leaves at radius r_2. The torque created is T and the rotor rotates at Ω radian/s.

Since the torque is equal to the rate of change of moment of momentum, this may be written in terms of the mass flow rate as

$$T = \dot{m}(r_2 V_{\theta 2} - r_1 V_{\theta 1}).\tag{9.1}$$

The power is then given by

$$\dot{W} = T\Omega = \dot{m}\Omega\left(r_2 V_{\theta 2} - r_1 V_{\theta 1}\right)$$
$$= \dot{m}\left(U_2 V_{\theta 2} - U_1 V_{\theta 1}\right), \tag{9.2}$$

where U_1 and U_2 are the speed of the blade row at inlet and outlet.

The power is also equal to mass flow rate times the change in stagnation enthalpy per unit mass, $\dot{W} = \dot{m}\,\Delta h_0$, and so by using Equation 9.2 we get

$$\Delta h_0 = U_2 V_{\theta 2} - U_1 V_{\theta 1}, \tag{9.3}$$

which is referred to as the Euler equation. In the case of a turbine rotor $V_{\theta 1} \gg V_{\theta 2}$ (i.e., the absolute tangential velocity is higher into the rotor than leaving it for a turbine) and the stagnation enthalpy falls, so the flow does work on the turbine. For a compressor $V_{\theta 1} \ll V_{\theta 2}$ and the stagnation enthalpy of the air rises so work is done by the compressor on the fluid. Very often an adequate approximation is to take $r_2 = r_1$ and this is adopted here. With this restriction one can write

$$\Delta h_0 = U\left(V_{\theta 2} - V_{\theta 1}\right), \tag{9.4}$$

which leads to the natural non-dimensional form for the *work coefficient*

$$\Delta h_0 / U^2 = V_{\theta 2} / U - V_{\theta 1} / U = \Delta V_\theta / U. \tag{9.5}$$

For the purpose of this very simplified treatment we will carry out calculations at the mean radius (half way between hub and casing). We will assume that this radius is constant across each compressor stage and across each turbine stage. For real machines the mean radius is not necessarily the same from front to back for a compressor or turbine as Figure 5.4 shows.

9.4 FLOW COEFFICIENT AND WORK COEFFICIENT

As the air is compressed the density increases and to maintain the axial velocity at an acceptably high value it is necessary to reduce the area – this is very clear in the cross-sectional drawings of engines such as Figure 5.4, which show span (i.e. the radial length) of the compressor blades decreasing progressively from front to back. Similarly in the turbine it is necessary to increase the area of the annulus as the gas expands and falls in density. There is no reason why the axial velocity should be precisely constant, but this is often not far from the truth at design point and it simplifies the present calculations. It has been found that compressors and turbines work most satisfactorily if the non-dimensional axial velocity, often called the *flow coefficient* V_x/U, is in a restricted range. For compressors the choice is normally $V_x/U \approx 0.4 - 0.75$, based on blade speed U at the mean radius. For turbines in the core $V_x/U \approx 0.5 - 0.65$ whilst for LP turbines $V_x/U \approx 0.7 - 1.1$.

Experience has allowed designers to choose combinations of work coefficient $\Delta h_0/U^2$ and flow coefficient V_x/U to give satisfactory performance. Figure 9.4 shows contours of efficiency for turbines with the work coefficient and flow coefficient as axes; superimposed are measured results

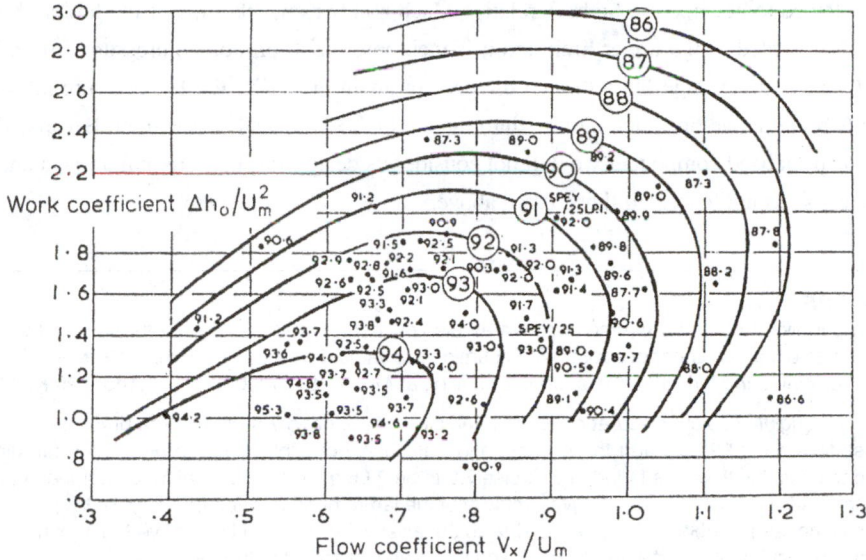

Figure 9.4. Variation of measured stage efficiency with stage loading and flow coefficient for axial-flow turbines (after Smith, 1965).

from actual machines. Very commonly the designer is unable to put the turbine working point at just the combination of V_x/U and $\Delta h_0/U^2$ which would give maximum efficiency for practical and/or geometric reasons and some compromise is accepted. In the case of compressors there is no diagram equivalent to Figure 9.4, but most stages are designed so that $\Delta h_0/U^2$ is in the range 0.35 to 0.5.

The temperature of the air increases through the compressor and so the speed of sound increases. As a result blade rows near the front of the compressor typically have a supersonic relative inlet flow whereas the stages near the rear are fully subsonic, even though the blade speed and flow velocity may be nearly equal at front and rear. If mean radius is held constant, and V_x/U is held constant from front to back, the axial Mach number must decrease through the compressor.

The fan is a very special compressor stage. It is invariably the case in modern civil engines that the fan has the rotor row as the first row and a stator row behind (i.e. no upstream stator row). The flow out of the fan divides, with most of it going down the bypass duct to a propulsive nozzle and a small fraction going into the core. The fan blades are long in relation to their axial extent; put another way, the ratio of the hub radius to the casing radius is small. A lower limit of about 0.25 can be put on this radius ratio, principally for mechanical reasons. The axial flow into the fan has a relatively high Mach number, about 0.6, and the mass flow per unit inlet area into the fan is about 85% of that required to choke the annulus were there no fan blades between hub and casing. Given the required mass flow of the engine the swallowing capacity of the fan determines diameter of the fan and effectively the size of the engine, as in Exercise 7.6. For reasons of efficiency, to prevent too much noise and to reduce the damage consequent on bird strike, the tip speed of the fan must not be allowed to become too high. A relative Mach number onto the tips of the fan of about 1.6 should be regarded as the upper limit and it is then possible to produce a pressure ratio of up to about 1.8 in a single stage fan with an efficiency of about 90%. For the fan pressure ratio of 1.5 chosen here a relative tip Mach number into the fan of 1.3 is plausibly assumed. Because the fan is such a specialised component we will not consider its design further here, but allow a choice of parameters so as not to exceed those listed above.

Exercises

9.1 a The intake around the engine decelerates the flow during cruise so that at the face of the fan the axial flow has a Mach number of about 0.6. Assuming this value, find the static temperature and the speed of sound at inlet to the fan when the aircraft is cruising at $M = 0.78$ at 35,000 ft.   (**Ans:** 229 K, 303 m/s)

b* The engine was specified at cruise in Exercises 7.2, 7.3 and 7.4. With a fan pressure ratio of 1.5 the mass flow was 392 kg/s and the bypass ratio was 11.2. For a hub-tip ratio of 0.3, the fan diameter required to pass this flow was found in Exercise 7.6 to be 2.66 m. Take the axial Mach number into fan to be uniform and equal to 0.6. The relative[2] velocity of the fan rotor tip is given by $\sqrt{(U_t^2 + V_x^2)}$ where $U_t = 2\pi\Omega_{LP}r_t$ is the fan rotor blade tip speed and V_x is the axial velocity. If the tip relative Mach number of 1.3 is plausible, calculate the blade tip speed U_t and the rotational speed of the fan, Ω_{LP}.
(**Ans:** $U_t = 350$ m/s, $\Omega_{LP} = 41.9$ rev/s)

c The average stagnation pressure ratio across the fan rotor for the core stream is 1.4, with efficiency 0.90. Find the stagnation pressure and temperature into the booster.   (**Ans:** 49.9 kPa, 273 K)

9.2 a* The hub radius increases through the fan and through the stator downstream of the fan. At the booster inlet the hub radius r_h and the tip radius r_t are 0.46 and 0.54 times the tip radius of the fan: the mean radius of the booster r_m is 0.5 times fan tip radius and this mean radius is constant from inlet to exit. The stators after the fan leave the flow direction at mean radius into the booster inclined at 20° from axial. The effective flow area at booster inlet is therefore $\pi(r_t^2-r_h^2)\cos(20°)$. The core mass flow can be found from Exercises 7.2 and 7.4 to be 32.1 kg/s. Use Equation 6.15 to confirm that the Mach number is equal to about 0.402. Hence find the resultant velocity at mid-span at fan core stator exit and thus the axial velocity. From this find the ratio of axial velocity to mean blade speed for the booster, assumed constant throughout.   (**Ans:** $V = 131$ m/s, $V_x = 123$ m/s, $V_x/U_m = 0.704$)

b Find the stagnation temperature and pressure at outlet from the booster if the pressure ratio for the core stream through the fan and booster is 2.5 and the efficiency is 90%. If the axial velocity is held constant through the booster, find the static properties and the air density at the booster exit. Hence find the flow area at exit from the booster and, if the area at booster exit is approximated by $2\pi r_m h$, find the blade height h.   (**Ans:** 327 K, 89.1 kPa, $\rho = 0.895$ kg/m^3, $A_{exit} = 0.291$ m^2, $h = 69.7$ mm)

c* The pressure ratio for the core flow through the fan is 1.4 so the pressure ratio across the booster is 2.5÷1.4. The stage loading coefficient $\Delta h_0/U^2$ for the booster is not to exceed 0.36. Find the minimum number of stages to give the required pressure ratio and the resulting value of loading coefficient, assumed equal for all stages.   (**Ans:** number of stages 5, $\Delta h_0/U^2 = 0.358$)

Note: The value of V_x/U_m is higher than would normally be allowed in HP compressors because the blade speed is low yet the axial velocity is relatively high. The low blade speed is because the LP shaft rotational speed is kept down by the constraint on fan tip Mach number. The high axial velocity is because the flow constraints in the layout keep the available flow area small.

9.3 a* The fan and LP compressor (the booster) give a pressure ratio of 2.5 to the core flow entering the HP compressor, with a combined efficiency of 90%. The flow at entry to the HP compressor is axial. Assume the inlet velocity is equal to 170 m/s (the true value will turn out to be close to this). Find the static

---

[2] Exercises 9.1b and 9.3b somewhat anticipate later parts of this chapter but all the necessary information is given in the questions.

temperature and pressure of the air at HP compressor entry and hence the density and flow Mach number. Supposing that the ratio of the hub radius to casing radius at inlet to the HP compressor is 0.60, find the mean radius at inlet and the hub and tip radii.

(**Ans:** $T = 313$ K, $p = 76.1$ kPa, $\rho = 0.848$ kg/m^3, $M = 0.480$, $r_m = 0.267$ m, $r_t = 0.333$ m, $r_h = 0.200$ m)

b* Throughout the HP compressor $V_x/U_m = 0.55$, where U_m is the blade speed at mid-span, and the axial velocity V_x is assumed to be uniform. Given that for the first stage the hub radius is 0.60 times the tip radius, show that the resultant relative velocity into the tip is $U_t (1 + 0.44^2)^{1/2}$, where U_t is the blade tip speed. If the relative Mach number into the tip is to be 1.2, find the resultant relative velocity at the tip of the first blade, the blade tip speed and the rotational speed of the HP shaft. What is the axial velocity given $V_x/U_m = 0.55$? (**Ans:** $V_{trel} = 425$ m/s, $U_t = 389$ m/s, $\Omega_{HP} = 186.0$ rev/s, $V_x = 171.3$ m/s)

9.4 a* At HP compressor exit the flow is purely axial and the mean radius r_m from Exercise 9.3a is equal to that at inlet. The pressure ratio of the HP compressor is 18 and the compressor efficiency is 90%. Find the static temperature and pressure at compressor exit and thence the density. Find the cross-sectional area at outlet from the compressor assuming that the axial velocity is constant throughout at 171 m/s. Then, assuming that the outlet area may be calculated using $2\pi r_m h$, where h is the blade height, find the blade height at outlet from the compressor. (**Ans:** $\rho = 6.72$ kg/m^3, $A = 0.0279$ m^2, $h = 16.7$ mm)

b* If the stage loading for the HP compressor $\Delta h_0/U^2 \le 0.45$, find the minimum number of stages and the corresponding value of stage loading. (**Ans:** 11 stages with average $\Delta h_0/U^2 = 0.439$)

Note: Because the turbines are more tolerant of operating conditions it is the fan and compressors which dominate the choice of rotational speed; here with Exercise 9.1b for the fan and Exercise 9.3 and 9.4 for the HP compressor. This is.

9.5 THE AXIAL TURBINE

The turbine can be visualised as a series of expansions. In each blade row the pressure and enthalpy fall whilst the velocity increases. If the high-pressure gas in the combustion chamber were expanded in a single expansion, a very high velocity indeed (about 1450 m/s) would be produced. Such a high velocity would be impossible to use efficiently. The 'trick' of the turbine is to make a series of smaller expansions, typically to velocities close the speed of sound and then, by changing the frame of reference, to apparently reduce velocity on entry to the next blade row. This can be seen for the stage in Figure 9.5: the velocity leaving the stator is high in the *absolute* frame of reference appropriate to the stator, but is much lower when seen by the rotor at entry. Likewise the velocity leaving the rotor is high in the *relative* frame of reference appropriate for it, but lower in the *absolute* frame of reference into the next stator. Each of the turbine blade rows takes in a flow which is not very far from axial and turns it towards the tangential, thereby reducing the flow area, increasing the velocity, reducing the static pressure and reducing static temperature.

The outlet conditions from the turbine stators, often called nozzles, become the inlet conditions to the rotor and so we need a simple way of going from absolute to relative and back again. The approach always adopted is to use *velocity triangles* and these are shown in Figure 9.5. Recall that we denote *absolute* velocities by V and *relative* velocities by V^{rel}. The flow enters stator row 1 with velocity V_1 inclined at angle α_1 to the *axial* direction. At this station the axial and tangential components are given by

$$V_{x1} = V_1 \cos \alpha_1 \text{ and } V_{\theta 1} = V_1 \sin \alpha_1, \text{ respectively.}$$

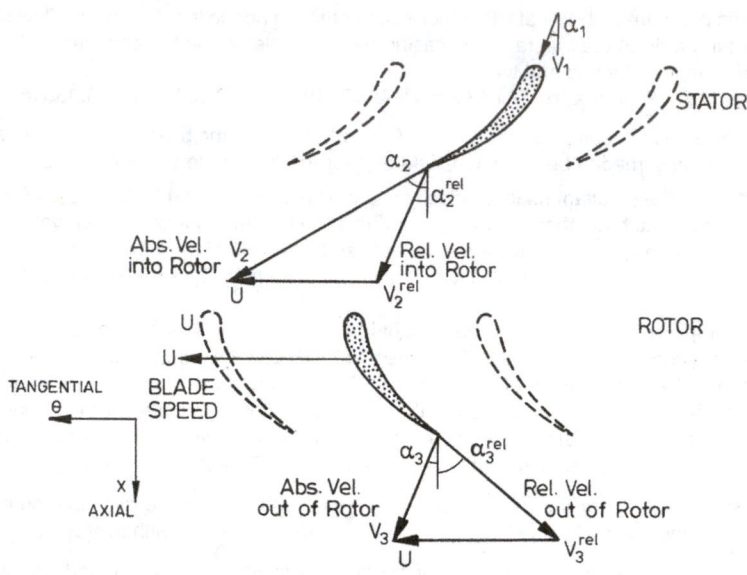

Figure 9.5. An axial turbine stage showing the velocity triangles into and out of the rotor.

At outlet from the stator the absolute velocity is denoted by V_2 and, just as for V_1, this can be resolved into its axial and tangential components. The relative velocity seen by an observer at inlet to the rotor, V_2^{rel}, may be obtained from the vector additions in the triangle drawn. As noted above, the *axial* velocity is equal in both the absolute and relative frames of reference, $V_{x2}^{\text{rel}} = V_{x2}$, but the tangential velocities are different so that

$$V_{\theta 2}^{\text{rel}} = V_{\theta 2} - U. \tag{9.6}$$

In the convention adopted here tangential velocity is positive if it is in the same direction as the rotor velocity U. As drawn in Figure 9.5, both the absolute swirl velocity $V_{\theta 2}$ and the relative swirl velocity $V_{\theta 2}^{\text{rel}}$ are positive. Flow angles are positive if they produce positive tangential velocity. From the components of velocity it is easy to obtain the absolute and relative flow angles so, for example,

$$\tan \alpha_2 = V_{\theta 2}/V_{x2} \quad \text{and} \quad \tan \alpha_2^{\text{rel}} = V_{\theta 2}^{\text{rel}}/V_{x2}$$

or
$$\cos \alpha_2 = V_{x2}/V_2 \quad \text{and} \quad \cos \alpha_2^{\text{rel}} = V_{x2}/V_2^{\text{rel}} \tag{9.7}$$

and so on for other possible combinations.

Conditions leaving the rotor are handled in the same way so, for example,

$$V_{\theta 3}^{\text{rel}} = V_{\theta 3} - U. \tag{9.8}$$

Inherent in Equations 9.6 and 9.8 is the simplification that the blade speed U is taken to be equal at inlet and outlet to the rotor, in other words the streamlines do not shift radially as they pass through the rotor blades. With this restriction to constant radius it is easy to show that the stagnation enthalpy is equal at inlet to and outlet from the rotor if the *relative* co-ordinate system is used, in other words

$$h_{02}^{\text{rel}} = h_{03}^{\text{rel}},$$

where

$$h_{02}^{\text{rel}} = h_2 + \left(V_2^{\text{rel}}\right)^2 / 2 \quad \text{and} \quad h_{03}^{\text{rel}} = h_3 + \left(V_3^{\text{rel}}\right)^2 / 2. \tag{9.9}$$

This is analogous to the situation in stator rows for which the stagnation enthalpy in *absolute* co-ordinates is conserved, that is $h_{02} = h_{01}$. (For the stator, however, this is true even if the streamline does not remain at the same radius.) In the absence of losses it would also be true that the relative stagnation pressure would be conserved in the rotor and the absolute stagnation pressure in the stator. In fact losses do occur and between 3% and 6% of the exit dynamic pressure is typically lost passing through a turbine blade row.

The Euler work equation shows that the drop in *absolute* stagnation enthalpy through the stage, is given for the stage in Figure 9.5 by

$$\Delta h_0 = U_2 V_{\theta 2} - U_3 V_{\theta 3}$$

$$= U \left(V_{\theta 2} - V_{\theta 3} \right) \tag{9.10}$$

with the restriction to streamlines at constant radius. Note that Δh_0 is negative for a turbine. This can be rewritten in terms of relative tangential velocity as

$$\Delta h_0 = U \left(V_{\theta 3}^{\text{rel}} - V_{\theta 2}^{\text{rel}} \right), \tag{9.11}$$

where V_θ and V_θ^{rel} are positive if in the same sense as the blade speed. If the axial velocity is chosen to be constant, Equation 9.10 then simplifies to

$$\Delta h_0 = U V_x \left(\tan \alpha_2 - \tan \alpha_3 \right), \tag{9.12}$$

which, it can be shown, is also equal to

$$\Delta h_0 = U V_x \left(\tan \alpha_3^{\text{rel}} - \tan \alpha_2^{\text{rel}} \right), \tag{9.13}$$

where α and α^{rel} are positive if they correspond to V_θ and V_θ^{rel} which are positive.

This work exchange takes place even when there are losses present. The effect of the losses is to make the pressure drop greater than it would be for the same temperature drop in an isentropic (loss-free) machine. The isentropic efficiency was defined in Section 4.2. If the design of the machine is unsatisfactory, however, these losses can become large, and it is for this reason that designs are normally restricted to ranges of $\Delta h_0 / U^2$ and V_x / U known to give satisfactory performance. Figure 9.4 gives an idea of the parameter range normally adopted. Particularly for the LP turbine, which tends to be big and heavy, there is a tendency for both flow and work coefficients to be at the higher end of what is allowable.

Figure 9.4 does not, however, give any guidance on the number of blades needed in a row or, more correctly, the ratio of the pitch of the blades to the chord. There are several reasons to want to reduce the number of turbine blades. Firstly they are costly and heavy. Secondly, where the blades must be cooled, more blades means more cooling air, which represents a source of loss in cycle (thermal) efficiency. (The same increase in cooling air comes if longer-chord blades are used.) More blades also means more blade surface area which increases the viscous drag. To be set against these incentives to reduce blade numbers and chord, there must be enough blades to produce the turning of the flow without boundary layer separation (after separation the losses are generally much greater[3]) and without too much disturbance of the flow on the endwalls (which can separate and lead to further sources of loss). One well known criterion for turbine blades is the Zweifel loading coefficient. This is the ratio of the tangential force per unit span on a blade, F_t, divided by a crude idealization of the maximum possible tangential force. The idealised force is the axial chord c_x multiplied by the difference in pressure between the *inlet stagnation* and the *exit static* pressure. This can be written for unit axial height as

$$Z_w = F_t / \{c_x \, (p_{0in} - p_{out})\}. \tag{9.14a}$$

If h is the blade height, $\dot{m}$ is the mass flow through one blade passage and ΔV_θ is the turning of the flow through the blade row the Zweifel coefficient can be rewritten as

$$Z_w = \dot{m}\Delta V_\theta / \{hc_x(p_{0in} - p_{out})\}. \tag{9.14b}$$

All the pressures and velocities, as well as the blade axial chord, vary along the span of the blade but it suffices for the purpose of this chapter to evaluate these at mid-height using average quantities. The mass flow through a blade passage is then proportional to blade height and average blade pitch, axial velocity and density, $\dot{m} \approx \rho h s V_x$.

If the flow can be approximated as locally two-dimensional, so V_x is equal at inlet and outlet, the change in tangential velocity can be written in terms of flow direction change as

$$\Delta V_\theta = V_x \, (\tan \alpha_{out} - \tan \alpha_{in}).$$

The term $p_{0in} - p_{out}$ in Equation 9.14 is equal to the dynamic pressure at outlet, if there are no losses. For the purpose of finding a criterion for blade design it suffices to use an incompressible expression, $p_{0in} - p_{out} = \frac{1}{2}\rho V_x^2 / \cos^2 \alpha_{out}$.

As stressed in Chapter 6 it is *not* correct to use an incompressible expression, but the error is absorbed into the chosen, arbitrary level set for an acceptable value of the coefficient. With these assumptions and approximations the Zweifel coefficient may be written

$$Z_w = (2s/c_x)\{\cos^2 \alpha_{out}(\tan \alpha_{out} - \tan \alpha_{in})\}. \tag{9.14c}$$

---

[3] Although the overall gradient of pressure in turbines is favourable strong adverse pressure gradients occur locally on the suction surface (i.e. the convex surface) of turbine blades and this can lead to separation.

Table 9.1   Approximate empirical guidelines for turbine design

Aspect ratio h/c	0.75–2.5	for LP turbines up to about 4
Pitch-chord ratio s/c	0.75–1.5	
Zweifel coefficient	< 0.8	for LP turbines up to 1.0
Flow coefficient V_x/U	0.5–0.65	for LP turbine 0.7–1.1
Work coefficient $\Delta h_0/U^2$	0.8–2.0	for LP turbine up to 2.4
Blade or vane exit Mach no.	0.85–1.1	for LP turbines down to 0.7

As a general design rule Z_w should not exceed 0.8, but in some cases, such as LP turbines, values up to about 1.1 are allowed. Note that as a rule α_{out} and α_{in} are of opposite sign so the tangents are additive in Equation 9.14c.

The results in Table 9.1 give some guidance in choosing the layout of axial turbines. These limits are not absolute but are evidence of where satisfactory performance has been achieved in the past. There are sometimes conflicting requirements, so that some "pushing at the boundaries" is allowed. Pitch-chord ratio follows from the reasoning of the Zweifel coefficient. The lower limit on aspect ratio comes from the effect of the endwalls, both the loss associated with viscous effects directly and loss associated with the clearance at each end of the blades. So, for example, a rule of thumb for unshrouded turbine blades is that $\Delta\eta/\eta = 1.5$ (tip clearance/blade span).

Exercises

Some of these do not have unique 'right' answers, they are open ended design exercises, for which one aims for the best solution.

9.5 The flow angle out of a turbine stator, which is the absolute flow angle into the rotor, is α_2 and the relative flow angle into the rotor is α_2^{rel}. Likewise out of the rotor and into the next stator the absolute and relative angles are α_3 and α_3^{rel} As noted, angles are taken as positive when they produce a swirl velocity in the same direction as the rotor. Show that if axial velocity is uniform through the stage:

$$V_x/U = 1/\left(\tan\alpha_2 - \tan\alpha_2^{rel}\right) = 1/\left(\tan\alpha_3 - \tan\alpha_3^{rel}\right)$$

and

$$\Delta h_0/U^2 = V_x/U\left(\tan\alpha_2 - \tan\alpha_3\right) = V_x/U\left(\tan\alpha_2 - \tan\alpha_3^{rel}\right) - 1.$$

Note: for a turbine it is normal for the signs of α_3 and α_3^{rel} to be the same, likewise α_2 and α_2^{rel}, so in the expression for V_x/U the tangent terms subtract. α_2 and α_3 are of opposite sign so the tangent terms add in the expressions for $\Delta h_0/U^2$.

9.6* The power from the HP turbine must equal the power into the HP compressor. Here we assume equal mass flow rate in compressor and turbine, neglecting cooling air and fuel mass flow rate. The enthalpy drop in the turbines is thus fixed by the pressure ratio of the compressors and the inlet temperature. Exercise 9.4 has fixed the rotational speed of the core shaft so the mean blade speed for the HP turbine depends only on the mean radius, which we take as equal for each stage of the HP turbine. Two stages will suffice and the flow leaving the second stage must be axial in the absolute frame because it will enter a duct on its way to the LP turbine. To maintain efficiency and to allow for the axial-flow constraint on the second stage, the loading coefficient, based on mean blade speed, $\Delta h_0/U_m^2$, is to be 2.2 for the first stage and 1.8

for the second stage. Find Δh_0 for the each HP turbine stage. Find the mean-height radius consistent with the prescribed loading coefficients and, based on a sketch of the engine, comment on its suitability.

(**Ans:** $\Delta h_{01} = 258$ kJ/kg, $\Delta h_{02} = 211$ kJ/kg, $r_m = 0.296$ m)

Note: The mean radius is similar to the mean radius of the HP compressor, so the outline shape of the engine core is promising.

9.7 Exercise 9.6 has fixed the non-dimensional work output from stages of the HP turbine at mid-radius and the mid-height radius is equal for both stages. Take $V_x/U_m = 0.50$ and assume that the axial velocity, as well as mean blade speed, is uniform from inlet to outlet.

a For stage 1 assume axial inlet flow to the stator and an absolute flow direction out of the rotor of $-30°$, that is absolute swirl in the *opposite* sense to the velocity of the blades. Determine the absolute flow direction out of the stator and the relative angles into and out of the rotor blades to achieve the desired loading and flow coefficients, $\Delta h_0/U_m^2$ and V_x/U_m. Sketch velocity triangles and the corresponding blade cross-sections. Assume that the flow direction is the same as the blade inlet and outlet direction, zero incidence and deviation.All angles are measured from the axial direction.

(**Ans:** $\alpha_2 = 75.3°$, $\alpha_2^{rel} = 61.2°$, $\alpha_3^{rel} = -68.8°$)

b For stage 2 assume the direction into the stator is obtained from Exercise 9.7a. The absolute flow direction out of the second stage rotor is purely axial. Determine the absolute and relative flow directions out of the second stator and the relative angle out of the rotor blades to achieve $\Delta h_0/U_m^2 = 1.8$ and $V_x/U_m = 0.55$. Sketch velocity triangles and the corresponding blade cross-sections. Again assume that the flow direction is the same as the blade outlet direction, i.e., zero incidence and deviation. All angles are measured from the axial direction.   (**Ans:** $\alpha_2 = 74.48°$, $\alpha_2^{rel} = 58.0°$, $\alpha_3^{rel} = -63.4°$ $\alpha_3 = 0°$)

9.8 a* The length of the HP turbine blades in the radial direction has not yet been chosen. Assume that the flow is choked at outlet from the stator blades so that $\overline{m} = 1.281$. Knowing the mass flow, the inlet stagnation temperature (which is 1500 K at cruise) and the pressure (assumed equal here to that at compressor exit, 1604 kPa), find the area of the flow passage at exit from the first row of stator blades (often called nozzle guide vanes, NGV). The flow is inclined at α_2 to the axial direction so the flow area A may be approximated by $A = 2\pi r_m h \cos \alpha_2$ where the blade height h is assumed short relative to the mean radius r_m. Use the mean radius from Exercise 9.6 to find h at NGV exit, which is assumed equal at inlet to height of the first rotor.   (**Ans:** $A = 0.0192$ m^2, $h = 41.2$ mm)

Note: This defines the radial length (span) of the first stage HP turbine rotor.

b* Repeat Exercise 9.7a for the second stage assuming that the NGV for this stage are also choked. (First find the stagnation temperature and pressure at inlet to the second stage of HP turbine. The first-stage temperature drop is known from the enthalpy drop and the pressure drop can be found when the efficiency is assumed to be 90%.)

(**Ans:** $T_0 = 1243$ K, $p_0 = 767$ kPa, $A = 0.0365$ m^2, $h = 74.2$ mm)

c For the core turbine the aspect ratio based on the *axial projection* of chord (aspect ratio is the span of the blade divided by its chord) should not be less than about one and should not exceed about 2.5. Sketch an outline of the core turbine in the axial–radial plane.

9.9 a* Find the enthalpy drop in the LP turbine using the temperature rise of the bypass stream and the bypass ratio and the temperature rise in fan root and booster.   (**Ans:** 459 kJ/kg)

b* The rotational speed of the LP shaft is fixed by the requirement of the fan tip speed in Exercise 9.1b to be 41.9 revs/s. If the LP turbine blade speed can be high, which means making the mean radius large, we can have fewer stages for the same loading coefficient. If one is not careful, however, the flow path needed to get the large mean radius becomes very unsatisfactory, both inside the turbine and in the bypass stream. For the LP turbine we must choose how many stages and up to seven would be acceptable.

For the LP turbine assume that $\Delta h_0/U_m^2 = 2.20$. Assume that the mean-height radius of the LP turbine is constant from front to back. For possible numbers of LP turbine stages find the required mean radius. Make some sketches of possible layouts, allowing the aspect ratio based on axial projections of chord to rise in this case to around 6.   (**Ans:** with 7 stages $r_m = 0.656$ m, with 6 stages $r_m = 0.709$ m)

Notes: The presumption of constant radius is to keep things simple and is adequate for the present purposes but this is not adhered to in real machines, see Figure 5.4. Similarly $\Delta h_0/U_m^2$ would not be identical for all stages.

The benefit of increased numbers of stage has been taken here as a reduction in radius, but, keeping radius constant, it could instead be taken as a reduction in $\Delta h_0/U_m^2$ and an increase in isentropic efficiency. For the present purposes we are not considering changes in loss and efficiency.

c The first and last stages are somewhat different from the embedded stages, but consider a stage near the middle of the LP turbine. At blade mid-span assume that the flow deflection is equal in rotor and stator, so the blades and the velocity triangles are like mirror images. For $V_x/U_m = 1.0$ sketch the blade sections and velocity triangles for a stage. Find the absolute flow direction into and out of the stator and the relative flow direction into and out of the rotor.

(**Ans:** $\alpha_2 = 58.0°$, $\alpha_2^{rel} = 31.0°$, $\alpha_3^{rel} = -58.0°$, $\alpha_3 = -31.0°$)

d If the Zweifel coefficient at mid-span for rotor and stator is to be 0.9, find the ratio of axial chord to blade pitch. For this part treat the flow as incompressible. (**Ans:** $c_x/s = 1.24$)

9.6 THE AXIAL CORE COMPRESSOR

Although compressors can be axial or radial, the latter often referred to as centrifugal compressors, the majority of turbofans use only axial compressors and this is always the case for large engines such as that which is the subject of this book. Figure 9.6 shows a compressor stage and the corresponding velocity triangles. At entry to the rotor the *absolute* velocity is V_1 inclined at α_1 to the axial. In the *relative* co-ordinates, corresponding to what an observer on the rotor would perceive, the velocity is V_1^{rel} which is inclined at α_1^{rel} to the axial direction.

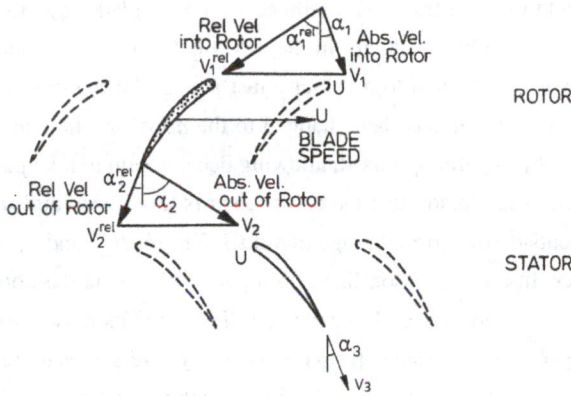

Figure 9.6. An axial compressor stage showing the velocity triangles into and out of the rotor.

For the flow into the rotor,

$$V_{x1} = V_{x1}^{rel} \quad \text{and} \quad V_{\theta 1} = V_{\theta 1}^{rel} + U \tag{9.15}$$

and likewise for the flow out of the rotor. The same trigonometric expressions as those for the turbine link the velocities, for example

$$\tan \alpha_1 = V_{\theta 1}/V_{x1} \quad \text{and} \quad \tan \alpha_1^{rel} = V_{\theta 1}^{rel}/V_{x1}.$$

In compressors V_θ and V_θ^{rel} are normally of opposite sign with the absolute positive and the relative negative; likewise α and α^{rel}. For the conditions in Figure 9.6 the work input, which is equal to the stagnation enthalpy rise per stage, is given by Euler's equation and is

$$\Delta h_0 = U_2 V_{\theta 2} - U_1 V_{\theta 1}$$

$$= U \left(V_{\theta 2} - V_{\theta 1} \right)$$

with the restriction to constant radius streamlines.

If the axial velocity is constant

$$\Delta h_0 = U V_x \left(\tan \alpha_2 - \tan \alpha_1 \right) \tag{9.16}$$

or equivalently

$$\Delta h_0 = U V_x \left(\tan \alpha_2^{\text{rel}} - \tan \alpha_1^{\text{rel}} \right). \tag{9.17}$$

Whereas for a turbine the flow deflection may be $90°$ or more, for a compressor it is rarely more than about $45°$. As a result $\Delta h_0 / U^2$ is several times smaller for a compressor stage than a turbine stage. In the frame of reference of a blade row the flow is decelerated when it is turned towards the axial direction so that $\left| \alpha_2^{\text{rel}} \right| < \left| \alpha_1^{\text{rel}} \right|$.

The velocity of a flow in a diffuser or compressor blade row can seldom be reduced below about 50% of its entry value because the boundary layers tend to separate. Once the flow has separated the 'wake' can expand to fill the area, thus effectively removing the increase of area implied by the diverging solid walls and preventing further reduction in velocity. In a compressor the 'trick' is to carry out the deceleration in a large number of steps, each one raising the pressure by a small amount, and to switch frame of reference after each blade row: the low velocity at exit from one blade row becomes high velocity in the next blade row downstream. Thus in the rotor in Figure 9.6 the flow is decelerated to a velocity just above that likely to cause boundary layer separation. The frame of reference is then changed to the *absolute* one appropriate for the stator and the velocity is thereby apparently raised, allowing deceleration to take place in the stator.

It has been remarked already that the compressor is less 'forgiving' than a turbine. With a turbine an excursion outside the normal range of values for $\Delta h_0 / U^2$ and V_x / U is likely to lead to some loss in efficiency. In a compressor, however, operation outside the normal bounds is likely to lead to the machine not working at all as expected. If the area increase is sufficient to cause the boundary layers to separate, the blockage from the boundary layers rises to diminish the amount of flow diffusion (and therefore pressure rise) and to increase the deviation steeply. The consequences of this are to be seen in Figure 9.1, which shows the fairly narrow range over which the compressor blade deflects the flow as intended and the rapid rise in deviation when this is exceeded. For the stator this would mean that α_1 and α_3 would be substantially larger than the design intent, and likewise for the rotor α_2^{rel} would be increased. As a result the work input drops sharply.

For compressors a common criterion used for preliminary assessment that a design specification is reasonable is the diffusion factor. The diffusion factor, written for a stator blade row but valid for a rotor row with change of symbols, is

$$DF = \left(1 - V_2 / V_1 \right) + \Delta V_\theta / \left(2 \sigma V_1 \right), \tag{9.18a}$$

Table 9.2   Approximate empirical guidelines for compressor design

Aspect ratio h/c	0.75–2.5, for fans up to about 4
Pitch-chord ratio s/c	0.6–1.5
Diffusion factor DF	< 0.5
Flow coefficient V_x/U	0.45–0.75, perhaps higher in LP compressor (booster)
Work coefficient $\Delta h_0/U^2$	0.35–0.5
Blade or vane inlet Mach no.	<1.0, but above $M_{rel} > 1$ at tips of fans and front stages

where V_1 and V_2 are inlet and outlet velocities, ΔV_θ is the change in tangential, or swirl, velocity across the blade row and $\sigma = c/s$, the ratio of blade chord over blade pitch. This is described in Cumpsty (2004). The first term is the one-dimensional deceleration, as in a straight diffuser, and this is affected by blade geometry only through the area ratio. The second term allows for the pressure gradients caused by turning the flow, which involves acceleration and deceleration. The magnitude of the pressure gradients to produce this turning depends on σ, the blade solidity. For a given level of diffusion factor the solidity must be higher, or pitch-chord ratio lower, if the blades change V_θ by a large amount. If the flow is at constant radius and constant axial velocity, the diffusion factor can be rewritten in terms of angles as

$$DF = \left(1 - \cos\alpha_1 / \cos\alpha_2\right) + \left(\tan\alpha_2 - \tan\alpha_1\right)\cos\alpha_1/2\sigma. \tag{9.18b}$$

For most designs the diffusion factor is kept to below about 0.45, though sometimes this is allowed to be a little higher. It should be realised that the diffusion factor was conceived as a design-point parameter, not a limit when the compressor is throttled towards stall. For the present purposes it is appropriate to use velocities at mid-span.

Since the diffusion factor was created in the 1950s there have been other approaches, but they are beyond the present scope. Diffusion factor would now be in industry used only for preliminary screening of designs and careful 3D calculations with computational fluid dynamics (CFD) would be used to verify the design and assess the efficiency. One of the key influences on efficiency and resistance to stall is the tip clearance but that is beyond the scope of this chapter.

Table 9.2 gives some parameters as guidance for compressor design including diffusion factor. The values are in no sense precise, but they reflect the knowledge base built up over years of designs which work satisfactorily. Even now, with CFD to support and guide designs, a parameter which diverged far from the values in the table would cause concern and some careful examination. Moreover, CFD is still rarely a direct route to design, but rather a method for interrogation of designs, so these simple parameters continue to have a place in guiding the initial phases of design such as engine layout.

In a turbine if the guide lines are exceeded by a wide margin there is likely to be a loss in efficiency, but this is not necessarily large. Compressor operation outside acceptable parameters can, however, lead to a steep loss in efficiency, with deficiencies compounding through the stages of a multistage compressor. As already noted, for a compressor there is a still more serious consequence of getting the design wrong: the compressor can stall and/or surge in the range where it is expected to operate. When a compressor stalls the pressure ratio and mass flow drop to low values and, this condition can sometimes be hard to get out of, leading to a large loss of thrust and potential serious damage to the engine. In surge the compressor flow oscillates; at one phase producing the full pressure ratio with a high mass flow rate and in another phase producing much reduced pressure ratio and reduced or even reversed mass flow. Both stall and surge are unacceptable conditions and much work goes into making sure they are avoided.

Exercises

Note: The flow which leaves the booster enters the HP compressor which has a smaller diameter and rotates significantly faster. There is a duct which takes the flow from booster exit to the HP compressor inlet. For structural reasons this should be as short as possible, but to reduce losses associated with flow separation it cannot be too short. The mean velocity, which has no tangential component, is taken to beconstant over the length of the duct. Some guidance for sketching the engine can be found from Figure 5.4b.

9.10 Sketch the velocity triangles for an HP compressor stage with constant axial velocity. The absolute flow angle out of a stator is written α_1 and the relative flow angle into the rotor is α_1^{rel}. Likewise out of the rotor and into the next stator the absolute and relative angles are α_2 and α_2^{rel}. Again angles are taken as positive when they produce a swirl velocity in the same direction as the rotor. Show that

$$V_x/U = 1/\left(\tan\alpha_1 - \tan\alpha_1^{rel}\right) = 1/\left(\tan\alpha_2 - \tan\alpha_2^{rel}\right)$$

$$\Delta h_0/U^2 = V_x/U\left(\tan\alpha_2 - \tan\alpha_1\right) = 1 - V_x/U\left(\tan\alpha_2^{rel} - \tan\alpha_1\right).$$

For the rotor

$$\Delta h/U^2 = 1 - \left(V_2^{rel}/V_1^{rel}\right)^2 = 1 - \left(\cos\alpha_1^{rel}/\cos\alpha_2^{rel}\right)^2 \quad \text{and}$$

for the stator

$$\Delta h/U^2 = 1 - \left(V_2/V_1\right)^2 = 1 - \left(\cos\alpha_1/\cos\alpha_2\right)^2.$$

Note: for a compressor it is normal for the signs of α_1 and α_{1rel} to be opposite; likewise for α_2 and α_{2rel}. As a result the tangent terms add in the expression for V_x/U whilst they subtract in the expression for $\Delta h_0/U^2$.

Also note that if the flow may be adequately approximated as incompressible and loss-free, the static pressure rise is given by $\Delta p/\rho U^2 = \Delta h/U^2$.

9.11 We assume that the axial velocity is uniform throughout the HP compressor and is equal to $0.55U_m$ and, from Exercise 9.4b, the loading coefficient $\Delta h_0/U_m^2 = 0.439$ for each stage. For a stage near the middle of the compressor assume that the static pressure rise in the rotor and stator are equal, so the rotor and stator velocity triangles and blades are "mirror images" of one another (a so-called 50% reaction stage). Sketch the velocity triangles and blades and find inlet (subscript 1) and exit angles (subscript 2) to the rotor. (These are equal, but of opposite sign, for rotor and stator.) Find the blade solidity if the diffusion factor is equal to 0.50 at mid-span.   **(Ans:** $\alpha_1 = 27.0°$, $\alpha_1^{rel} = -52.6°$, $\alpha_2^{rel} = -27.0°$, $\alpha_2 = 52.6°$, $\sigma = 1.3$)

Note: There needs to be a row of stators upstream of the first rotor for a compressor like this to impart swirl to the flow into the first rotor; they are referred to as inlet guide vanes. "Mirror image" blades at mid-span, usually referred to as a 50% reaction stage, is an acceptable choice and a good starting point for design, but real machines are not obliged to adopt this.

9.12 For a stage part way along the compressor the blade span is 65 mm. If the aspect ratio (span divided by chord) is 1.5, use the solidity obtained in Exercise 9.11 to find the number of blades in the row.

(Ans: 52)

Note: The number of rotor and stator blades are never chosen to be equal to avoid vibration and noise interactions. The number found here could be adjusted up or down to avoid this.

9.7 SPANWISE EFFECTS

The treatment of turbomachinery so far in this chapter has either been one-dimensional, looking at average quantities, or else has been carried out at mid-span, often referred to as meanline. If the effect of the blade shape is taken into account in a two-dimensional calculation on a constant radius surface it is referred to as a blade-to-blade treatment. A proper treatment which allows for the spanwise, or radial, effects is beyond the scope of this book, but it is covered for compressors in Cumpsty (2004). In the early days of turbomachinery there were attempts to treat the blade-to-blade and the spanwise effects analytically, but it is now realised that to do it properly requires full three-dimensional computational fluid dynamics – and even then some empiricism is currently still required. The CFD does not yet lend itself to direct design so approximate methods are still used. All that will be done here is to indicate the trends which occur because of spanwise variation and indicate the approach adopted in the design methods. In multistage compressors and turbines it is normal to design so that the stagnation temperature is maintained approximately uniform in the radial direction. Expressed differently, the change in stagnation enthalpy Δh_0 across a blade row or stage is approximately uniform in the radial direction. If the losses were uniform in the radial direction then uniform stagnation temperature would correspond to uniform stagnation pressure. In fact losses are not uniform and roughly half the losses are produced in the endwall regions. Moreover, in regions where relative Mach numbers are high the losses can be larger. In compressors in particular there is a tendency to increase the work input in regions of high loss to maintain stagnation pressure roughly constant along the span. These are complications the designer has to allow for but which, for simplicity, we will ignore here. The underlying problem can be approached by considering the loading coefficient, $\Delta h_0/U^2$, which was introduced earlier. Since Δh_0 is to be uniform in the spanwise direction and since the blade speed U is proportional to radius, it follows that $\Delta h_0/U^2$ is inversely proportional to the square of radius. It will be recalled that $\Delta h_0 = U \Delta V_\theta$, the product of blade speed and the change in tangential velocity across the rotor. So changes in radius result in changes in tangential velocity and maintaining uniform stagnation temperature therefore requires

$$r\Delta V_\theta = \text{constant.}$$

One way that this equation can be satisfied is if the swirl velocity V_θ is inversely proportional to radius both upstream and downstream of each rotor. This can be written as

$$rV_\theta = \text{constant,}$$

which is the equation for a free vortex. Rather confusingly the spanwise distribution of swirl velocity is frequently referred to as the vortex distribution. In general the free vortex form is not rigorously followed, but $rV_\theta = \text{constant}$ form does provide a convenient starting point. One of the particular advantages with this distribution is that in the special case of radially uniform Δh_0

and radially uniform loss production the axial velocity is also radially uniform; in general axial velocity in non-uniform in the spanwise direction. As noted above, because losses are known to be non-uniform the work distribution is often adjusted to compensate, with the aim of keeping stagnation pressure uniform, but the simple uniformity of axial velocity is lost. This is best handled by iterative numerical methods and is not appropriate for the simple treatment here.

The basis of a design method can be summarised as follows. A possible compressor or turbine will be designed along the lines of earlier exercises in this book at mid-radius. If this meanline design looks promising, distributions of V_θ will be created after each blade row, agreeing with the meanline value at mid-radius. A calculation method will then be used which treats the flow as axisymmetric (i.e. as if there were no blade-to-blade variation or an infinite number of blades) to calculate the flow between hub and casing. (The common generic term for this axisymmetric calculation is a streamline curvature method because that approach to solution has become so common.) The axisymmetric calculation allows the flow into and out of the blades to be found at different radii, most importantly near the hub and near the casing, and these may reveal that the blades will be incapable of satisfactory operation. In that case a redesign has to take place. When spanwise variation in loss is included some variation in work will be needed, which means a variation in ΔV_θ, and then the calculation needs to be repeated.

Exercises

9.13 a Take the flow angles from Exercise 9.11, which are at mid-span, and calculated the *absolute* tangential velocity V_θ upstream and downstream of the first rotor of the HP compressor at this radius. (Use subscript 1 and 2 to be upstream and downstream of the rotor respectively and subscript m to denote mean radius.)   (**Ans:** $V_{\theta 1m} = 87.3$ m/s, $V_{\theta 2m} = 224.1$ m/s)

b Assume that the tangential velocity is described by $rV_\theta =$ constant both upstream and downstream of the rotor. Find the tangential velocity at hub and tip, when the ratio of hub to tip diameter is 0.6.
(**Ans:** $V_{\theta 1h} = 116$ m/s, $V_{\theta 2h} = 299$ m/s; $V_{\theta 1t} = 69.9$ m/s, $V_{\theta 2t} = 179$ m/s)

c Assuming radially uniform losses, so that $V_x = 171$ m/s $=$ constant, find the absolute flow angles at hub and tip.   (**Ans:** $\alpha_{1h} = 34.2°$, $\alpha_{2h} = 60.2°$; $\alpha_{1t} = 22.3°$, $\alpha_{2t} = 46.3°$)

d Find the diffusion factor at the rotor and stator hub and tip. Assume that the blade chord is radially uniform so that the solidity $\sigma = c/s$, found in Exercise 9.11, is equal for rotor and stator and is inversely proportional to radius.   (**Ans:** rotor hub 0.364, rotor tip 0.394; stator hub 0.547, stator tip 0.461)

9.14 Sketch a longitudinal section of the LP and HP core compressors from inlet to outlet, holding the mean radius constant in each. The aspect ratio (based on the blade span and the projection of the chord in the axial direction) can be up to 3 for the first stage, but should decrease uniformly to around 1 at the last stage – this is known to give a reasonable compromise between length, robustness, good aerodynamics and a modest number of blades.

SUMMARY OF CHAPTER 9

Compressors and turbines consist of stages made of rows of stationary blades (stators) and rotating rows (rotors). The pressure rise of a compressor stage is much less than the pressure fall of a turbine stage because of the generally favourable pressure gradient in the turbine and the adverse pressure gradient in the compressor.

Compressor blades generally operate satisfactorily over a narrower range of incidence; when the incidence becomes too large, massive boundary layer separation can lead to a large increase in loss and a reduction in turning (i.e. deviation). It can also lead to stall or surge.

Satisfactory operation of both compressors and turbines is possible only in a limited range of V_x/U and $\Delta h_0/U^2$. Acceptable values of these non-dimensional parameters are frequently given in terms of the conditions at mean height along the blades. Practical constraints can make it impossible to stay within the desired limits and because the turbine is more 'forgiving' it is generally in this component where the compromises are most evident. Although there is no need to maintain the axial velocity exactly constant through a multistage compressor or turbine, this is a reasonable first approximation. This requires the blade height to reduce along the length of the compressor and to increase along the turbine.

The work exchange in a turbine or compressor is given by the Euler work equation

$$\Delta h_0 = U_2 V_{\theta 2} - U_1 V_{\theta 1}$$

and for two-dimensional flows at constant radius, $U_2 = U_1$.

The practical way to consider turbine or compressor blade rows is to adopt a frame of reference fixed to the blade row being studied, so for the rotor the *relative* frame of reference is used and for the stator the *absolute* frame is used. The easy way to carry this out is to use velocity triangles, and it is strongly recommended that these be drawn whenever change in frame of reference is carried out.

Considerable simplification is possible if the blade speed is equal at inlet and outlet to the rotor (i.e. the streamline does not shift radially) and also if the axial velocity is equal at inlet and outlet of the rotor and across the stage. Both these approximations are reasonably close to typical design choices.

For a turbine one method of assessing the capability of the blades to turn and accelerate the flow as intended (i.e. to deliver the power output) is the Zweifel coefficient, defined in Equation 9.14. For most designs Z_w should be less than about 0.8.

For a compressor the designs can be assessed for feasibility using diffusion factor defined in Equations 9.18. Diffusion factor is normally kept below 0.45.

For a turbine blade row the deviation, the difference between flow outlet angle and blade (metal) outlet angle, is small and normally does not vary much with flow coefficient (that is with incidence). For a compressor blade row with old designs of blade shape the deviation angle may be 20% of the turning (camber) of the blades but with modern designs of compressor blades it is typically only 1 or 2 degrees. For compressors deviation is inclined to rise sharply if the compressor is asked to operate beyond the normal incidence and for the high speeds typical or aero engine compressors the allowable incidence is small; for preliminary design a choice of zero incidence is sensible.

Conditions vary along the span of the blades. The correct description is fully three-dimensional but some simple approximations can be made to show features of the flow. To achieve radially uniform work or pressure rise flow deflection must be greater near the hub than the tip and one simple description is to have tangential velocity inversely proportional to radius at each location. With this, and for idealised conditions with spanwise-uniform loss, the axial velocity is uniform in the radial direction.

CHAPTER 10 OVERVIEW OF THE CIVIL ENGINE DESIGN

The emphasis of Part 1 of the book has been overwhelmingly towards the aerodynamic and thermodynamic aspects of a jet engine. These are important, but must not be allowed to obscure the obvious importance of a wide range of mechanical and materials related issues. In terms of time, cost and number of people, mechanical aspects of design consume more than those which are aerodynamic or thermodynamic. Nevertheless this book is concerned with the aerodynamic and thermodynamic aspects and it is these which play a large part in determining what are the *desired* features and layout of the engine. Clearly, an aerodynamic specification which called for rotational speed beyond what was possible, or temperatures beyond those that materials could cope with, would be of no practical use.

An aircraft engine simultaneously calls for high speeds and temperatures, light weight and phenomenal reliability; each of these factors is pulling in a different direction and compromises have to be made. Ultimately an operator of jet engines, or a passenger, cares less about the efficiency of an engine than that it should not fall apart. Engines are now operating for times in excess of 20,000 hours between major overhauls (at which point they must be removed from the wing), and this may entail upward of 10,000 take-off and landing cycles. In-flight engine shut-downs are now rare and the rate for the fleet of modern civil aircraft is one shut-down in about 250,000 flying hours. As a result most pilots will never experience a compulsory engine shut-down during their whole careers.

The speeds of some of the components of an engine are comparable with the shells from some guns with the potential to do terrible damage to the pressurized aircraft fuselage and the passengers. The fan blades are the largest rotating items and the fan is surrounded by a containment ring so that in the event of a fan blade detaching the parts will not travel sideways to hit the fuselage. Much engineering effort is expended in design and testing of the containment ring around the fan and this ring adds considerable extra weight to the engine. Other rotating items are not designed to be contained but, instead, enormous care is taken in the manufacture of the discs to ensure failure is very rare.

Any design is a compromise. The aerodynamicist would like thin blades, but for mechanical reasons they may have to be thicker than he would choose. The mechanical designer has to find a way to hold the bearings and to supply and remove the lubricating oil – these tend to give fairly massive structures which conflict with the aerodynamicist's wishes. This is a conflict that the mechanical designer is likely to win! The positioning of the bearings has a marked effect on the strength and integrity of the engine and on its weight – bearings which are not in the right axial place need heavy structural elements to transmit the loads to them. Moreover, the mechanical design has a major influence on the aerodynamic performance. One issue, for example; how round and concentric do the engine components remain when knocked about in service?

A huge amount of work has gone into the specification of the layout of engines and the mechanical design which follows. If you look at drawings for engines of different companies you will see the differences in such important things as the positions of the bearings and the load bearing struts. Different companies have arrived at families of engine styles for which they have evolved workable solutions to the problems inherent in the design. By keeping to a familiar style it is possible to avoid some of the problems of a new design. One of the fascinating aspects of this is that a very elegant solution to one problem, for example where to put a bearing, may cause a problem in another aspect, for example in the aerodynamic specification of the HP turbine Decisions may be taken at an early stage without fully appreciating some of the consequences which can make other things more difficult at a later stage. As so often there is not a 'right' solution – there are a range of solutions, some of which are more elegant and more satisfactory than others. All of these considerations favour a "family" style for each company.

The first ten chapters have taken the simplest possible approach to the design of civil transport engines. By sketching the design, as suggested in Exercise 10.1, a strong similarity will emerge with the two-shaft engine shown in Figure 5.4b. This, of course, is no coincidence, since the assumptions and empiricism used are close to those adopted by engine manufacturers. The treatment here has considered the engine operating only at the design point, chosen to be cruise. The significance of three shafts, as in Figure 5.4a, emerges with off-design performance, including starting. Off-design operation is addressed in the following two chapters. Chapter 11 looks at the performance of the components of the engine, mainly the compressor and turbine, including these operating off-design. Chapter 12 examines how the engines behave when they go off-design, for example when T_{04}/T_{02} is greater or less than the value at the design point.

The early civil engines evolved out of military engines intended for fighter aircraft. Over the years the designs of military and civil engines have diverged. Civil engines all have high bypass ratios with low fan pressure ratio and low specific thrust, whereas military combat engines are now generally low bypass ratio turbofans. Chapters 13, 14 and 15 look at the requirements for military combat engines which are more varied and complicated than those for the civil aircraft. Chapter 16 considers the design-point operation of the engine for a possible new fighter aircraft and Chapter 17 looks at the off-design operation. The study of military engines closes in Chapter 18 with a short section on the turbomachinery requirements specific to combat engines.

In considering the military engine several of the complicating factors are included which were omitted in Chapters 1 to 10. Chapter 19 therefore returns to the engine for the New Efficient Aircraft with these additional effects included. Chapter 20 is a brief overview of the whole scene, emphasising that any straightforward treatment outlined in a short book is likely to be a serious oversimplification of what goes on in practice.

Exercise

10.1 Sketch an engine, having in mind the core compressor and turbine you have specified. An important component which we have neglected in this course is the combustor. Use the engine cross-sections in Figure 5.4 to obtain some idea of this.

Try to avoid making the flow go through tortuous ducts to change radius. Try to put in bearing supports. To get everything to match is difficult and you will not have the time to try many schemes. The main thing is to get a feel for the difficulty involved and the scope for different solutions. (It is recommended that this be done on squared paper.)

Part 2

Engine Component Characteristics and Engine Matching

CHAPTER 11 COMPONENT CHARACTERISTICS

11.0 INTRODUCTION

Up to this point consideration has been given only to the design point of the engine. This is clearly not adequate for a variety of reasons. Engines sometimes have to give less than their maximum thrust to make the aircraft controllable and to maintain an adequate life for the components. Furthermore all engines have to be started, and this requires the engine to accelerate from very low speeds achieved by the starter motor. The inlet temperature and pressure vary with altitude, climate, weather and forward speed and these need to be allowed for.

To be able to predict the off-design performance it is necessary to have some understanding of the way the various components behave and this forms the topic of the present chapter. It is fortunate that to understand off-design operation and to make reasonably accurate predictions of trends it is possible to approximate some aspects of component performance. The most useful of these approximations is that the turbines and the final propulsive nozzle are perceived by the flow upstream of them as choked. Another useful approximation is that turbine blades operate well over a wide range of incidence so that it is possible to assume a constant value of turbine efficiency independent of operating point. These approximations make it possible to consider the matching of a gas turbine jet engine and to see how the various components operate together at the conditions for which they are designed (the design point) and at off-design conditions. Off-design performancewill form the topic of Chapter 12.

The present chapter will consider only the major components: the fan and compressor, the combustor, the turbine and the propulsive nozzle. Because of its simplicity it is convenient to begin by considering the nozzle, but prior to this the issue of gas properties will be addressed.

11.1 GAS PROPERTIES IN THE AIRCRAFT GAS TURBINE

In the treatment of the engines for the New Efficient Aircraft in Chapters 1–10 the specific heat capacity of the gas at constant pressure c_p and ratio of specific heat capacities $\gamma = c_p/c_v$ were assumed to be equal for the air and for the products of combustion and to be constant regardless of temperature and pressure. This is a major over-simplification which will be corrected somewhat in the present and later chapters. A very extensive collection of thermodynamic properties relevant to the gas turbine is given in tabular form in Banes, McIntyre and Sims (1967) for the range of interest for gas turbines, extending from about 216 K to about 2300 K and from about 20 kPa to

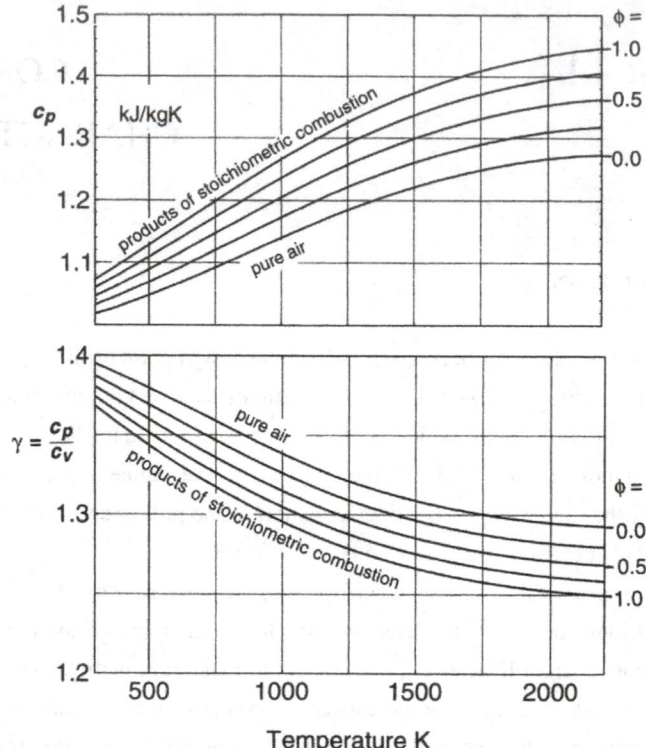

Figure 11.1. Variation in specific heat at constant pressure c_p and specific heat ratio γ with temperature for air and for combustion products of kerosene C_nH_{2n}. ϕ is the equivalence ratio.

45 MPa respectively. In this range the effect of pressure on the values of c_p and γ is of the order of 0.1% and is therefore negligible for the level of precision required here. The effect of temperature and composition is not negligible, as Figure 11.1 shows. Curves are shown for different levels of equivalence ratio ϕ, which is the ratio of the fuel–air ratio to the fuel–air ratio for stoichiometric combustion. If all of the reactants are consumed in the combustion and there is no excess fuel or oxygen, the combustion is said to be stoichiometric and the equivalence ratio is unity. For a kerosene fuel which is typical of what might be burned in an aircraft gas turbine, with the empirical formula C_nH_{2n}, the mass of fuel per unit mass of air in stoichiometric combustion is 0.0676. For a gas turbine combustor at maximum thrust the equivalence ratio is typically about 0.3 for a civil engine, indicating that only 30% of the oxygen is burned, whereas for an afterburner on a military engine at maximum thrust the equivalence ratio is about unity.

The curves in Figure 11.1 were derived assuming that the component gases behave as perfect gases (i.e. there is no pressure dependence) and dissociation of the gas molecules as well as the formation of additional species, such as oxides of nitrogen, are neglected. These assumptions are reasonably good at the pressures and maximum temperatures involved: at 2000 K and a pressure of 100 kPa the molar concentration of atomic oxygen at equilibrium is about three orders of magnitude below that of molecular oxygen. The concentration of atomic nitrogen is many orders of magnitude below that of atomic oxygen. Of the oxides of nitrogen the highest concentration is that of NO at about 0.8%, smaller than the concentration of argon. Although such a small concentration of NO is not significant to the energy release in the combustion process, it is highly significant as a pollutant.

The curves in Figure 11.1 show that any simple choice of gas properties is going to lead to inaccuracy if applied to compressions and expansions in which there are substantial temperature changes. For the air in the compressor, γ falls from just below 1.40 to about 1.30 over the range of temperature involved. With $\phi = 0.3$, γ at entry to the HP turbine is about 1.28 and γ increases to about 1.33 at exit from the LP turbine. For simplicity and consistency the values for γ of 1.40 and 1.30 will be used for the compressor and turbine respectively from this chapter forwards. In Chapter 19 an empirical relationship derived by Cumpsty and Marquis (2014) is used, which gives much greater accuracy.

Exercise

11.1 a Pure air consists of 78.03% by volume of nitrogen (molar mass 28.01), 29.05% by volume of oxygen (molar mass 32.00) and 0.98% by volume of argon (molar mass 39.95). Confirm that the molar mass of air is 28.97 and gas constant is 287.0 J kg^{-1}K^{-1}. (Take $\mathcal{R} = 8.314$ kJ kmol^{-1}K^{-1})

b Find the molar mass of the products of stoichiometric combustion of a fuel representative of kerosene with the formula C_2H_{2n}. Hence find the gas constant for the combustion products.

(**Ans:** 28.91; 287.6 J kg^{-1}K^{-1})

Note: It is convenient that molecular mass is almost equal for air and the combustion products of a fuel like kerosene (C_nH_{2n}). As a result $R = 287$ J kg^{-1}K^{-1} is a good approximation for both air and for products of combustion of kerosene. This is not true for all fuels, so for stoichiometric combustion of natural gas (CH_4), the molar mass is lower, 27.74, and $R = 299.7$ J kg^{-1}K^{-1}.

11.2 THE PROPULSIVE NOZZLE

The bypass and core streams may have separate nozzles or the two flows may be mixed prior to the contraction forming the nozzle. In either case the flow is assumed to be uniform when it enters the nozzle; non-uniformity contributes to the discharge coefficient being less than unity. (Discharge coefficient is the ratio of the actual mass flow to the mass flow assuming an expansion to the exit static pressure with no loss and with uniform flow over an area equal to the geometric area of the nozzle.) For a nozzle which is just choked with a 5° convergence, the discharge coefficient is about 0.97, whilst the velocity coefficient is about 0.998, indicating that the velocity is within 0.2% of that calculated assuming the flow is ideal, a level of imperfection small enough to be neglected in the present treatment. For high-speed propulsion it is not uncommon to use a convergent–divergent nozzle, but for subsonic aircraft the cost and weight is not justified.

The variation in non-dimensional mass flow through a convergent nozzle with pressure ratio (upstream stagnation pressure to downstream static pressure) across the nozzle is shown in Figure 11.2 for air with $\gamma = 1.40$ and for exhaust gases, for which $\gamma = 1.30$ is more appropriate.

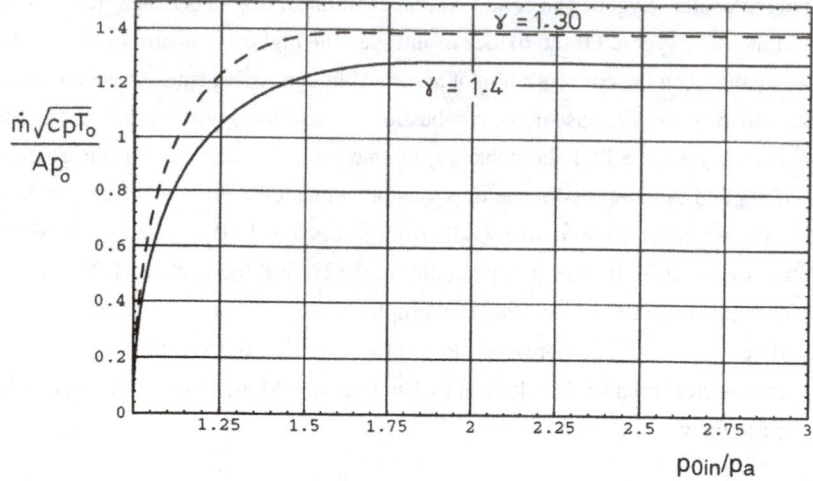

Figure 11.2. Variation in non-dimensional mass flow rate through a convergent nozzle with ratio of inlet stagnation pressure to atmospheric pressure at outlet.

It was assumed here that the flow is reversible to the throat, which is the exit plane in the case of a simple convergent nozzle. The mass flow increases with pressure ratio until the pressure ratio reaches the value at which the nozzle chokes. As is shown in Chapter 6, choking of a convergent nozzle occurs when the inlet stagnation pressure is equal to at least $((\gamma + 1)/2)^{\gamma/\gamma-1}$ times the downstream ambient static pressure, or 1.89 in the case when $\gamma = 1.40$. When the nozzle is choked, the Mach number is unity at the throat and the non-dimensional mass flow rate is constant. In other words, repeating Equation 6.15,

$$\overline{m}_{choke} = \left(\frac{\dot{m}\sqrt{c_p T_0}}{A p_0}\right)_{M=1.0} = \frac{\gamma}{\sqrt{\gamma - 1}}\left(\frac{\gamma + 1}{2}\right)^{-\frac{\gamma+1}{2(\gamma-1)}}.$$

For pure air, such as through the bypass nozzle, taking $\gamma = 1.4$ gives

$$\overline{m}_{choke} = 1.281.$$

For the flow in a turbine or in the core nozzle, taking $\gamma = 1.3$, one obtains

$$\overline{m}_{choke} = 1.389.$$

The mass flow and all conditions *upstream* of the throat are unaffected by conditions downstream of the throat once the flow is choked. The relationship between $\overline{m}$ and p_0/p shown in Figure 11.2 also applies to a convergent–divergent nozzle where A is then the throat area and p is the pressure at the throat. (The variation in area for an isentropic convergent–divergent nozzle as a function of Mach number may be inferred from Figure 6.2; using the curves of p/p_0, also in Figure 6.2, the dependence of area on local static pressure can be deduced.) In many cases of interest for aircraft propulsive nozzle will be choked.

Although conditions downstream of a choked throat do not affect the conditions upstream in the engine this does not mean that what happens downstream of the throat is unimportant, since it greatly affects the thrust. This is illustrated in Exercise 11.2.

Exercise

11.2 Consider a nozzle discharging into an atmosphere with ambient static pressure p_a. The flow at nozzle exit plane (exit area A) has uniform static pressure and velocity, p_9 and V_9; in general p_9 is not equal to p_a. Some distance downstream of the exit plane the pressure in the jet is equal to p_a and at that condition the jet velocity is V_j. The mass flow in the jet remains effectively constant during this process and is equal to $\dot{m}$. By considering a suitable control volume and applying conservation of momentum, show that the gross thrust is given by

$$F_G = \dot{m}V_j = \dot{m}V_9 + (p_9 - p_a)\, A.$$

For which of the following conditions is the second term on the right hand side of this equation non-zero: an unchoked nozzle, a convergent nozzle which is just choked, a convergent nozzle which has a pressure ratio significantly greater than that to choke it, and a fully expanded convergent–divergent nozzle? Show that the gross thrust may be written in non-dimensional form as

$$\frac{F_G}{\dot{m}\sqrt{c_p T_{09}}} = \frac{V_9}{\sqrt{c_p T_{09}}} + (p_9/p_{09} - p_a/p_{09})\left(\frac{\dot{m}\sqrt{c_p T_{09}}}{A p_{09}}\right)^{-1}.$$

Noting that the conditions at the nozzle exit plane are fixed for a choked convergent nozzle (p_9/p_{09}, the non-dimensional velocity and the non-dimensional mass flow are all constant) explain why the rate of increase in thrust with p_{09}/p_a is so small for large values of pressure ratio.

Note: The very weak dependence of $F_G/\dot{m}\sqrt{c_p T_{09}}$ on p_{09}/p_a for large values of p_{09}/p_a means that at high speeds, ($M > 1.8$, say) when the inlet ram pressure is high, there is little benefit in having a large pressure ratio in the engine. Instead the designer should go for getting the highest mass flow through the engine possible, and this strategy is adopted for some modern high-performance combat engines.

Figure 11.3, based on Exercise 11.2 shows the non-dimensional gross thrust $F_G/\dot{m}\sqrt{c_p T_{09}}$ versus pressure ratio for two nozzles in the case of air ($\gamma = 1.40$). One is a simple convergent nozzle, the other a series of convergent–divergent nozzles each of sufficient exit area to allow full reversible expansion at each pressure ratio. The expansion is assumed reversible inside the convergent nozzle but not downstream of its exit plane. The nozzle is choked for pressure ratios $p_{09}/p_a > 1.89$, but the effect of irreversibility in the flow downstream of the convergent nozzle does not become apparent until the pressure ratio exceeds about 3. This is substantially higher than is common for modern civil transport engines, but much lower than that for high-speed engines.

A modern fighter engine may have a pressure ratio of 16 across the nozzle, and at this condition the loss in thrust associated with a convergent nozzle is more than ten per cent. From a plot like Figure 6.2 it can be seen that the exit areas can become large for a fully reversible expansion inside the nozzle if the pressure ratio is large; with a pressure ratio $p_{09}/p_a = 16$, for example, the exit area would be more than 2.5 times the throat area.

For test purposes the nozzle may be replaced by a variable area throttle, as for example, during the testing of a compressor. Again the non-dimensional mass flow rate is independent of

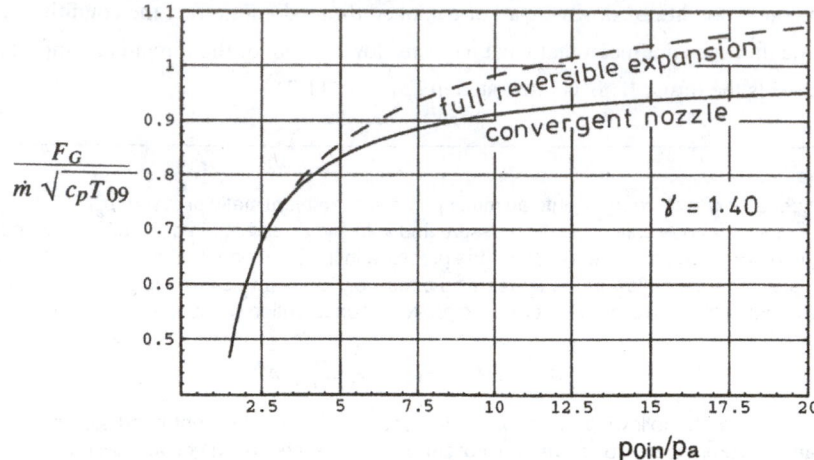

Figure 11.3. Variation in non-dimensional gross thrust from nozzles as a function of the ratio of inlet stagnation pressure to atmospheric pressure at outlet.

pressure ratio once the pressure ratio is large enough to choke the throat or narrowest part. In this case, however, the flow is strongly non-uniform across the area of the throttle, so $\overline{m}_{choke}$ based on the throttle open area would not have the same magnitude as that for a well-shaped choked nozzle with nearly uniform flow. So far as the compressor is concerned, however, the throttle behaves just like a nozzle.

11.3 THE FAN

The fan is simply a specialised form of compressor. For civil applications it is a single stage producing a pressure ratio of no more than about 1.8. In military applications the pressure ratio may be as high as 4, with two or three stages.

      The fan should pass as much flow per unit area as possible, so the blades are long and the hub radius small; at fan entry it is typically less than 0.3 times the tip radius. In modern civil fans there are no stator blades upstream of the rotor, but for military fans upstream stators, usually referred to as inlet guide vanes (IGVs) are commonly used. An important reason for avoiding inlet guide vanes for engines on civil aircraft is to reduce the noise from the fan. Figure 11.4 shows the characteristics for the bypass stream of a civil fan plotted in the conventional way, which is the ratio of outlet to inlet stagnation pressure, p_{013}/p_{02}, versus non-dimensional mass flow rate for different non-dimensional speeds. (The numbering system for stations used throughout and illustrated in Figure 7.1 is retained here; p_{013} is the average pressure of the bypass stream whilst $\dot{m}$ is the mass flow rate of the core and bypass streams entering the fan.) The characteristic map of a military fan is generally similar to Figure 11.4, though the maximum pressure ratio is likely to be about 4 and the efficiencies are significantly lower.

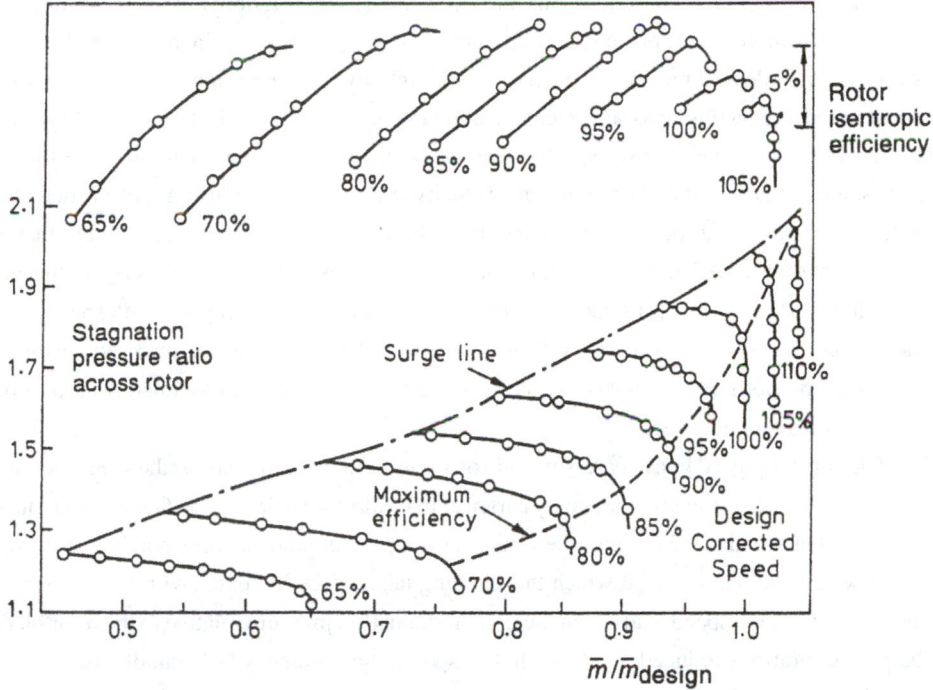

Figure 11.4. Characteristics of a civil fan showing pressure ratio and isentropic efficiency versus mass flow rate (non-dimensionalised by inlet conditions and normalised by design mass flow) for lines of constant non-dimensional rotational speed.

As noted in Chapter 8 it is common to drop from the non-dimensional terms those quantities which are constant for a given machine, typically properties of air (R, c_p and γ) and area A. The mass flow is then expressed as

$$\frac{\dot{m}\sqrt{T_{02}}}{p_{02}}$$

and the rotational speed as

$$\frac{N}{\sqrt{T_{02}}}.$$

Results are often presented in terms of *corrected* mass flow and speed, where $\dot{m}$ and N are referred to standard inlet conditions using $\delta = p_{02}/p_{02\text{ref}}$ and $\theta = T_{02}/T_{02\text{ref}}$; typically $p_{02\text{ref}} = 1.01$ bar and $T_{02\text{ref}} = 288$ K.

The corrected mass flow and speed are then

$$\dot{m}_{corr} = \frac{\dot{m}\sqrt{\theta}}{\delta} \quad \text{and} \quad N_{corr} = N/\sqrt{\theta}.$$

In this system the units are retained (for example, kg/s and rev/min) and since the terms δ and θ are often near to unity the physical magnitudes of corrected quantities are often helpful.

The curves for pressure ratio versus mass flow at constant rotational speed show properties common to all compressors. As the mass flow is reduced the pressure rise increases. Reducing the mass flow is equivalent to reducing the axial velocity into the fan or compressor and for constant rotational speed a reduction in axial velocity gives an increase in incidence. The effect of increased incidence can be appreciated in several different but compatible ways: an increase in the force on the blades; an increase in the change in whirl velocity ΔV_θ across the rotor; a greater increase in streamtube area passing through the rotor and therefore a larger decrease in velocity and increase in static pressure rise. All these are compatible with the pressure ratio increasing as the mass flow is reduced. Both the pressure ratio and the mass flow rate increase rapidly with speed, so the constant speed lines are well separated. If incidence were held constant the mass flow would be approximately proportional to rotational speed and the rise in pressure approximately proportional to the square of rotational speed.

The constant speed lines of pressure ratio on Figure 11.4 vary in shape as the speed increases. At low rotational speeds the lines gradually curl over to be almost horizontal as flow rate is reduced, but for the higher speeds the lines become vertical at low pressure ratios, corresponding to the rotor blades choking. The mass flow at which this choking takes place increases with rotational speed because the increase in speed leads to an increase in stagnation pressure relative to the rotor blades. As the pressure ratio is reduced for these choked speeds the efficiency falls rapidly. Efficiency is strongly dependent on both mass flow and speed; alternatively on pressure ratio and speed.

On the plot of pressure ratio against mass flow in Figure 11.4 are two lines in addition to the lines of constant rotational speed. The highest is the stall or surge line, which denotes the maximum pressure rise which the fan can produce at any rotational speed; attempts to operate above and to the left of this line result in one of the following: a collapse of pressure ratio (stall); a violent oscillatory flow (surge) or an aero-elastic phenomenon called flutter. Any of these unsteady phenomena is unacceptable. The other line is the locus of maximum efficiency as rotational speed is altered. Not shown on Figure 11.4, but derived in Exercise 11.4, is the *working line*, sometimes called *operating line*, produced by a nozzle at the rear of the bypass duct. The working line is roughly parallel to the surge line and tends to move away from the locus of maximum efficiency as speed is reduced. For recent civil fans with pressure ratio about 1.5 the bypass nozzle is not choked for static operation but for cruise conditions the high inlet stagnation pressure to the fan, relative to the ambient static, does give a choked bypass nozzle and the working line then corresponds to a constant value of non-dimensional mass flow at exit from the fan. This idea is explored in Exercises 11.4, 11.5 and 11.6.

Also shown in Figure 11.4 are curves of isentropic efficiency versus mass flow rate for different speeds. The isentropic efficiency for a compressor or fan is given by

$$\eta_{\text{isen}} = \frac{T_{013\text{is}} - T_{02}}{T_{013} - T_{02}}, \tag{11.1}$$

where $T_{013\text{is}}/T_{02} = (p_{013}/p_{02})^{\gamma-1/\gamma}$.

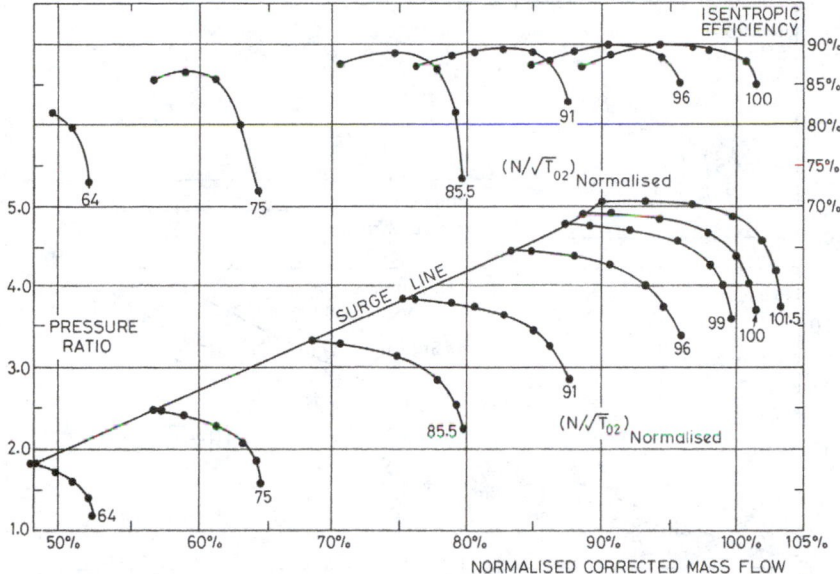

Figure 11.5. Characteristics of a modern compressor showing pressure ratio and isentropic efficiency versus corrected mass flow rate for lines of constant non-dimensional rotational speed.

Figure 11.4 shows the conventional form for displaying compressor or fan performance, but comparable information would be conveyed if the temperature ratio T_{013}/T_{02} or the ratio of temperature rise to inlet temperature $(T_{013} - T_{02})/T_{02}$ were shown instead of pressure ratio.

Exercise

11.3 For the fan in Figure 11.4 draw a working line (pressure ratio versus mass flow) for a choked nozzle; three points will define the line adequately. The line should give $pr = 1.5$ at a rotational speed around that for peak efficiency. For simplicity, take the efficiency to be 0.90 all along the working line. To achieve the choking of the nozzle assume that the stagnation pressure at inlet $p_{02} = 1.5p_a$.

Note: the working line does not follow the line of peak efficiency at all closely.

11.4 THE CORE COMPRESSOR

Figure 11.5 shows the performance map of a multi-stage compressor in the form of pressure ratio p_{03}/p_{023} versus non-dimensional mass flow, $\overline{m}$, based on inlet conditions to the core compressor,

$$\overline{m}_{23} \frac{\dot{m}\sqrt{c_p T_{023}}}{A p_{023}},$$

using the numbering system of Figure 7.1. The non-dimensional mass flow has been expressed as a percentage of the design value. The maximum pressure ratio of this compressor is around 5 and

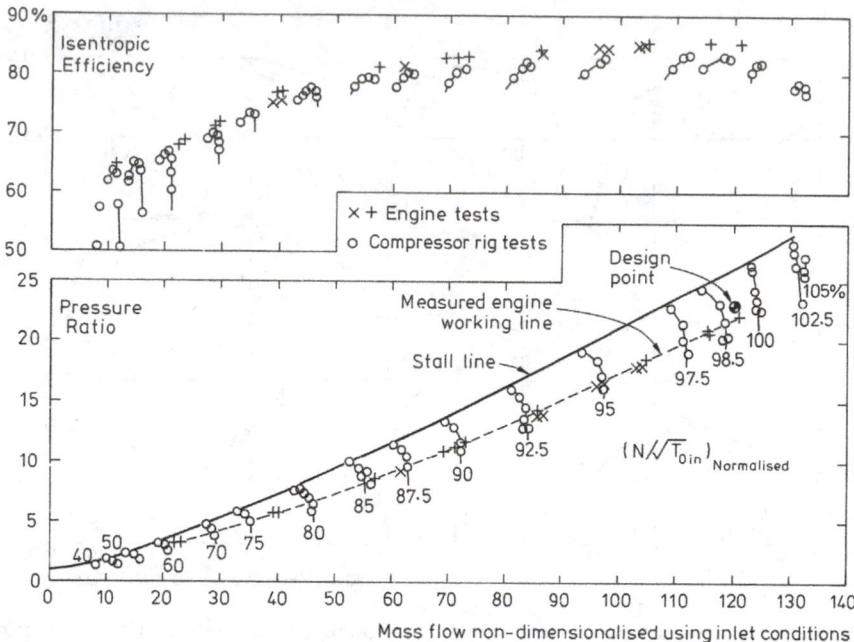

Figure 11.6. Characteristics of the GE E^3 compressor showing pressure ratio and isentropic efficiency versus mass flow rate (non-dimensionalised by inlet conditions) for lines of constant non-dimensional rotational speed.

would be suitable for a three-shaft high bypass ratio engine with no variable stators. A pressure ratio of 7 is approaching the upper limit which can be achieved without incorporating variable stators for reasons which are discussed later in this section. Variable stators are arranged so that as the speed of the compressor falls the stators in the front few rows of the machine are turned to be more nearly tangential, i.e. their stagger angle is increased.

Figure 11.6 is plotted in similar variables for a compressor with a pressure ratio of about 25. This is the General Electric E^3 (Energy Efficiency Engine) compressor, which, in developed and modified form, is in the core of the GE90 and GEnx engines. For this compressor there are variable stators on the first 6 rows of stator blades (including the inlet guide vanes ahead of the first rotor).

Notice again that at the highest speeds the compressor speed lines show choking at lower pressure ratios. Two types of data are shown on this figure, results from a compressor tested on a rig with an external motor and variable throttle (shown with open symbols) and data obtained in engines (shown with crosses). The data obtained on the rig is that which makes possible the speed lines (loci of constant speed operating lines). Obtaining speed lines on an engine requires the engine geometry to be modified between points and this is costly and time consuming so for the engine the data is measured along the working line.

Definitions of efficiency – isentropic and polytropic efficiency

In Figures 11.4, 11.5 and 11.6 the efficiencies shown are the *isentropic* efficiencies, defined in Equation 11.1, and sometimes called overall or adiabatic efficiency. In dealing with multi-stage compressors and turbines it is useful to introduce an efficiency defined slightly differently, the *small-stage* or *polytropic* efficiency η_p. Using polytropic efficiency makes some algebra easier, but it also removes a bias in the isentropic efficiency when comparing machines of different pressure ratio.

Consider a small *stagnation* pressure rise δp_0 which will be accompanied by a stagnation temperature and enthalpy rise δT_0 and δh_0. The familiar thermodynamic equation states that

$$T_0 \delta s_0 = \delta h_0 - \delta p_0 / \rho_0,$$

where $\delta h_0 = c_p \delta T_0$ and by definition $\delta s_0 = \delta s$. (The change in entropy is the same for stagnation and static property definitions.) For an ideal isentropic compression process, for which $\delta s = 0$, the enthalpy and pressure changes are related by

$$\delta h_{0i} = c_p \delta T_{0i} = \frac{\delta p_0}{\rho_0} = \frac{RT_0}{p_0} \delta p_0,$$

which may be integrated between states 1 and 2, using $c_p = \gamma R / (\gamma - 1)$ to give the familiar

$$T_{02i}/T_{01} = (p_{02}/p_{01})^{(\gamma-1)/\gamma}.$$

The actual temperature rise for a real compression process will be larger than the ideal and can be written in terms of the polytropic efficiency η_p as

$$\delta h_0 = c_p \delta T_0 = \frac{\delta p_0}{\eta_p \rho_0} = \frac{RT_0}{\eta_p p_0} \delta p_0,$$

which integrates to give

$$\frac{T_{02}}{T_{01}} = \left(\frac{p_{02}}{p_{01}} \right)^{\frac{(\gamma-1)}{\eta_p \gamma}} \quad \text{for a compressor} \tag{11.2}$$

and by similar arguments
$$\frac{T_{02}}{T_{01}} = \left(\frac{p_{02}}{p_{01}} \right)^{\frac{\eta_p (\gamma-1)}{\gamma}} \quad \text{for a turbine.} \tag{11.3}$$

Exercise

11.4 Show that the polytropic or small-stage efficiency for a compression process between stagnation states 1 and 2 can be written as

$$\eta_p = \frac{\gamma - 1}{\gamma} \frac{\ln(p_{02}/p_{01})}{\ln(T_{02}/T_{01})} \tag{11.4}$$

and that the overall, adiabatic or isentropic efficiency is given by

$$\eta_{\text{isen}} = \frac{(p_{02}/p_{01})^{(\gamma-1)\gamma} - 1}{(p_{02}/p_{01})^K - 1} \quad \text{where } K = \frac{\gamma - 1}{\eta_p \gamma}. \tag{11.5}$$

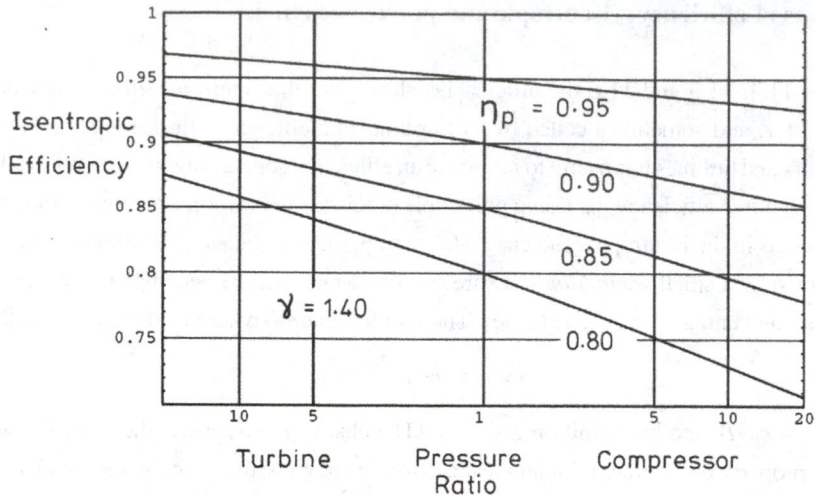

Figure 11.7. Variation in *isentropic* efficiency with pressure ratio for various levels of *polytropic* efficiency.

Using Equation 11.2 the *polytropic* efficiency may be related to the *isentropic* efficiency and this is shown in Figure 11.7 for air for both a compressor and turbine. For compressors the overall isentropic efficiency is always lower than the polytropic efficiency, whereas for turbines it is always higher; in turbines this effect is referred to as the *reheat factor.*

The reason for the lower isentropic efficiency for the compression process can be explained with the aid of the temperature–entropy diagram in Figure 11.8, which shows a compression between stagnation states 01 and 03 carried out two ways: in one step or in two smaller equal steps.

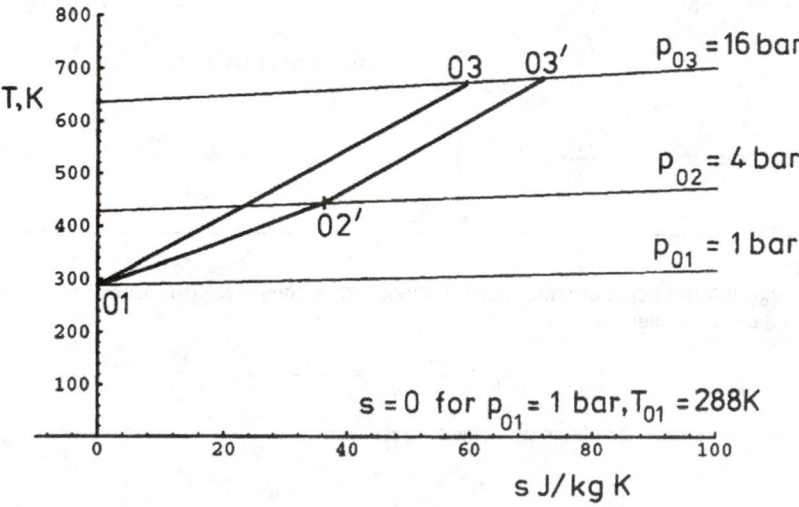

Figure 11.8. Temperature–entropy diagram for compression between 1 and 16 bar achieved in two ways: process 01–03 is in a single step with $\eta_{isen} = 0.9$; process 01–02'–03' is in two steps each with $\eta_{isen} = 0.9$.

As drawn the overall isentropic efficiency for the single-step compression is 90%; an efficiency of 90% has also been used for each step of the two-step compression and in this case the final temperature $T_{03'}$ can be seen to be higher, implying a lower overall efficiency. In other words when the compression process is broken down into steps or stages the isentropic efficiency of each step must be higher than that over the whole. This is because the temperature of the gas is raised in the first step before it enters the next and as a result the temperature rise in the second step is higher than in the first even if the pressure ratio and the efficiency were equal for both steps.

In the limit, when the increment in pressure rise across a step tends to zero, the isentropic efficiency of the step approaches the polytropic limit. For axial compressors the pressure rise across each stage is normally small and in this case the stage isentropic efficiency and stage polytropic efficiency are almost equal; furthermore if all the stages had equal efficiency the polytropic efficiency of the overall machine would equal that of each stage. This is illustrated in Exercise 11.4. The polytropic efficiency is therefore not only more convenient algebraically in many instances, but it offers a better comparative measure of the performance of machines, both compressors and turbines, having different pressure ratio.

Exercises

11.5 a Suppose that a compressor has ten stages each with a pressure ratio of 1.3. Find the overall pressure ratio. If the polytropic efficiency of each stage is 90%, find the overall temperature ratio across the whole machine. Confirm that the polytropic efficiency across the whole machine is 90% and find the isentropic efficiency across a stage, $\eta_{isen,st}$, and across the whole machine, $\eta_{isen,ov}$.

$$\text{(\textbf{Ans}: (a) 13.78, 2.300, } \eta_{p,ov} = 0.90, \ \eta_{isen,ov} = 0.858, \ \eta_{isen,st} = 0.896)$$

b Suppose that the stage pressure ratios remain 1.3, but that for the first five stages the efficiency is 90% and for the last five stages it is 80%. Find the overall temperature ratio and thence the polytropic and isentropic efficiencies for the whole machine.   (**Ans**: (b) 13.78, 2.423, $\eta_{p,ov} = 0.846$, $\eta_{isen,ov} = 0.784$)

Non-dimensional mass flow based on outlet conditions

In the conventional presentation of compressor performance the non-dimensional mass flow used is that based on inlet conditions, $\overline{m}_{23}$ for the core compressor. Results can, however, be presented in terms of p_{03} and T_{03}, the outlet pressure and temperature

$$\overline{m}_3 = \frac{\dot{m}\sqrt{c_p T_{03}}}{A_{out} p_{03}} = \frac{\dot{m}\sqrt{c_p T_{023}}}{A_{in} p_{023}} \frac{p_{023}}{p_{03}} \frac{\sqrt{T_{03}}}{\sqrt{T_{023}}} \frac{A_{in}}{A_{out}}$$

$$= \overline{m}_{23} \left(\frac{p_{023}}{p_{03}} \right)^L \frac{A_{in}}{A_{out}},$$

where the index $L = 1 - (\gamma - 1)/2\eta_p\gamma$ and η_p is the polytropic efficiency. For reasonable values of efficiency, say $\eta_p = 0.9$, and taking $\gamma = 1.4$ one obtains $L = 0.84$. Since the value of L is not far from unity it shows that the ratio of non-dimensional mass flow rates is almost proportional to the pressure ratio. (The manipulation carried out above illustrates how convenient the polytropic efficiency can be.)

Exercises

11.6 For the compressors shown in Figures 11.5 and 11.6 calculate the polytropic efficiency corresponding to the highest isentropic efficiency at the design (100%) speed for each machine.   **(Ans:** 0.92, 0.88)
          The compressors were each tested using a downstream throttle. Assuming that the throttle is choked and its area is constant (i.e. constant corrected mass flow through the throttle), draw a working line on Figure 11.6. The working line should pass through a pressure ratio of 22.5 on the 100% speed line.

11.7 For the compressor in Figure 11.6 redraw the 100%, 90% and 80% lines for constant speed in terms of the non-dimensional mass flow based on *outlet* conditions. Assume a constant value of polytropic efficiency on each speed line in calculating the outlet temperature – a simplification which will not seriously alter the form of the characteristic. Use the value for η_p derived in Exercise 11.5 for 100% speed, and take $\eta_p = 0.86$ at 90% speed and $\eta_p = 0.83$ at 80% speed.

Note: The non-dimensional mass flow rate based on outlet conditions can be equal at the different speeds. This is the case, for example, at a pressure ratio of 22.5 at 100% speed and a pressure ratio of about 12 and 7 for 90% and 80% respectively, corresponding to the working line drawn in Exercise 11.6.

Off-design operation of multi-stage compressors

It was remarked in Chapter 9 that the design of a compressor is inherently difficult because the pressure is increasing in the flow direction. If the pressure rise for a stage or for a complete compressor becomes too large for the speed of rotation or for the design the compressor can either surge or go into rotating stall. Shown on Figures 11.5 and 11.6 are surge lines for the two compressors. (It is common to refer to the boundary where the flow breaks down into surge or rotating stall as the surge line regardless of which form of breakdown actually occurs.) Surge is an oscillatory motion of the air, which is usually violent; rotating stall is a non-uniform pattern with reduced flow rate and pressure rise. Either surge or rotating stall is an unacceptable operating condition which it is important to avoid, but this becomes progressively more difficult as the overall pressure ratio of the compressor is increased. This section attempts to explain why. Stall and surge in multi-stage compressors are more complicated than in a single-stage fan because the different stages from front to back may be operating under very different conditions at the same time. For example, the front stage may be close to stall whilst the back stage is choked, and this is a consequence of the matching of the stages at off-design conditions.
          To understand the behaviour better it is helpful to consider the schematic example in Figure 11.9, where the overall pressure ratio versus non-dimensional inlet mass flow $\overline{m}_2$ is shown. The compressor is assumed here to be made up of many stages of identical performance. For each stage the non-dimensional pressure rise $\Delta p_0 / \rho U^2$ is the same unique function of the flow coefficient V_x / U (U being the mean rotor blade speed and V_x the mean axial velocity[1]) and this

---

[1] It can be shown that Vx/U is proportional to $(\dot{m}\sqrt{c_p T_0}/A_0) \div (N)/\sqrt{T_0}$, so for constant non-dimensional rotational speed a reduction in Vx/U is equivalent to a reduction in non-dimensional mass flow.

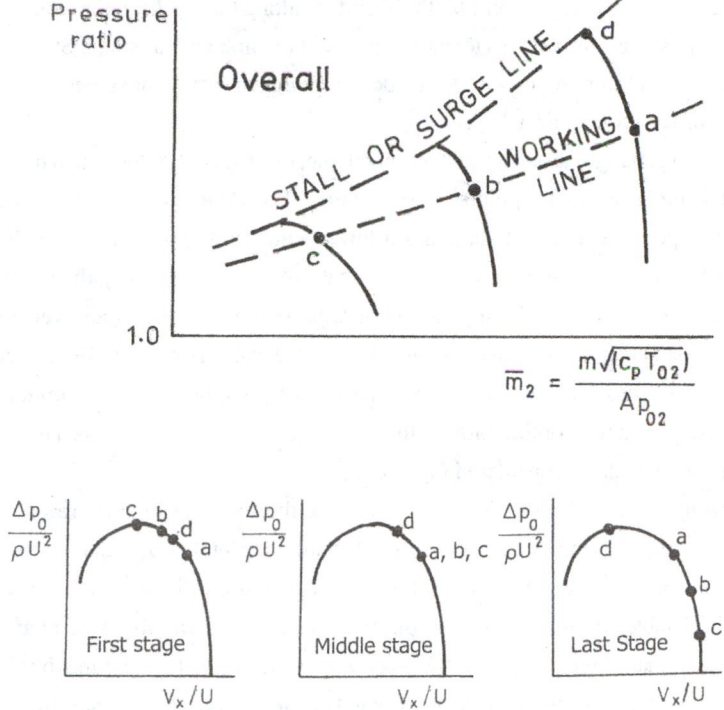

Figure 11.9. Schematic overall and stage characteristics for a multi-stage compressor to illustrate different behaviour of front and back stages.

relationship, for the first, a mid-point and the last stage is also shown in the lower part of Figure 11.9. Reducing V_x/U leads to an increase in stagnation pressure rise until the stage starts to stall, when further reduction in flow leads to a reduction in pressure rise. At the design point, shown as 'a', all stages operate at the same V_x/U which is possible because the annulus height is reduced from front to back of the compressor to compensate for the rise in density and keep V_x constant.

On the pressure ratio versus $\overline{m}_2$ plot a working line is drawn through the design point 'a' and this working line might be produced by a throttle on a test bed or by the turbine of an engine. Reducing the rotational speed, but staying on the working line through the design point, takes the compressor first to point 'b' and then to 'c'. To understand what happens at the level of the individual stages we must look again at the stage characteristics. As the rotational speed is reduced the pressure rise from each stage is reduced and so too is the density rise. The annulus height decreases from inlet to outlet of the compressor to accommodate the increasing density and keep the axial velocity constant but the dimensions are calculated for the design condition 'a'. For 'b' and even more so for 'c', the annulus height decreases too quickly and as a result V_x/U increases stage by stage through the compressor and the stage pressure rise falls. This is illustrated for points 'b' and 'c' in Figure 11.9. Eventually, the latter stages become choked or virtually choked and this constrains the mass flow through the entire compressor with the result that V_x and consequently V_x/U are reduced at the inlet. Reducing V_x/U drives a stage towards the point of surge and hence the overall effect of moving from point 'a' to 'b' and 'c' along the working line can, as illustrated in the figure, be to push the early stages towards surge and to choke a rear stage. Stages in the middle of the compressor tend to remain close to the design point. In fact front stages very often do have some rotating stall at low rotational speeds.

Considering changes along a constant speed line, the effect of reduction in mass flow relative to the working line is shown by points 'd' on Figure 11.9. Point 'd' is on the design speed line and compared to point 'a' the first stage has a lower value of V_x/U. This lower flow coefficient results in a higher pressure rise and therefore a greater increase in density than was envisaged at design. The greater increase in density means that the second stage has an even lower value of V_x/U than the first stage, with an even greater value of density rise. The effect is cumulative, so the last stage approaches stall whilst the front stage is only slightly altered. Similar effects occur at reduced speed, where the combination of throttling and the effect of speed changes described in the paragraph above must be considered together.

The strong effects of reduction in speed along the working line are revealed in Exercise 11.8. Although some rotating stall in the front stages is acceptable at reduced speed, the extent or degree of stall becomes too large when the design pressure ratio across the machine exceeds about 7 unless variable stators are fitted to the front stages. The variable stators are arranged for the front stages so that at reduced rotational speed the swirl velocity (i.e. tangential flow velocity in the direction of rotor motion) is raised at entry to the each rotor. This reduces the rotor incidence and allows the stages to be unstalled and to produce a pressure rise. Even with several rows of variable stators the mismatch at low speeds becomes so severe that pressure ratios exceeding 20 are unusual.

As noted above, the matching problem at speeds below design arises from the tendency for the rear stage to choke. This can be alleviated at low speeds where efficiency is not crucial by bleeding off some of the air around the middle of the compressor and dumping it into the bypass duct. This has the effect of increasing the flow through the front stages, tending to unstall these stages, thereby improving their performance. If the front stages operate more efficiently (more rise in pressure, less rise in temperature) it is possible to pass more mass flow through the rear stages at the same value of corrected mass flow. All multi-stage compressors for aircraft engines use bleed for some low-speed conditions, including starting.

Exercise

11.8 For the E^3 compressor of Figure 11.6 the non-dimensional mass flow along the working line will remain constant, based on compressor exit conditions, when tested with a choked throttle of constant area. The excursion in mass flow (and non-dimensional mass flow based on inlet conditions) is almost proportional to the pressure ratio at design to pressure ratio off-design. Suppose for simplicity that at design speed the *absolute* flow into the first rotor is axial, the *relative* inlet flow angle into the first rotor row is 60° at the mean radius and the incidence there is zero. If the axial velocity is proportional to mass flow at these conditions, what would be the relative flow direction and incidence at mean radius on the working line at 60% speed in the unlikely situation were there were no variable stators?

(**Ans:** $\beta = 81°$, $i = 21°$ – note this incidence is too large for the blades to work satisfactorily)

11.5 THE COMBUSTOR

The combustor is required to convert the chemical energy of the fuel into thermal energy with the smallest possible pressure loss and with the least emission of undesirable pollutants, such as products of incomplete combustion or oxides of nitrogen. To achieve this in a small volume it is necessary for the flow to be highly turbulent – the flow in a combustor is dominated by complex turbulent motions which still cannot be fully described quantitatively. Because the behaviour of the burning process depends on the turbulence, whilst at the same time the energy release brings about large alterations in the turbulence properties, computations for the three-dimensional flow in the combustors are more challenging than for other components of the engine where they are becoming fairly routine. The design of the combustor therefore continues to rely on experience and experimental data, though this is now augmented by the insights provided by highly complex numerical calculations. For this book this means that the combustor does not lend itself to simple, useful calculations in the same way that the compressor and turbine do. The combustor is therefore treated rather differently here than the other components, with only one numerical exercise which does not address the major problems of combustor design.

The treatment is necessarily brief and more information is readily available elsewhere: Hill and Peterson (1992) give a very clear account of the fluid dynamic and thermodynamic aspects of combustion, whilst Kerrebrock (1992) gives a particularly good introductory account of the chemistry and of the historical progress to reduce pollution. For a much more detailed treatment the reader is referred to LeFebvre and Ballal (2010) and Bahr and Dodds (1990). An excellent discussion of practical issues is given in *The Jet Engine* (2005).

Historically the combustor has always been one of the hardest components to get right, and problems with combustion of kerosene held up the Whittle engine for many months; von Ohain avoided the problems in the engine flown in 1939 by burning hydrogen instead of kerosene. Burning hydrogen remains an attractive option so far as the design and operation of the combustor is concerned, since many problems are removed, but it is currently ruled out as a fuel by its cost and practical problems of its use. Hydrogen would need to be liquefied (at around $-250°C$) which would require a thick layer of insulation around fuel tanks, which in turn would make the wings unsuitable for storing fuel. Finally the density of liquid hydrogen is low so that the energy stored per unit volume is only about one third that of kerosene.

Chemical energy release

Inside the combustor of a gas turbine there is a very large energy release rate per unit volume, typically 100 times that for the boiler in a large steam power station. The combustor for a jet engine needs to be small to fit between the compressor and turbine without making the shaft and engine unnecessarily long, since extra length would add to the weight and introduce problems of mechanical stiffness. The stiffness of the engine is very important and the outer walls of the combustor play a crucial role in shielding the load bearing outer casing from the extreme temperatures of the combustion process. The high energy release in a small volume is made possible by the high pressures (at least 40 bar for a recent large civil engine at sea-level take-off) and by the very high level of turbulence created in the combustor which mixes the fuel and air. It is also noteworthy that the length in the flow direction of combustors increases much less with engine thrust than other components. This is because the length of the combustor determines the time for the chemical process of combustion to take place and the time required is fixed by the chemical reaction rates.

Fuel–air ratio and turbine inlet temperature

The combustor delivers hot gas to the turbine at an average temperature that the turbine can tolerate with a level of non-uniformity that is tolerable. The crucial temperature is that to which the first turbine rotor is exposed, the stator outlet temperature (SOT) from the first row of nozzles. SOT is denoted here as T_{04} and is the temperature of the flow into the rotor after hypothetical mixing of the cooling air used in the first row of nozzles with the hot gases. For a modern large civil engine T_{04} is not more that about 1700 K in normal operation for take-off, but for combat engines being developed for entry into service in a few years time, this temperature could be as high as 2300 K. (For comparison it can be noted with stoichiometric combustion with inlet air at the elevated temperature of compressor exit gives a temperature of about 2600 K.) As discussed in Section 11.1, the fuel–air ratio (the mass flow rate of fuel for a given mass flow rate of air through the core) for stoichiometric combustion is about 0.0676. For a compressor delivery temperature of 918 K and turbine entry temperature of 1700 K it can be shown in Exercise 11.9 that the fuel-air ratio is only 0.0236. The maximum fuel–air ratio would be somewhat higher for a military engine at low flight Mach numbers because the temperature leaving the combustor is somewhat higher whilst the temperature of the air entering the combustor is lower in military engines because the pressure ratio is lower.

A difficulty with hydrocarbon fuels is that they will not burn if the fuel–air ratio is far below stoichiometric value of about 0.0676. The way around this is illustrated in Figure 11.10. Fuel is injected with some air through the injector. However, most of the air from the compressor is diverted to avoid the region where the fuel is injected so that combustion starts in a relatively rich primary region with a fuel-air ratio of about 0.25 (i.e. about 3.7 times stoichiometric) for take-off and about 0.1 at idle. Additional air is then fed in through holes in the combustor lining to complete the combustion process and, at the same time, reduce the temperature to a level acceptable to the turbine. After dilution the effective overall fuel-air ratio is about 0.03 at take-off power (i.e. less than half stoichiometric). The air entering the dilution region is also used to create a layer of relatively cool air on the walls and to modify the exit temperature radial profile to be suitable for entry into the turbine. The stress in the turbine rotor blades is highest near the root and to minimise creep it

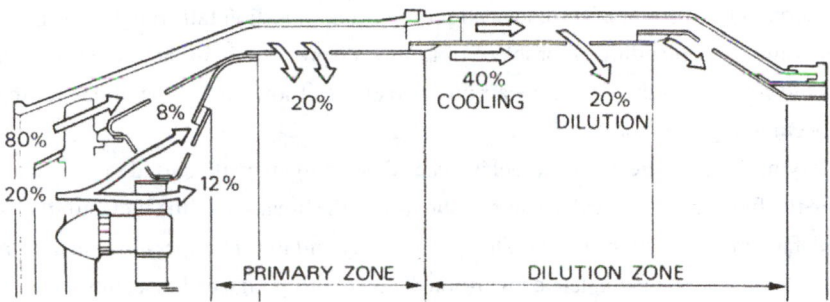

Figure 11.10. A schematic of a combustor showing the apportioning of airflow.
From *The Jet Engine* (1986).

is desirable to have lower temperature near the root than the tip. It is also desirable to have lower temperature at the turbine rotor tips to avoid them being damaged.

Flame speed and the stabilisation of flames

Air leaves the compressor at a speed on the order of 180 m/s, but this is reduced to around 50 m/s before entering the combustor. Laminar flames for hydrocarbon fuels, however, cannot travel at more than about 0.3 m/s, though this can be raised to around 5–8 m/s in turbulent flow. To prevent the flame from being blown away some means must be found to stabilise the reaction within the combustor by creating local regions with low velocity. One solution is to have the flame in the wake of a solid object, but it is now more common to set up a recirculation zone. The recirculation is normally created by swirling the flow around the fuel spray so that an enclosed region is formed. To create the swirl which will produce the recirculation, and to generate the high turbulence to increase the flame speed, there has to be a drop in stagnation pressure in the combustor. The actual drop in pressure depends on the type of combustor, but a loss of 4% of the absolute stagnation pressure entering the combustor is representative.

Fuel injection, combustion rate and combustion efficiency

Liquid fuel is supplied to injectors and a flow of air passing through the fuel injector breaks up the fuel into small droplets. The need to have a substantial air velocity to break up the fuel, typically around 40 m/s, requires a drop in pressure between the compressor and the centre of the burning regions and sets one lower limit on the pressure drop in the combustor.

Burning liquid droplets, even ones as small as 30×10^{-3} mm, require the liquid to evaporate and then for the fuel and air to diffuse together to form locally a near-stoichiometric mixture. This process is normally much slower than the main chemical process of combustion which normally takes place quickly. The rate of the chemical process of combustion falls roughly as the square of the air pressure, but an additional problem from low air pressure is in the failure to break-up of liquid fuel to form small droplets. Low air pressure arises at low power conditions for the engine and can occur at high altitude.

Varying the pressure at which fuel is injected is the main method used to vary the fuel flow rate; when the fuel flow is low and the air pressure is low the break-up of the fuel into small droplets is less satisfactory than at high powers. The coarser spray and lower temperatures in the combustor at low thrust can lead to incomplete combustion. This in turn produces low combustion efficiency with high levels of carbon monoxide and unburned hydrocarbons in the exhaust. The efficiency of combustion (defined as the actual temperature rise divided by that for complete combustion) for a civil airliner is very close to 100% for high thrust levels at sea level but decreases to around 99% at idle conditions. After starting and during the run up to idle conditions the combustion efficiency can be significantly lower. When the engine is throttled back to give small thrust at the start of descent, but still at the normal cruise altitude, the combustion efficiency can become so low that the process is no longer self-sustaining and the flame can be extinguished.

An important requirement, which has to be designed for, is the ability to re-light the combustor at altitude, even when the compressor is producing very little pressure rise. Combustion efficiency at starting is important for altitude restarts, requiring enough residence time for fuel and air to react sufficiently. Enabling this, so that the engine can pull away to idle conditions, tends to fix the minimum volume of the combustor itself.

Wall cooling and the use of annular combustors

The temperatures of the gas in the combustion region are high enough that rapid failure of the combustor walls would occur if the hot gas came in contact with them. Modern large engines are expected to have a combustor life of at least 20,000 hours. Moreover, thermal stress and fatigue are critical and the combustor should be able to tolerate not less than about 5000 cycles of take-off, climb, cruise, landing and taxiing. The combustor walls are therefore shielded and cooled with compressor delivery air, as illustrated in Figure 11.10. There is an obvious advantage in reducing the amount of surface area in relation to the flow area. Early engines arranged the combustor as a series of discrete cans or tubes (referred to as can-annular or tubo-annular combustors) but large engines designed since the late 1960s have used an annular geometry, as illustrated in Figure 11.11, with corresponding saving in surface area. The fuel is still supplied through a number of discrete injectors.

Modern combustors normally have an inner wall lined with a thermal barrier coating. These coatings reduce the metal temperatures, for the same gas temperatures, and also inhibit oxidation of the surfaces. The modern combustor often has many small cooling holes, similar to the surface of turbine blades, but an alternative is to use metallic tiles, which are cooled from behind. The tiles,

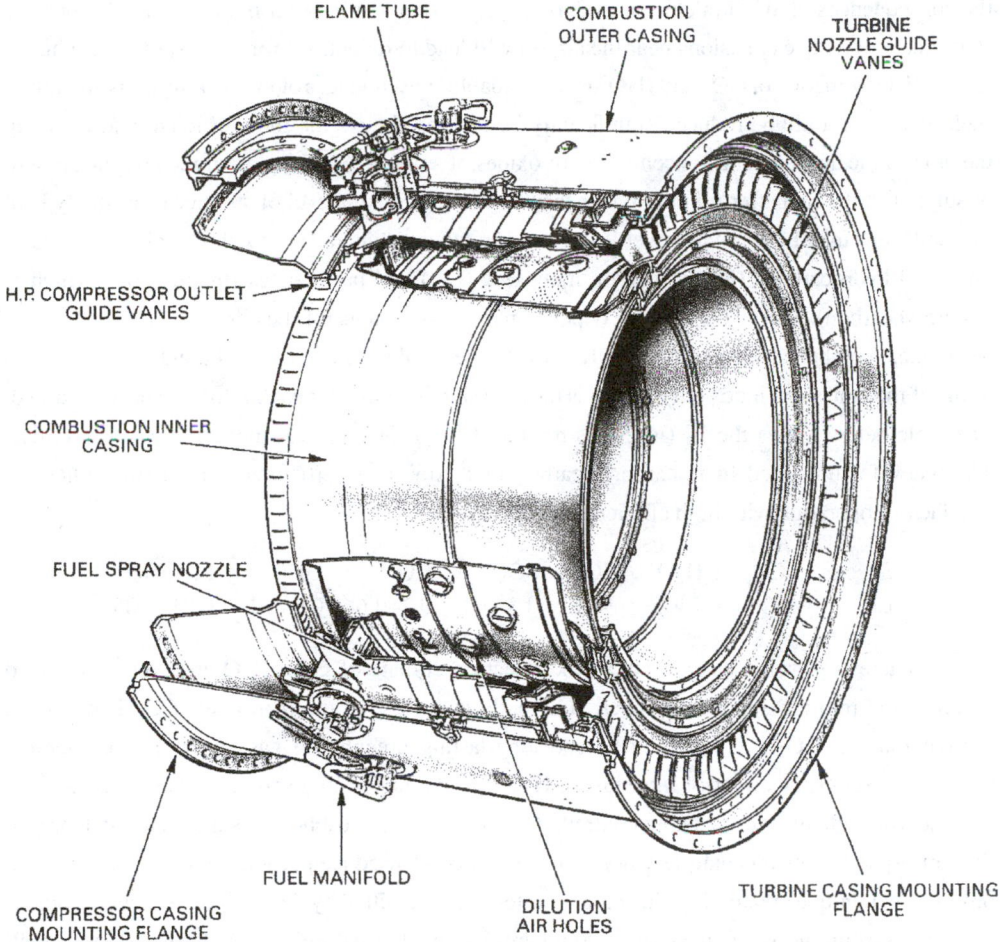

FLAME TUBE

COMBUSTION
OUTER CASING

TURBINE
NOZZLE GUIDE
VANES

H.P. COMPRESSOR OUTLET
GUIDE VANES

COMBUSTION INNER
CASING

FUEL SPRAY NOZZLE

FUEL MANIFOLD

COMPRESSOR CASING
MOUNTING FLANGE

DILUTION
AIR HOLES

TURBINE CASING MOUNTING
FLANGE

Figure 11.11. An annular combustor. From *The Jet Engine* (1986).

which carry very little load and can be made of high temperature cast alloys, can therefore operate at higher temperature than the outer combustion walls.

Emissions – formation, regulation and control

Emissions, the creation of harmful or toxic gases during combustion, can be considered important for two different points of view. One relates to the effect on the global environment, such as global warming, climate change and ozone changes, and is primarily a problem during cruise when most of the fuel is burned. The other relates to the immediate surroundings of the airport and it was the effect of aircraft emissions during starting, taxiing, taking off and landing which were first apparent and gave rise to the first protests and then to legislation. At present legislation applies only near the airport and is based on simulations of landing and take-off operations. This is despite the fact that the consequences of aviation emissions during cruise are potentially far more serious; fortunately steps taken to reduce emissions near the airport will lead to reductions for the rest of the flight.

The formation of CO_2 and H_2O are unavoidable consequences of the burning of hydrocarbon fuel, which can only be reduced significantly by making the engine more efficient and reducing the drag of the aircraft. The concentration of oxides of sulphur (SO_x) is determined by the amount of sulphur in the fuel after refining rather than by the engine; the level of sulphur is normally kept low and steps may be taken to lower it further. The other pollutants are oxides of nitrogen (NO_x), unburned hydrocarbons (UHC), carbon monoxide (CO) and particulates (mainly soot, which is unburned carbon). The levels of these depend on the performance of the combustion chamber and for an ideal combustor would be virtually zero. The level of emission of a pollutant is expressed in terms of the emissions index (EI), which is the emission in grams for each kilogram of fuel burned. The table below shows the EI (in grams per kg of fuel) for a typical modern aircraft at cruise. The index for unburned hydrocarbons, carbon monoxide and particulates are small, indicating satisfactory operation with high efficiency.

Species	CO_2	H_2O	NO_x	SO_x	CO	UHC	Particulate
EI	3200	1300	9–15	0.3–0.8	0.2–0.6	0.0–0.1	0.01–0.05

There is growing concern about the effect on climate of CO_2, H_2O and NO_x introduced by air traffic into the upper atmosphere. Currently about 2% of the man-made CO_2 is produced by aviation and CO_2 is a well understood long-lasting greenhouse gas. There is considerable uncertainty about the effect of introducing H_2O into the upper atmosphere; either introducing it into the troposphere, where visible contrails can form if the atmosphere is super-saturated, or into the stratosphere, which is naturally dry. The mass of condensed ice particles in contrails is several orders of magnitude greater than the mass of water vapour emitted by the engines. The water vapour in the exhaust plume first condenses to liquid droplets and these droplets then freeze to form small ice particles. These ice particles form nucleation sites which enables the super-saturated vapour in the ambient air to condense. The effect of NO_x is also complicated, affecting ozone and the greenhouse effect. All of this was discussed at considerable length in the special report of the IPCC On Aviation and the Global Atmosphere, published in 1999.

Kerrebrock (1992) explains how until the late 1970s the designers of combustors were so fully occupied with making the combustion stable and efficient, whilst avoiding the burning of the walls, that they could do little for the pollution aspects. The first target pressed by the US Environmental Protection Agency was to make the engines produce no visible smoke (unburned carbon particles) in the phase of operation below 3000 ft altitude that had a direct impact on the environment around the airport. As a result smoke had been greatly reduced before 1981 when the ICAO[2] regulations for new aircraft types were issued, applying to carbon monoxide, unburned

---

[2] ICAO, the International Civil Aviation Organisation, is the United Nations specialised agency that has global responsibility for the establishment of standards, recommended practices, and guidance on various aspects of international civil aviation, including environmental protection.

hydrocarbon and oxides of nitrogen, none of which had been addressed before. The aviation industry at first resisted the proposed regulations for reducing pollutants on the grounds of impracticality, but since then there has been considerable progress and compliance with the limits became obligatory. The ICAO regulations for CO, UHC, NO_x and smoke lay down procedures as well as actual levels. (For smoke the limit is set at the maximum level to ensure invisibility.) For each species the mass generated is summed for measurements made of the engine on a sea-level static test bed when operating over a standard landing and take-off (LTO) cycle: 42 seconds at 100% thrust, 2.2 minutes at 85% of maximum thrust (to simulate climb to 3000 feet), 4 minutes at 30% thrust (to simulate the approach) and 26 minutes at 7% thrust to allow for taxiing and idle on the ground. This mass Dp of a pollutant in grammes is then divided by maximum rated thrust F_{00} in kilo newton at standard sea-level conditions; the quantity is denoted in the regulations by Dp/F_{00}. (Dp is from the French '*des pollutants*' and F_{00} stands for maximum thrust at sea level on a static test bed.) For unburned hydrocarbons and carbon monoxide the allowable level has remained constant at the original ICAO level of 19.6 g/kN and 118.0 g/kN respectively, summed over the LTO cycle using the maximum rated thrust.

The first regulations, the ones issued in 1981, are usually referred to as the ICAO regulations, but the later regulations are known by the committee of ICAO which now considers them, the Committee for Aviation Environmental Protection (CAEP). The first regulations to come from CAEP, known as CAEP2, were implemented in 1986 and included a production cut-off regulation (effectively stopping production after a fixed future date of engines already in production which were not modified to the new standard). The ICAO/CAEP regulations for NO_x up to CAEP8 are summarised in the appendix to this chapter.

Whereas NO_x and smoke are the main emission of concern at high thrust conditions (when the fuel flow to the combustor is high and the temperatures are also high) the emission of CO and UHC, tend to be greatest during the taxi and idle conditions. At least conceptually, the removal of CO, UHC and smoke is straightforward: the combustion should be prolonged for as long as possible at high temperature in the presence of ample excess oxygen. This also has the effect of increasing the combustion efficiency, though it may not necessarily assist high-altitude re-light ignition capability. The requirements for reducing NO_x are much more subtle. The rate of formation of NO_x increases rapidly with temperature but these chemical reactions are much slower than those leading to the formation of CO_2 and H_2O. Because of the *comparatively* slow rate of formation of NO_x, the amount created depends on both the temperature and the *residence time* at that temperature. Unfortunately the long residence time at high temperature which would reduce CO, UHC and smoke would favour the formation of NO_x. When temperature is subsequently reduced the breakdown of NO_x is significantly slower than its formation. Figure 11.12 shows a cross-section through a modern combustor together with an inset plot showing schematically the production and consumption of NO_x and smoke.

The picture of the combustor in Figure 11.12 illustrates how complicated the designs have become to achieve satisfactory emissions, with carefully placed chutes to direct the mixing

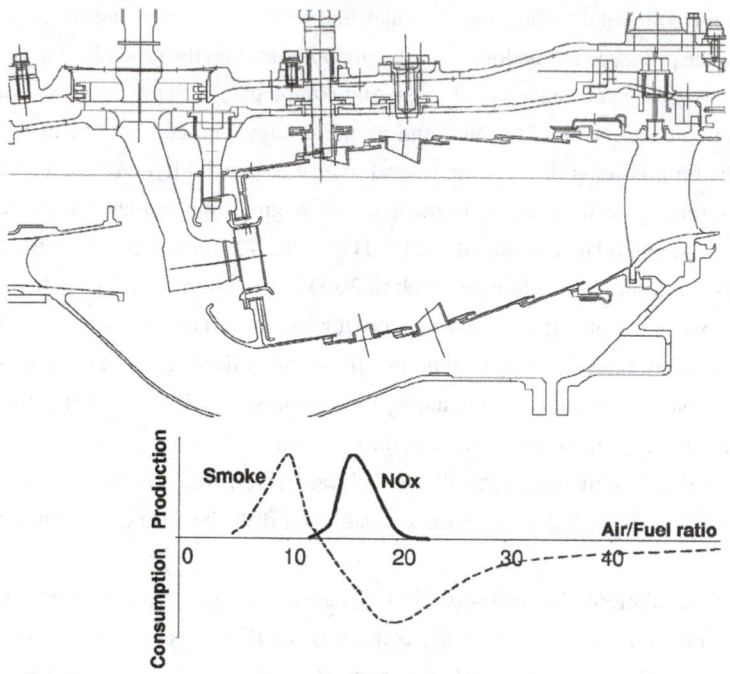

Figure 11.12. A cross-section through the combustor of a modern combustor with a sketch of the smoke and NO$_x$ formation.

air where it is most needed. The diagram shows that soot is formed where the mixture is rich (and oxygen correspondingly scarce), but since oxygen is scarce little NO$_x$ is formed at first; the NO$_x$ forms rapidly when more air is added and temperatures are locally close to stoichiometric. What happens downstream is a balance between maintaining the temperature high enough in the presence of excess air to burn the soot whilst cooling the gas sufficiently to avoid forming high levels of NO$_x$. The standard approach to reducing NO$_x$ is to minimise the residence time at high temperature as much as possible, having in mind the need to burn off soot, UHC and CO, and also the need to keep an acceptable level of combustion efficiency at high altitude and low fuel flow rate. The very hot gases, which locally contain pockets of stoichiomentric combustion, are therefore rapidly quenched by mixing with cool air to drop the temperature below that at which the NO$_x$ formation is significant. The need to reduce residence time to avoid high levels of NO$_x$ is a powerful driver to the reduction in size of the combustor in relation to the flow rates of air and fuel.

As already noted, to keep NO$_x$ low at high thrust (requiring a short residence time at high temperature) and to keep CO and UHC levels down at low thrust (requiring a long residence time) are fundamentally contradictory. For very low levels of emissions (or for acceptable levels at very high combustor inlet and outlet temperatures) the solution may lie with some form of *staged* combustor. For staged combustors different injectors and/or different regions of the combustor are

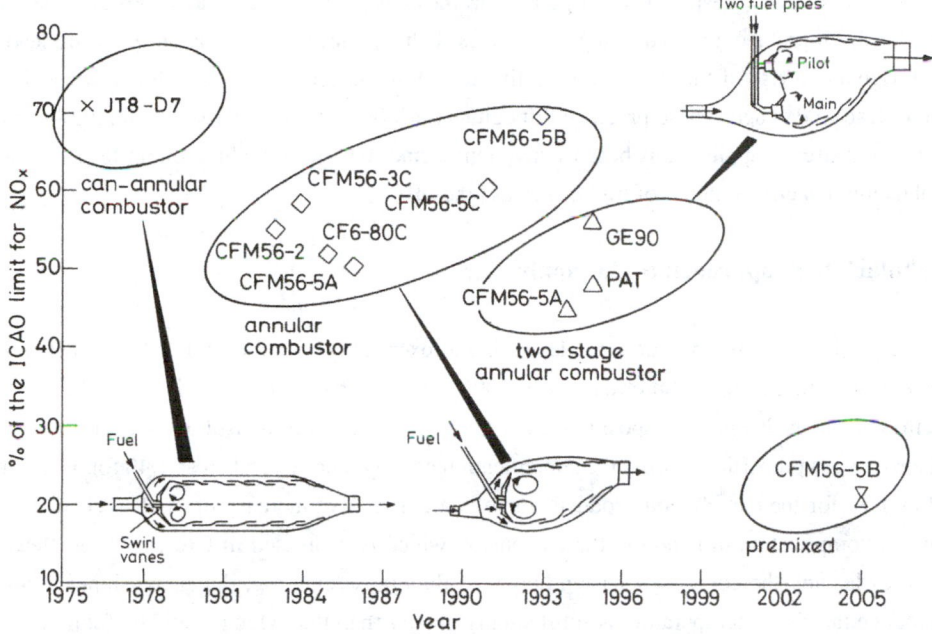

Figure 11.13. The variation in NO$_x$ with date. (Based on figure supplied by SNECMA.) Also shown are schematics of three types of combustor.

used for low thrust and high thrust; for idle and low power there is a pilot stage and for high power there is a separate stage. The use of staged combustion introduces extra cost and complexity compared with a combustor like the one from a Rolls-Royce Trent shown in Figure 11.12. The CFM56 was offered with a two-stage combustor but due to poor sales and combustor trade-offs this variant on the CFM56 is no longer available. The experience with the CFM56 was, however, a step towards the General Electric GEnx lean staged combustor that is now in service.

The progress that has been made in the case of NO$_x$ is illustrated in Figure 11.13. The ordinate is the percentage of the limit laid down in the original ICAO regulation for the engines shown. Although the levels have, fallen, the extent of the fall to date is not enormous. What is certain, however, is that without the improvements in design to lower NO$_x$ the level would have been higher, in part because of the higher compressor delivery temperature. The graph also shows sketches of three types of combustor: the old can-annular (or tubo-annular) type, the annular combustor which is the main type in service now and the two-stage annular combustor. Attention in the companies is now focused on developing a lean burn combustor in a single annular architecture.

If even lower levels of NO$_x$ are to be achieved, especially if it becomes necessary to limit NO$_x$ production at cruise, then a premixed arrangement will probably be necessary; here the fuel will be mixed with the air and partially vaporised before entering the combustion region. An advantage of premixing is that it avoids the near-stoichiometric burning that takes place in most present-day combustors as fuel and air diffuse together and burn with high local temperatures. Premixed combustion is now widespread for large gas turbines designed for land-based power generation but it has been problematic to develop, sometimes with high levels of oscillatory pressure leading to damage and a risk of the flame being extinguished or 'flashing back' to stabilise where it can rapidly lead to damage. These practical difficulties must be overcome before premixing becomes feasible for aircraft applications because a system cannot be allowed which might lead to loss of combustion in a critical phase of flight, such as take-off.

The 'black box' approach to the combustor

For the purpose of this book and the design based exercises the combustor will be treated as a 'black box' with a combustion efficiency of 100%. In using the 'black box' it should be recalled that the change in chemical composition and in temperature in the combustion process brings about a change in gas properties. Within the simplification adopted here γ is taken as 1.40 for air at entry and as 1.30 for the combustion products; there is a consequent increase in c_p. The change in c_p causes a complication in handling the combustor, which is addressed in Exercise 11.9. Because the gases leaving the combustor have a higher c_p than those entering, the energy input required to produce the rise in temperature is substantially greater than that which would be found using a constant value of c_p.

Exercise

11.9 Write down the energy balance for a calorimeter with inlet and outlet flows at 298 K for which the energy removed per kg of fuel is the lower calorific value LCV. (LCV is defined at 25°C and not 288 K.) Then consider the adiabatic combustor with inlet temperature T_{03} and outlet temperature T_{04},

$$\dot{m}_f LCV = \left(\dot{m}_{air} + \dot{m}_f\right)\left(h_{p04} - h_{p298}\right) - \dot{m}_{air}\left(h_{03} - h_{298}\right),$$

where h_{p04} and h_{p298} refer to enthalpy of the products at turbine entry temperature and the reference temperature 298 K whilst h_{03} and h_{298} refer to the air at T_{03} and 298 K. If c_p and c_{pe} are the specific heat capacity of the gas at entry and exit, assumed constant from 298 K to T, show that

$$\dot{m}_f LCV = \left(\dot{m}_{air} + \dot{m}_f\right) c_{pe}\left(T_{04} - 298\right) - \dot{m}_{air} c_p \left(T_{03} - 298\right), \tag{11.6}$$

where $\dot{m}_f$ is the mass flow rate of fuel and $\dot{m}_{air}$ the mass flow of air entering the combustor.

Taking $LCV = 43$ MJ/kg, $c_p = 1005$ and $c_{pe} = 1244$ J kg^{-1}K^{-1}, find the mass flow of fuel needed per unit mass flow of air into the combustor, to create $T_{04} = 1700$ K when $T_{03} = 918$ K (from Exercise 4.1).

(Ans: 0.0272 kg/s)

Repeat the calculation taking $c_{pe} = c_p = 1005$ J kg^{-1}K^{-1}, which corresponds to assuming $\gamma = 1.4$ throughout.

(Ans: 0.0186 kg/s)

Notes: (1) Calculations have also been performed for the combustor using enthalpy tabulated against temperature to calculate fuel flow, an approach which is essentially exact. Using this method the mass flow of fuel for $T_{03} = 918$ K and $T_{04} = 1700$ K was found to be 0.0236 kg/s per kg of air.

(2) In practice a substantial amount of air is used to cool the nozzle guide vanes to the HP turbine and by the normal convention the turbine inlet temperature is the mixed-out temperature at the exit from these stators. This is the temperature used here, T_{04}.

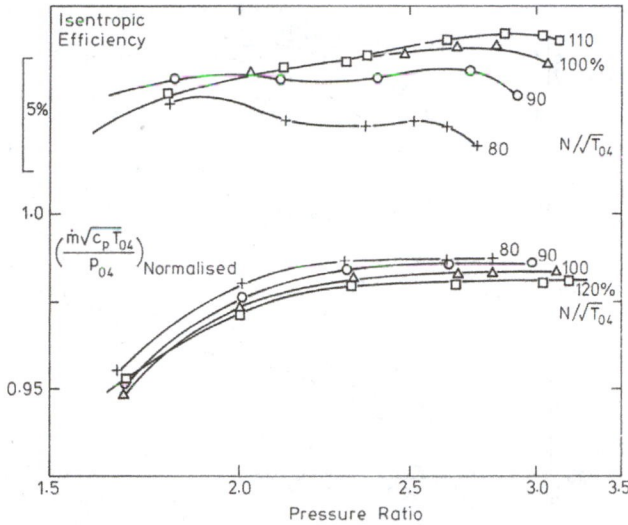

Figure 11.14. Characteristics of a modern HP turbine showing mass flow
(non-dimensionalised by inlet conditions) and isentropic efficiency versus pressure
ratio for lines of constant non-dimensional rotational speed.

11.6 THE TURBINE

The pressure ratio, p_{04}/p_{05} of an HP turbine stage is plotted against the non-dimensional mass flow
in Figure 11.14. In this case the non-dimensional mass flow is evaluated using the inlet stagnation
pressure and temperature, p_{04} and T_{04}. The results shown are for various non-dimensional speeds,
$N/\sqrt{c_p T_{04}}$, but the variation in non-dimensional mass flow with pressure ratio is almost independent
of the speed over the range shown. In fact the turbine behaves to the upstream flow like a choked
nozzle for all but the lowest speeds; because the choking flow is almost independent of speed this
indicates that it is the nozzle row which gives most of the choking effect. Although the turbine has
a mass flow/pressure ratio variation like a choked nozzle, most turbines are not actually choked,
though the maximum average Mach numbers are close to unity. The combination of several rows
of blades, each nearly choked, simulates a truly choked row.

A second important property of the turbine shown in Figure 11.14 is the efficiency. The
changes in efficiency are small enough to be neglected in the analyses which will be carried out
here, in which the turbine is not far removed from its design condition. As a result the temperature
ratio and non-dimensional power output will be assumed to be fixed only by the pressure ratio
across the turbine.

Figure 11.15 shows the corresponding flow rate and efficiency curves for an LP turbine for
a high bypass ratio engine. The LP turbine rotates comparatively slowly, because the fan which it
drives cannot rotate very fast, and as a result the Mach numbers in such a turbine are lower than in

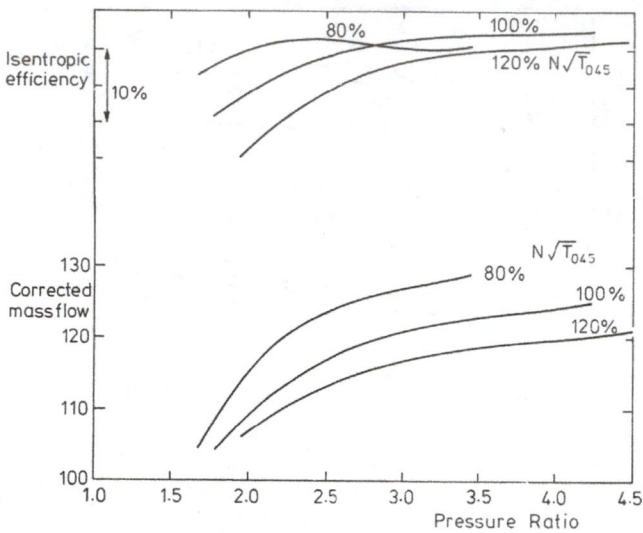

Figure 11.15. Characteristics of a modern LP turbine showing corrected mass flow and isentropic efficiency versus pressure ratio for lines of constant non-dimensional rotational speed.

the HP. To get the large work output the flow turning in the LP turbine blades is large and there are generally several stages, sometimes up to seven (see Figure 5.4b). The effect of these is to give only a small variation in non-dimensional flow rate and efficiency with pressure ratio over a substantial range of rotational speed. Again, so far as mass flow rate is concerned, the LP turbine behaves very much like a choked nozzle and the dependence of efficiency on speed and pressure ratio is small enough to be neglected in the approximate analysis here. (Much less than for compressors.)

Figure 11.15 is for the LP turbine of a civil engine, where rotational speeds are low. Military LP turbines are more like HP turbines with relatively high rotational speeds in relation to gas speed. Moreover, most military LP turbines, unlike civil LP turbines have some cooled blade rows. One of the radical alterations in engine architecture which may become commonplace for civil engines is the introduction of a gear box between the LP turbine and the fan. If this happens the LP turbine will be able to rotate faster, near its optimum speed for the flow conditions, and the number of stages will be significantly lower (see Figure 5.4c). The pressure ratio across each stage will consequently increase and the turbine will again appear to the upstream flow to have characteristics like a choked nozzle.

Exercise

11.10 Replot the pressure ratio–mass flow characteristic of the turbine shown in Figure 11.15 at 100% $N/\sqrt{T_{04}}$ in terms of the non-dimensional mass flow based on *outlet* conditions. Take the polytropic efficiency to be constant at 90%.

Turbine cooling

As was briefly discussed in Section 5.3, many turbines operate with gas temperatures in excess of the metal melting value and an essential requirement is turbine cooling using air. For modern engines the HP turbine stages are always cooled and very often some stages of the IP or LP turbine are cooled as well. Typically 20% of the air entering the core compressors may be directed around the combustor to cool the turbines and, as gas temperatures go up, there is a tendency for the amount of cooling to rise too. Typically half the cooling air is used in the HP turbine nozzle guide vanes. Because this air mixes with the combustion products *before* entering the rotor it does not affect the turbine work and the effect of HP nozzle vane cooling is the same as the dilution air in the combustor. By choosing to denote turbine inlet temperature T_{04} to be the temperature downstream of HP nozzle guide vanes this cooling air is automatically accounted for. Cooling air introduced further downstream in the turbine, such as in the HP rotor, does have an effect on the turbine work and the cycle efficiency and needs to be explicitly considered. Likewise cooling air to nozzles other than the first nozzle guide vane row does introduce a cycle penalty.

Cooling air can affect the flow and aerodynamic loss in the turbine, but the main effect, and the only one to be considered here, is the effect on the thermodynamic cycle. Work has been done in compressing the air which is used for cooling, but the potential work is not fully realized when it is used for cooling. The penalty in cycle efficiency is reduced if the cooling air is bled from the compressor at the lowest possible pressure to enter the turbine flow (meaning extracting the cooling air as far upstream in the compressor as possible). For the simple analysis in Chapters 12, 16 and 17 it will be assumed that all the cooling air is taken at compressor deliver pressure p_{03} with temperature T_{03}. The cooling air is assumed to mix with turbine main flow at the local pressure in the turbine; for the HP turbine rotor the mixing is assumed to occur just downstream of the rotor with no work extracted from the cooling air in that blade row. If T_{045} is the temperature which would emerge from the HP rotor without cooling and $T_{045'}$ is the temperature after mixing then, with an approximation for the specific heats, one can write

$$\left(\dot{m}_{a45} + \dot{m}_f\right) c_{pe} T_{45'} = \left(\dot{m}_{a4} + \dot{m}_f\right) c_{pe} T_{045} + \dot{m}_c + \dot{m}_c c_p T_{03}, \qquad (11.7)$$

where $\dot{m}_{a4}$ is the air mass flow at turbine nozzle inlet, $\dot{m}_{a45}$ is mass flow of air at HP turbine exit after mixing of the cooling air, $\dot{m}_f$ is the fuel mass flow rate and $\dot{m}_c = \dot{m}_{a45} - \dot{m}_{a4}$ is the cooling mass flow. The cooling air specific heat capacity is denoted by c_p whereas that of the turbine main stream is c_{pe} into the HP turbine and $c_{pe'}$ after mixing with the cooling air. The mixed gas at pressure p_{045} and temperature $T_{045'}$ then enters the next row of nozzle vanes.

The turbine polytropic efficiency of the HP turbine can be calculated using the polytropic relation

$$p_{045}/p_{04} = (T_{045}/T_{04})^{\gamma/(\gamma-1)\eta},$$

specifically based on the temperature *before* mixing the cooling air, T_{045}. The ratio of specific heats is given by $\gamma_e = c_{pe}/(c_{pe} - R)$.

Exercise

11.11 The HP turbine nozzle guide vanes are cooled with 10% of air compressed in the HP compressor but this is assumed fully mixed out at rotor inlet to give a stagnation temperature $T_{04} = 1500$ K and therefore is accounted for. If the stagnation pressure downstream of the rotor is given by $p_{045} = 0.226 p_{04}$, find the temperature T_{045} when the turbine polytropic efficiency is 0.9 and then calculate the adiabatic efficiency.

The HP turbine rotor is cooled with 8% of the air entering the core. This air is taken from compressor at exit with a stagnation temperature of 826 K. The rotor cooling air is assumed to mix out at rotor exit without having done work in the rotor, so its temperature is that of the compressor exit, whilst the turbine gas temperature is that found above. If c_{pe} for the gas leaving the combustor is approximated by 1244 kJ kg^{-1}K^{-1} (corresponding to $\gamma = 1.30$) and the cooling air by $c_p = 1005$ kJ kg^{-1}K^{-1} (corresponding to $\gamma = 1.40$) find $c_{pe'}$ for the mixed-out gas, using a mass weighting. Assume that the mass flow of fuel is equal to 0.0225 times the mass flow entering the core compressor.

Then use Equation 11.7 to find the mixed-out temperature from the HP turbine before the gas enters the next turbine nozzle row.

(**Ans:** $T_{045} = 1101.4$ K, $\eta_{ad} = 0.915$, $c_{pe'} = 1223.3$ kJ kg^{-1}K^{-1}, $T_{045'} = 1083.2$ K)

Summary of chapter 11

For an ideal convergent nozzle there is a unique relation between non-dimensional mass flow and the ratio of inlet stagnation pressure to exit static pressure. For a ratio of inlet stagnation pressure to outlet static pressure exceeding 1.89 the nozzle is choked in the case of air at near ambient temperatures (i.e. when $\gamma = 1.4$) and the non-dimensional mass flow does not increase further if pressure ratio is increased. For a convergent–divergent nozzle the choking relation is also valid, but based on the area and static pressure at the throat in place of exit conditions.

For a fan or compressor the pressure ratio increases at a given rotational speed as the mass flow rate is reduced. A reduction in mass flow leads to pressure rise increasing as axial velocity falls and incidence increases. In non-dimensional form the pressure ratio and non-dimensional mass flow are strong functions of the non-dimensional rotational speed: pressure rise is roughly proportional to the square of rotational speed; at low speed the mass flow increases approximately linearly with rotational speed, but when choking becomes important the rate of increase can be much smaller.

Although pressure rise and pressure ratio for a fan or compressor increase as the mass flow rate is reduced at fixed speed, there is a limit to this. The limit is marked on the performance map as the surge or stall line; attempts to operate above and to the left of this line result in either rotating stall (with a large drop in pressure rise) or surge (a violent pulsation of the entire flow). The working line, sometimes called operating line, of the fan or compressor is determined by the downstream components, which may be a nozzle, a throttle or the other engine components. The working line through the design point lies so that it is likely to cross the surge line at low rotational speeds.

The crucial requirement for a multistage compressor is the matching of the stages so that the annulus area gives the required axial velocity into each stage by allowing for the increase in density. If this is correct at design it normally cannot be correct at off-design conditions because the density variation will be different. A serious problem of mismatching arises in compressors with a large design pressure ratio. When the speed is reduced there is a tendency for the rear stages to choke and the front stages to stall. This is alleviated to some extent by having variable stagger stators in the front stages (arranged to become more nearly tangential as the rotational speed falls) and having bleed ports part way along the compressor which open at low speeds. Even with these the design of compressors for pressure ratios above about 20 is difficult.

Combustors are complicated components in which the designer has simultaneously to meet conflicting requirements. Not only do requirements for low pollution conflict with some operability aspects, but techniques for reducing carbon monoxide and unburned hydrocarbons can lead to an increase in oxides of nitrogen. There have been very remarkable improvements over the last few years to meet the regulations. For the elementary treatment of engine matching it suffices to assume that the combustion efficiency is 100% and that there is a loss in stagnation pressure equal to 4% of the inlet stagnation pressure.

Turbines have pressure ratio versus non-dimensional mass flow characteristics very similar to those of a nozzle, and for most conditions relevant to the jet engine the turbine behaves as if it were choked. The mass flow rate is barely affected by the rotational speed of the turbine. Turbine efficiency is dependent on non-dimensional rotational speed as well as pressure ratio, but much less than the efficiency of a compressor and to a sufficiently small extent that it may be neglected in the cycle analyses to be performed here. With this approximation the temperature ratio across the turbine is therefore fixed by the pressure ratio alone.

As an alternative to the isentropic efficiency the polytropic efficiency can be defined. This allows compressors or turbines of different overall pressure ratio to be compared without the bias introduced in the definition of isentropic efficiency. It also allows algebraic simplification in some cases.

APPENDIX: NO_x EMISSIONS AND REGULATIONS

New regulations for emissions are drawn up by the Committee for Aviation Environmental Protection (CAEP). After agreement is reached in CAEP the proposals have to be agreed by ICAO and then ratified by the governments of all the member countries, processes which take three or more years. The first regulations of CAEP (known as CAEP2) were implemented in 1986 and included a production cut-off regulation (effectively stopping production after a fixed future date of engines already in production which were not modified to the new standard). Subsequently CAEP4, CAEP6 and CAEP8 have tightened the NO_x regulation by implementing new-type standards (i.e. specifying the stricter standard which engines first entering service must achieve) and providing production cut-offs associated with those regulations. These are shown for NO_x in Figure 11.16 and Table 11.1.

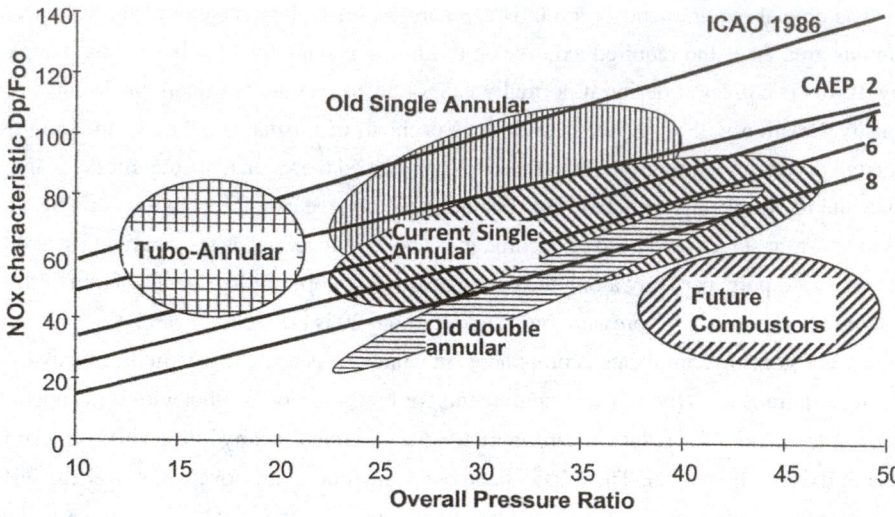

Figure 11.16. The ICAO regulations versus pressure ratio with the regions of expected performance for different types of combustor.

To compensate for the rise over time of the overall engine pressure ratio, and consequently the compressor delivery temperature and stoichiometric temperature, the allowable level of NO_x is allowed to increase in proportion to pressure ratio. The original ICAO limit was set as $40 + 2p_{03}/p_{02}$. This simple proportionality to the overall pressure ratio was smaller for CAEP2 regulations but was restored in later regulations. Figure 11.16 shows the original ICAO levels for NO_x in the landing and take-off (LTO) cycle and the more stringent regulations from CAEP2 down to CAEP8.

To visualise the tightening of standards which has occurred it is helpful to consider how the regulations would apply to succession of large engines all with a constant overall pressure ratio of 40; the impact of the regulations on allowable NO_x is summarised in Table 11.1.

Table 11.1 Dp/F_{oo} for NO_x levels allowed for an engine with $opr = 40$

Title	Dp/F_{oo} (g/kN)	Date of agreement	Date of implementation	
			New types	In production
ICAO	120	1981	1986	1986
CAEP2	96	1991	1996	2000
CAEP4	87	1998	2004	N/A
CAEP6	79	2004	2008	2013
CAEP8	70	2010	2014	not yet agreed

CHAPTER 12 ENGINE MATCHING OFF-DESIGN

12.0 INTRODUCTION

In Chapter 11 the performance of the main aerodynamic and thermodynamic components of the engine was considered. In earlier chapters the design condition of a high bypass ratio engine had been specified and a design arrived at for this condition. At the design point all the component performances would ideally fit together and only the specification of their performance at this particular condition would be required. Unfortunately engine components never exactly meet their aerodynamic design specification and we need to be able to assess what effect these discrepancies have. Furthermore engines do not only operate at one non-dimensional condition, but over a range of power settings (for take-off, climb, cruise and descent) and there is great concern that the performance of the engine should be satisfactory and safe at all off-design conditions. For the engines intended for subsonic civil transport the range of critical operating conditions is relatively small, but for engines intended for high-speed propulsion and for combat aircraft, performance may be critical at several widely separated operating points. Although the chapter is predominantly aimed at the engine for the New Efficient Aircraft, the chapter also lays the ground for the off-design behaviour and treatment of combat aircraft.

The treatment in this chapter is deliberately approximate and lends itself to very simple estimates of performance without the need for large computers or even for much detail about the component performance. The ideas which underpin the approach adopted are physically sound and the approximations are sufficiently good that the correct trends can be predicted; if greater precision is required the method for obtaining this, and the information needed about component performance, should be clear. The programme GasTurb is widely available and allows for more precise modelling of engine behaviour.

12.1 ASSUMPTIONS AND SIMPLIFICATIONS

Engine performance is primarily determined by the inlet air stagnation pressure and temperature and by the fuel flow. It may also be affected by the ambient static pressure, as discussed in Chapter 8. Of these inputs only the fuel flow may be treated as the control or independent variable. Depending on the fuel flow are the thrust of the engine, the mass flow of air, the rotational speeds of the shafts and the temperatures and pressures inside the machine. If the engine operates at a constant condition the appropriate non-dimensional values of these quantities must be constant. The problem of determining these quantities can be set as a list of constraints and for this we assume that the engine is a multi-shaft machine.

1. The rotational speed of the compressor and turbine must be equal on each shaft.
2. The mass flow through the compressor and turbine must be equal (neglecting in this chapter the mass flow removed in bleeds and the small mass flow of fuel).
3. The power output of the turbine must equal the power input into the compressor on the same shaft (neglecting the small power losses in the bearings and windage and the sometimes substantial power off-take to supply electrical and hydraulic power to the aircraft).
4. The pressure rise in the compression processes (including the intake) must equal the pressure drops in the expansion process, including the combustor, turbines and propulsive nozzle(s).

In general matching these constraints is a process which involves iteration, since the measured performance characteristics of the fan, compressors and turbines have to be used such as those in Figures 11.6 and 11.14. In a calculation carried out for which detailed and accurate predictions are necessary, the pressure loss in the combustor, the bleed flows (for cooling the blades, de-icing the aircraft and cabin pressurisation) and the power off-take would all be included. For the present purpose, which is to show the trends, these complications can be neglected.

Two very important simplifications are possible as a result of the turbine operating characteristic. As was shown in Chapter 11, the dependence of non-dimensional mass flow on pressure ratio for the turbine is almost independent of rotational speed at pressure ratios likely to be encountered above idle conditions. Using the stagnation temperature and pressure at *inlet* to the turbine the form of this dependence is effectively identical to that for a choked convergent nozzle. In other words, for each turbine in a multi-shaft engine

$$\frac{\dot{m}\sqrt{c_p T_{0in}}}{A_{in}p_{0in}} = \overline{m}_{in} = \text{constant},\tag{12.1}$$

where the area A is an appropriate area such as the throat area of the turbine nozzle guide vanes. Taking $\gamma = 1.30$ to be representative of the gas through the turbine the value $\overline{m}_{in} = 1.389$.

The second great simplification for the turbine, which follows from the great tolerance of well designed turbine blades to incidence, is that the efficiency is little affected by the rotational speed over the range of speeds experienced above the idle condition. It therefore becomes possible to derive the ratio of stagnation temperature from inlet to outlet of a turbine stage from the pressure ratio and the polytropic efficieny. From the temperature ratio and inlet temperature follows the temperature drop and thence the power per unit mass flow:

$$\frac{T_{0out}}{T_{0in}} = \left(\frac{p_{0out}}{p_{0in}}\right)^{\frac{\eta_p(\gamma-1)}{\gamma}}.\tag{12.2}$$

The significance of the approximations applicable to the turbine, that it is effectively choked and that its efficiency is constant and independent of rotational speed, becomes apparent when there are two turbines in series, as in a two-shaft engine, or when a turbine operates upstream of a choked propulsive nozzle.

In contrast to the turbine, the compressor performance is, as shown in Chapter 11, strongly dependent on rotational speed and it is the compressor characteristics which largely determine at what speeds the engine shafts rotate.

We also need to consider the non-dimensional mass flow through the propulsive nozzle. In this case the area is either the throat or the minimum area at exit. Sometimes the nozzle is choked, in which case $\overline{m}_{in} =$ constant, but this is not always the case. For modern engines with low fan pressure ratio the bypass nozzle is choked at cruise but not at take-off and the core nozzle is never choked. Choke occurs for $p_{0in}/p \geq 1.893$ when $\gamma = 1.40$ and for $p_{0in}/p \geq 1.832$ when $\gamma = 1.30$. The choking value of $\overline{m}$ is equal to 1.281 for $\gamma = 1.40$ and 1.389 when $\gamma = 1.30$. For the unchoked case $\overline{m}$ is a function of the ratio of stagnation pressure at inlet to the nozzle to static pressure at outlet, the latter being ambient pressure. The relevant equation was given as Equation 6.18 and is

$$\frac{\dot{m}\sqrt{c_p T_0}}{A p_0} = \overline{m}\left(\frac{p_{ex}}{p_{0in}}, \gamma\right) = \frac{\gamma}{\gamma - 1}\sqrt{2\left[\left(\frac{p_{ex}}{p_{0in}}\right)^{\frac{2}{\gamma}} - \left(\frac{p_{ex}}{p_{0in}}\right)^{\frac{(\gamma+1)}{\gamma}}\right]}. \tag{12.3}$$

This equation should be used when p_{0in}/p is not large enough to choke the nozzle.

The engines of current interest are the modern two- and three-shaft engines. Even if they are of the high-speed military type, these engines usually now have a bypass stream. The multiple shafts and the bypass stream complicate the treatment so, as an introduction to the approach, it is worthwhile to look first at a single-shaft turbojet engine which was first designed in the 1950s, the Rolls-Royce Viper illustrated in Figure 5.1. For the main conditions of interest in flight the turbine is effectively choked and the propulsive nozzle is also choked. Because this simple configuration shows the important off-design effects it is treated in some detail, including a graphical explanation of the behaviour.

12.2 A SINGLE-SHAFT TURBOJET ENGINE

A Viper turbojet engine is shown schematically in Figure 12.1a, using the standard numbering system. As noted above, the final propulsive nozzle and the turbine are choked at and near the design point. The information in Table 12.1 is for the engine at design point on a stationary sea-level test bed

The Viper compressor has eight stages and the turbine considered here has one stage. The polytropic efficiency of the compressor may be assumed to be about 0.90 and of the turbine about

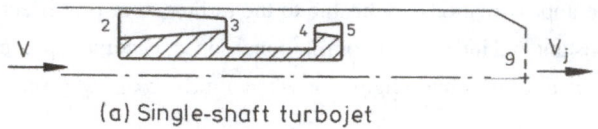

(a) Single-shaft turbojet

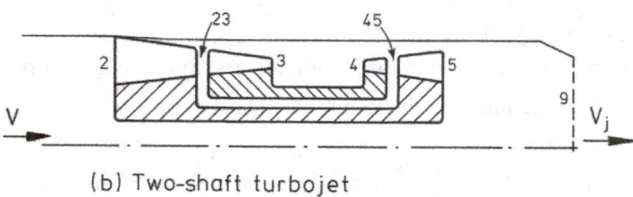

(b) Two-shaft turbojet

Figure 12.1. Schematic representation of turbojet engines.

0.85; the efficiency of the turbine is realistically put lower because, with only one stage, it is relatively highly loaded. (Later versions of the Viper engine, such as the one shown in Figure 5.1, had two turbine stages.)

From the thrust and mass flow in Table 12.1 the jet velocity $V_j = F_G/\dot{m} = 637$ m/s.

Table 12.1   Parameters for the Viper on a sea-level test bed

Gross thrust (static)	$= 15{,}167$ N
Specific fuel consumption sfc	$= 0.993$ kg h^{-1}kg^{-1}
Air mass flow $\dot{m}_{air}$	$= 23.81$ kg/s
Fuel flow $\dot{m}_f$	$= 0.4267$ kg/s
Stagnation pressure ratio p_{03}/p_{02}	$= 5.5$

Note: $T_{02} = T_a = 288$ K, $p_{02} = p_a = 101$ kPa.

Basic equations and approach

The energy release during combustion is found, as in Exercise 11.9, using the approximation

$$\dot{m}_f LCV = (\dot{m}_{air} + \dot{m}_f)c_{pe}(T_{04} - 298) - \dot{m}_{air}\,c_p(T_{03} - 298). \tag{12.4}$$

For the simplified treatment here the pressure drop in the combustor is neglected and as a result the pressure out of the compressor is taken to be the pressure into the turbine,

$$p_{04} = p_{03}.$$

The sum of the rise in pressure in the intake from atmospheric static, $p_{02} - p_a$, and the rise in the compressor, $p_{03} - p_{02}$, must then be equal to the drop in pressure in the turbine, $p_{04} - p_{05}$, plus the drop in pressure in the nozzle back to ambient static, $p_{05} - p_a$. It is assumed throughout that in the jet pipe between the turbine and the propulsive nozzle there is no drop in pressure, so that $p_{05} - p_{09}$ and no heat transfer, so that $T_{05} = T_{09}$.

Exercise

12.1 a Use the design pressure ratio across the Viper for operation on a sea-level test bed with standard atmospheric conditions, Table 12.1, to find the temperature ratio across the compressor and the stagnation temperature at compressor discharge, T_{03}. From the temperature rise across the compressor find the temperature drop across the turbine assuming $\gamma = 1.30$

(**Ans:** 1.72, $T_{03} = 495.1$ K, $\Delta T_{0c} = 206.9$ K, $\Delta T_{0t} = 167.2$ K)

b From the given air and fuel flow rates find the energy release in the combustor per kg of air flow. Take $LCV = 43$ MJ/kg. Hence find the turbine inlet stagnation temperature T_{04} and, downstream of the turbine, the stagnation temperature T_{05} and the stagnation pressure p_{05} downstream of the turbine. Treating the nozzle as isentropic, find the jet velocity and compare this with the value derived from thrust in Table 12.1.

(**Ans:** 771 kJ/kg, $T_{04} = 1063$ K, $T_{05} = 896$ K, $p_{05} = 233$ kPa, $V_j = 624$ m/s)

c Use the design condition information to determine the area of the turbine nozzles and the area of the propulsive nozzle, both of which are choked.   (**Ans:** 0.354 m², 0.777 m²)

Note: a discrepancy of 13 m/s in jet velocity relative to the value in Table 12.1 is fortuitously small having in mind the simplicity of the approach, the neglect of pressure loss in the combustor, the assumptions for efficiency and the simple assumptions for gas properties.

Mass flow through the engine

The mass flow into the turbine and into the final propulsive nozzle will be equal. As noted, for the simple turbojet the turbine nozzle guide vanes will be effectively choked for all cases of interest so for $\gamma = 1.30$ we obtain $\overline{m}_4 = 1.389$. The final or propulsive nozzle will be effectively choked over much of the operating range of the engine but we retain the more general $\overline{m}_9 \, (p_{09}/p_a)$.

It is assumed that there is no loss in stagnation pressure nor any external heat transfer between the turbine exit and propulsive nozzle so that propulsive nozzle entry conditions correspond to turbine exit, $T_{09} = T_{05}$ and $p_{09} = p_{05}$. It is now possible to divide the non-dimensional mass flow into the turbine $\overline{m}_4$ and into the propulsive nozzle $\overline{m}_9$ to give

$$\frac{\overline{m}_4}{\overline{m}_9 \, (p_{09}/p_a)} = \frac{A_9}{A_4} \frac{p_{05}}{p_{04}} \sqrt{\frac{T_{04}}{T_{05}}}$$

And on re-arranging

$$\left(\frac{p_{05}}{p_{04}}\right) = \left(\frac{A_4}{A_9}\right) \times \left(\frac{T_{05}}{T_{04}}\right)^{1/2} \times \left(\frac{\overline{m}_4}{\overline{m}_9 \, (p_{09}/p_a)}\right). \tag{12.5}$$

Expansion: pressure ratio and temperature through the turbine

The temperatures and pressures upstream and downstream of the turbine are also related by the polytropic efficiency, so that it is possible to write Equation 12.2 as

$$\left(\frac{p_{05}}{p_{04}}\right)^{\frac{\eta_p(\eta-1)}{\eta}} = \frac{T_{05}}{T_{04}}.$$

Equations 12.2 may be introduced into Equation 12.5 to remove temperature ratio across the turbine thus

$$\left(\frac{p_{05}}{p_{04}}\right) = \left(\frac{A_4}{A_9}\right) \times \left(\frac{\overline{m}_4}{\overline{m}_9}\right) \times \left(\frac{p_{05}}{p_{04}}\right)^{\frac{\eta(\gamma-1)}{2\gamma}}$$

or

$$\left(\frac{p_{05}}{p_{04}}\right)^{1-\frac{\eta(\gamma-1)}{2\gamma}} = \left(\frac{A_4}{A_5}\right) \times \left(\frac{\overline{m}_4}{\overline{m}_9\,(p_{09}/p_a)}\right). \qquad (12.6)$$

The special case of the choked propulsive nozzle

For the special case when the propulsive nozzle is choked (as well as the turbine) $\overline{m}_4 = \overline{m}_9$ and the relation for turbine pressure ratio, Equation 12.6, reduces to

$$\frac{p_{05}}{p_{04}} = \left(\frac{A_4}{A_9}\right)^{\frac{2\gamma}{2\gamma-\eta(\gamma-1)}}, \qquad (12.7a)$$

whilst the temperature ratio becomes

$$\frac{T_{05}}{T_{04}} = \left(\frac{A_4}{A_9}\right)^{\frac{2\eta(\gamma-1)}{2\gamma-\eta(\gamma-1)}}. \qquad (12.7b)$$

Equations 12.7 show that both the temperature ratio and pressure ratio across the turbine are uniquely determined by area ratio because the turbine is operating between two choked nozzles, the turbine inlet and the propulsive nozzle. From the temperature ratio and the turbine inlet temperature the turbine work per unit mass flow is given, so the turbine power are determined by area ratio too. Once the turbine and nozzle are choked the pressure ratio and temperature ratio across the turbine can *only* be changed by altering one of the areas and these are used in practice as the means of matching engines during their development.

It is revealing to put numbers to the powers to which area ratio in Equations 12.7 is raised. For $\gamma = 1.30$ and $\eta_p = 0.85$, applicable to the Viper engine, Equations 12.7 can be written as

$$\frac{p_{05}}{p_{04}} = \left(\frac{A_4}{A_9}\right)^{1.109} \qquad (12.8a)$$

leading to

$$\frac{T_{05}}{T_{04}} = \left(\frac{A_4}{A_9}\right)^{0.217}. \tag{12.8b}$$

Reducing the propulsive nozzle area A_9 gives an *almost* proportional increase in p_{05}/p_{04}, that is it reduces the pressure ratio across the turbine. With the assumption of constant turbine efficiency the reduction in pressure ratio across the turbine corresponds to a reduction in power output. Similarly, increasing A_4, the area of turbine nozzle guide vanes (i.e. the stator blades), also reduces turbine pressure ratio and power output. The areas can be obtained from on-design calculations. Once the areas are known the solution is simple with choked turbine and propulsive nozzle.

When the propulsive nozzle is unchoked a numerical iteration is required, using, for example, Equation 12.6. As a starting point it is convenient to assume the propulsive nozzle is choked from which one can get a first estimate for turbine pressure ratio p_{05}/p_{04} and thence, from overall pressure ratio, one obtains p_{09}/p_a. Knowing p_{09}/p_a one can find $\bar{m}_9 (p_{09}/p_a)$ using Equation 12.3 which can be introduced into Equation 12.6 to improve the estimate of pressure ratios – the procedure is repeated until adequate convergence is achieved.

Compressor and turbine power with choked propulsive nozzle

If the temperature ratio across the turbine, T_{05}/T_{04}, is a constant wholly determined by the area ratio, Equation 12.7b, it may be shown that the temperature drop in the turbine is proportional to the turbine inlet temperature, and given by

$$T_{04} - T_{05} = T_{04} (1 - T_{05}/T_{04}) = k_H T_{04}. \tag{12.9a}$$

In other words the turbine power per unit mass flow rate is proportional to the turbine inlet temperature and given by $c_{pe} k_H T_{04}$. With the numerical approximations adopted ($\gamma = 1.30$, $\eta_p = 0.85$),

$$k_H = \left(1 - (A_4/A_9)^{0.217}\right). \tag{12.9b}$$

The power produced by the turbine must be equal to the power into the compressor. Neglecting the increase in mass flow through the turbine because of the fuel and the flow bled off the compressor for cooling and other purposes, this balance in power can be expressed as

$$c_p (T_{03} - T_{02}) = c_{pe} (T_{04} - T_{05})$$
$$= c_{pe} k_H T_{04}.$$
$$\text{or} \quad T_{03}/T_{02} = 1 + (c_{pe}/c_p) k_H (T_{04}/T_{02}). \tag{12.10}$$

From Equation 12.10 the temperature ratio in the compressor T_{03}/T_{02} is known once the ratio of turbine inlet temperature to engine inlet temperature is specified. If the compressor polytropic efficiency η_p can be taken as constant (which presumes that the engine can match at a condition in which the compressor is operating efficiently) this allows the pressure ratio to be found,

$$\frac{p_{03}}{p_{02}} = \left(\frac{T_{03}}{T_{02}}\right)^{\eta_p(\gamma/\gamma-1)}$$

so that

$$\frac{p_{03}}{p_{02}} = \left(1 + k_H \frac{c_{pe}T_{04}}{c_p T_{02}}\right)^{\eta_p \gamma/(\gamma-1)} \tag{12.11}$$

If the engine overall pressure ratio and turbine inlet temperature are known at one condition, such as the design point, the value of k_H can be determined. Then, knowing k_H, the overall pressure ratio can be found at other conditions, i.e. at other values of T_{04}/T_{02}.

It is worth reiterating that within the approximations adopted here, with a choked propulsive nozzle the pressure ratio of the compressor is wholly determined by the ratio of turbine inlet area to nozzle area and the turbine inlet temperature ratio T_{04}/T_{02}.

A graphical view of the turbine and nozzle in series

Figure 12.2 shows side by side the characteristics of the turbine, such as that for the Viper engine, on the left and the final propulsive nozzle on the right, both plotted as normalised mass flow versus pressure ratio. The normalised mass flow function, $\dot{m}\sqrt{c_p T_0}/p_0 = A\,\overline{m}$, with units of area, is proportional to the non-dimensional mass flow $\overline{m}$, and is convenient for comparing the two graphs. For the turbine the pressure ratio is p_{04}/p_{05} and for the propulsive nozzle the pressure ratio is, p_{05}/p_9, where $p_{05} = p_{09}$ and $p_9 = p_{ex}$, the static pressure at nozzle exit. All turbines of interest to us are effectively choked at inlet.

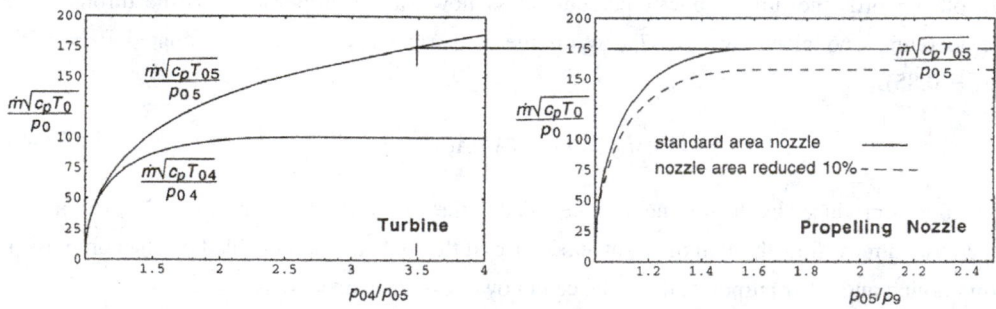

Figure 12.2. A graphical representation of the matching of a turbine with a downstream propulsive nozzle. For the turbine, mass flow function is shown at turbine inlet and turbine exit. (The horizontal line from the choked propulsive nozzle the fixes the mass flow function at turbine exit.)

The characteristic for the turbine on the left in Figure 12.2 is shown in two ways: the conventional way, where the stagnation pressure and temperature are the *inlet* values p_{04} and T_{04}, and also in a less familiar way based on the *exit* conditions p_{05} and T_{05}. The normalised mass flow function for the turbine in terms of *inlet* conditions becomes constant when the turbine chokes: the normalised mass flow based on *exit* conditions, $\dot{m}\sqrt{c_p T_{05}}/p_{05}$, does *not* become constant when the turbine chokes, but increases continuously as pressure ratio is increased because p_{05} falls. (In calculating the normalised mass flow in terms of exit conditions for Figure 12.2 the polytropic efficiency of the turbine has been assumed to remain constant at 0.85, believed appropriate for the rather old Viper.) The important feature of the turbine exit properties is that they form the inlet conditions to the propulsive nozzle.

If the exit mass flow function of the turbine is equal to the inlet mass flow function of the propulsive nozzle there can be a unique value of mass flow function $\dot{m}\sqrt{c_p T_{05}}/p_{05}$ between turbine exit and nozzle inlet if the propulsive nozzle is choked. To indicate this a horizontal line has been drawn on Figure 12.2 from the choked part of the propulsive nozzle to the *exit* flow function of the turbine. The characteristic of the turbine in terms of the exit flow function fixes the pressure ratio across the turbine at about 3.5 for the example in Figure 12.2. Because the turbine is choked the turbine pressure ratio does not alter the inlet mass flow function to it.

If the propulsive nozzle area is reduced (indicated by the dotted curve in Figure 12.2) there will be a proportional drop in the mass flow function at which the nozzle is choked so the turbine exit mass flow function must reduce and the pressure ratio across the turbine will also necessarily reduce. A reduction in propulsive nozzle area therefore leads to reduced power output from the turbine because of the lower pressure ratio and thus lower temperature ratio. Reduced power from the turbine leads to less power to the compressor, which decreases the pressure rise in the compressor and in turn decreases the mass flow rate swallowed by the compressor. A reduced mass flow and reduced pressure ratio requires a decrease in compressor rotational speed.

The propulsive nozzle ceases to be choked if the pressure ratio across it falls below 1.832, as happens for low fuel flows and low rotational speeds. In this case the horizontal line from the nozzle to the turbine would be at a level depending on the nozzle pressure ratio. This new horizontal line will intersect the line for turbine exit mass flow function at a lower turbine pressure ratio than was the case when the propulsive nozzle was choked. Only if this line is so low that the turbine is also no longer effectively choked does significant complication arise.

A practical numerical solution to the off-design condition

The graphical presentation points to a practical and convenient way of proceeding to compute off-design behaviour even when this is done algebraically. The design point calculation determines pressure and temperature into the turbine and into the propulsive nozzle so, knowing the mass flow through the engine, the corresponding areas A_4 and A_9 can be found from the values of $\overline{m}$.

For off-design operation the turbine is assumed to remain choked, so $\overline{m}_4 = 1.389$, and this may be true for the propulsive nozzle too; in any case, assuming both turbine and propulsive nozzle are choked gives a good starting point. The nozzle areas are obtained from an on-design calculation. Suppose that turbine inlet temperature is reduced below the design value. A procedure for off-design calculation is as follows:

i. assuming both the turbine at inlet and, initially, the propulsive nozzle remain choked, find the turbine pressure and temperature ratio using Equations 12.7.

ii. calculate engine performance with Equations 12.10 and 12.11, including overall pressure ratio, and thence calculate pressure ratio across the propulsive nozzle, p_{09}/p_a.

iii. if the nozzle is choked the engine performance is known.

iv. if propulsive nozzle is unchoked calculate $\overline{m}_9(p_{09}/p_a)$ using Equation 12.3 and the pressure ratio p_{09}/p_a calculated in part (ii) as an initial estimate. Use $\overline{m}_9$ with Equation 12.6 to find the revised turbine pressure ratio and temperature ratio.

v. knowing turbine temperature ratio find turbine power which is equal to compressor power. Hence find compressor pressure ratio, overall pressure ratio and pressure ratio across propulsive nozzle p_{09}/p_a. Use this to recalculate $\overline{m}_9(p_{09}/p_a)$.

vi. if $\overline{m}_9$ computed in (v) agrees with the estimate used in (iv) the engine properties are found. Otherwise step (iv) must be repeated with a new value of p_{09}/p_a.

It is, of course, implicit in this that the turbine and compressor efficiencies remain constant. If this is not true then additional cycles of iteration are required to correct the values of efficiency.

Exercise

12.2 Use the design point conditions to determine k_H for the engine of Exercise 12.1, taking the propulsive nozzle to be choked. When the turbine inlet temperature is reduced to 900 K findthe temperature drop in the turbine and thence the temperature rise in the compressor. Assuming that the polytropic efficiencies of the compressor and turbine do not alter from the values assumed at design point, find the pressure ratio across the compressor and the stagnation pressure in the jet pipe. Treating the flow in the nozzle as isentropic, find the jet velocity.   (**Ans:** $k_H = 0.157$, 141.4 K, 175.1 K, $pr = 4.46$, 189 kPa, 503 m/s)

The compressor working line for the viper

Although the pressure ratio across the compressor is fixed for the choked nozzle case by the ratio of turbine inlet area to propulsive nozzle area, the mass flow of air through the engine is not yet determined. To get this it is necessary to again consider the choked condition at turbine inlet

$$\overline{m}_4 = \frac{\dot{m}\sqrt{c_{pe}T_{04}}}{A_4 p_{04}} = 1.389,$$

because the compressor will be forced to supply a pressure ratio and mass flow to satisfy this. We can assume for the present that there is a negligible loss in pressure across the combustor, so

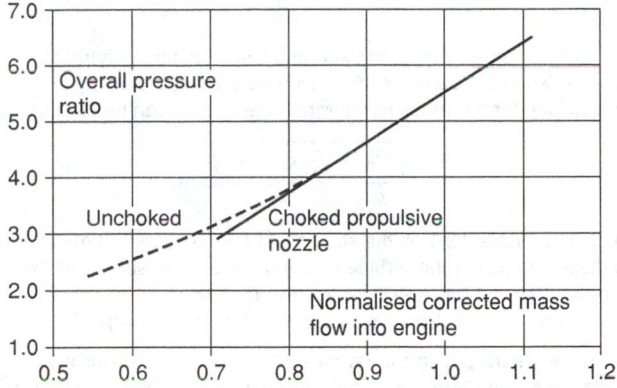

Figure 12.3. The pressure ratio versus non-dimensional mass flow (working line) for the Viper. The solid line corresponds to treating the propulsive nozzle as choked; the broken line considers correctly the effect of unchoking the nozzle.

$p_{03} = p_{04}$. The non-dimensional mass flow for the compressor, based on *inlet* conditions follows from

$$\bar{m}_2 = \frac{\dot{m}\sqrt{c_p T_{02}}}{A_2 p_{02}}$$

$$= \bar{m}_4 \frac{A_4}{A_2} \frac{p_{03}}{p_{02}} \frac{\sqrt{c_p T_{02}}}{\sqrt{c_{pe} T_{04}}}. \qquad (12.12)$$

Since the pressure ratio p_{03}/p_{02} can be found from Equation 12.11 as a function of T_{04}/T_{02}, it follows that $\bar{m}_2$, is a function of T_{04}/T_{02}. By varying T_{04}/T_{02} it is then possible to plot the compressor working line on axes of compressor pressure ratio and non-dimensional mass flow based on compressor inlet conditions. This has been used to calculate working lines for the Viper engine which are shown as Figure 12.3. The working line has been chosen so as to pass through the design point and the values of mass flow shown in Figure 12.3 are normalised by the value at the design point. Two lines are shown; for the solid line both nozzles are assumed to remain choked to the lowest pressure ratios, with pressure and temperature ratio across the turbine constant according to Equations 8a and 8b. For the broken line allowance is made for the unchoking of the propulsive nozzle. The propulsive nozzle becomes unchoked at values of T_{04}/T_{02} below 3.08, compared with the design value of about 3.7. The working line moves upwards when allowance is made for unchoking. The working line does *not* require the compressor rotational speed to be selected; rather, given the working line, the compressor 'chooses' the rotational speed which gives the necessary pressure ratio and mass flow. Alternatively the compressor may be imagined to select the speed which absorbs the turbine power output and passes the flow necessary to choke the turbine inlet guide vanes. To predict the off-design rotational speed it is therefore necessary to have the measured or predicted pressure ratio versus non-dimensional mass flow characteristics along constant speedlines of the compressor.

Exercises

12.3 With the turbine and propulsive nozzle choked and with constant polytropic efficiency in the turbine show that the temperature rise in the compressor of a single-spool turbojet $T_{03} - T_{02}$ is proportional to T_{04}. Hence show that the condition for the mass flow through the engine can be rewritten as

$$\frac{\dot{m}\sqrt{c_p(T_{03} - T_{02})}}{p_{03}} = \text{constant}.$$

12.4 a If for the engine of Exercise 12.1 on the static test bed the corrected mass flow at the design point is 23.8 kg/s, find the mass flow when the turbine inlet temperature is 900 K. At this temperature find the gross thrust, the fuel mass flow and the specific fuel consumption.

         (**Ans:** 21.0 kg/s, $F_G = 10.6$ kN, $\dot{m}_f = 0.290$ kg/s, $sfc = 27.4$ g s^{-1}kN^{-1} $= 0.968$ kg h^{-1}kg^{-1})

    b Find the minimum pressure p_{05} in the jet pipe to choke the propulsive nozzle of the Viper on a sea-level test bed. Find the turbine inlet temperature at which this occurs, assuming (correctly) that the turbine remains choked. Confirm that in flight at $M = 0.78$ the propulsive nozzle will be choked even at $T_{04} = 600$ K.   (**Ans:** $p_{05} = 185$ kPa $T_{04} = 888$ K)

Note: For the static test bed the static pressure at nozzle exhaust is equal to the stagnation pressure at engine inlet. In flight the stagnation pressure at inlet is higher than ambient static, so the pressures inside the engine are higher in the ratio p_{02}/p_a and the pressure ratio across the final nozzle will increase by an amount similar to this.

It is a necessary assumption in the approach used here that the working line remains in a part of the compressor map such that it does not stall and that it is reasonable to assume that the compressor efficiency remains approximately constant. If the working line takes the compressor into areas of the map where the efficiency varies significantly then an iterative calculation is necessary, which requires measured or estimated values of the compressor performance. If the turbine or the nozzle cease to be choked, an iteration is also necessary. Codes used in industry, such as GasTurb, would certainly use iterative methods to include these effects.

The results of the present simple calculation assuming choked nozzles for the compressor of the Viper engine are compared with measurements in Figure 12.4. In Figure 12.4a the working line calculated with a normal sized propulsive nozzle and the normal turbine nozzle guide vane area is superimposed on a plot of the measured working lines. The calculated and measured working lines move apart at low rotational speeds, which is in line with the way the working line diverged when the propulsive nozzle became unchoked at low pressure ratios. In addition some divergence is possible because the efficiency of the compressor starts to fall. Figure 12.4b shows the working line measured with the turbine nozzle guide vane area reduced by 20% and the propulsive nozzle area reduced by 11%; compared to the nominal ratio, A_4/A_9 has been reduced to 0.90 times the nominal value. According to Equation 12.9 this increases the turbine temperature ratio T_{05}/T_{04} (and therefore reduces the power output). Superimposed on the measurements of Figure 12.4b are the working lines from the above analysis. The agreement is reasonably good, particularly at the higher pressure ratios, showing the correct trend with mass flow for each working line and the correct effect of the change in areas. The discrepancy at

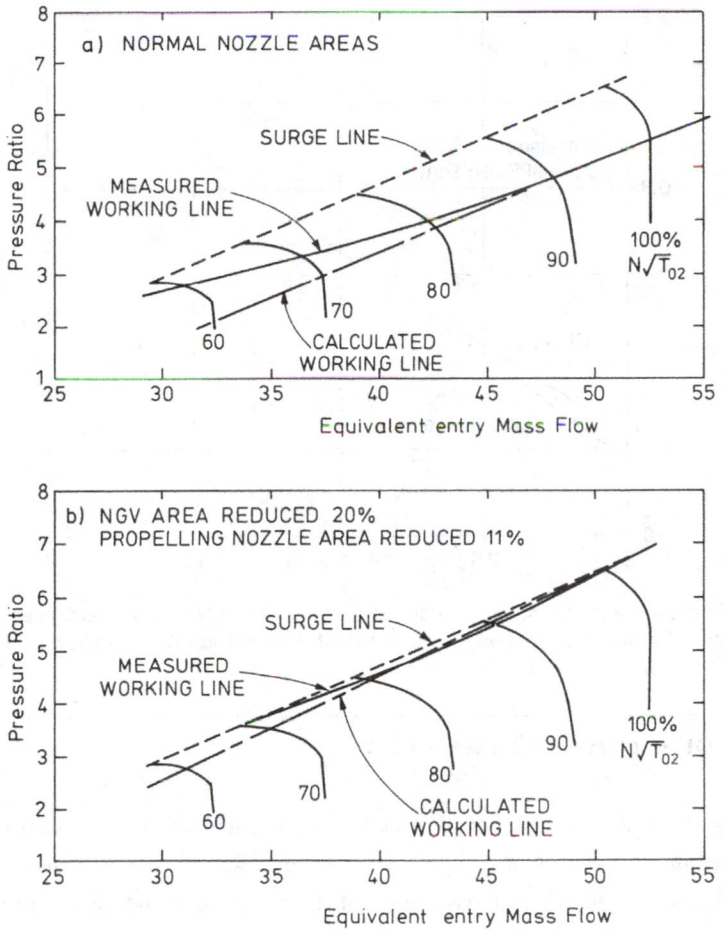

Figure 12.4. The predicted working line for the Viper superimposed on measured characteristics
for the compressor. Case (a) with the normal NGV and propulsive nozzle areas;
case (b) with reduced areas. All assuming choked propulsive nozzle.

the lower speeds and pressure ratios is again probably because the propulsive nozzle will have
unchoked.

Figure 12.4 allows the rotational speed to be estimated for the reduced operating temperatures. At $T_{04} = 900$ K, for example, the pressure ratio was found in Exercise 12.2 to be 4.46, and
Figure 12.4a shows the rotational speed would be about 86% of the design value.

Shown in Figure 12.5 is the variation in gross thrust with temperature ratio. Of particular
note is the very rapid fall in the thrust with T_{04}/T_{02}. Again results are shown as a solid line for the
choked propulsive nozzle down to low values of T_{04}/T_{02} together with the more realistic unchoked
assumption of unchoking for the nozzle as a broken line.

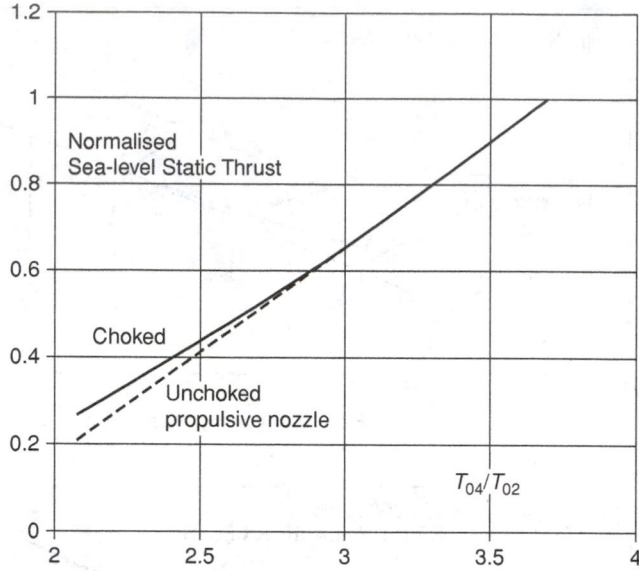

Figure 12.5. Predicted gross thrust for the Viper on a sea-level test bed versus turbine inlet temperature. Calculations shown with propulsive nozzle assumed choked and when it is allowed to unchoke.

SUMMARY OF SECTIONS 12.1 AND 12.2

The matching of conditions for the single-spool turbojet requires that the rotational speed, mass flow and power are the same for the compressor and turbine, with the pressure changes compatible. In general this requires the characteristic maps of the compressor and turbine to be used in an iterative calculation.

Enormous simplification is possible in approximate calculations when the special nature of the turbine characteristic is utilised. The approximate calculations are capable of indicating the correct trends. In terms of its upstream effect the turbine behaves as a choked nozzle; moreover the efficiency of the turbine is only a weak function of rotational speed and pressure ratio. For a simple turbojet the propulsive nozzle is choked for many conditions of interest, which means the non-dimensional mass flow is constant at inlet and outlet of the turbine. Under this constraint, and with the turbine efficiency constant, the turbine is forced to operate at a constant stagnation pressure ratio and stagnation temperature ratio. Turbine pressure ratio and temperature ratio are then fixed by the ratio of the turbine nozzle guide vane throat area and the propulsive nozzle area. It follows immediately that the turbine enthalpy drop is proportional to the stagnation temperature at turbine inlet.

The calculations are more involved when the propulsive nozzle is unchoked because an iterative calculation is required; non-dimensional mass flow into the propulsive nozzle, $\bar{m}_9$, must be calculated using the pressure ratio across the nozzle, p_{09}/p_a, and the pressure ratios adjusted so that the mass flow rate through the nozzle is equal to that through the HP turbine inlet guide vanes.

With a choked propulsive nozzle the turbine power per unit mass flow rate, and therefore the compressor power, is proportional to the turbine inlet temperature. From this the pressure ratio across the compressor becomes a function of turbine entry temperature ratio T_{04}/T_{02} if the compressor efficiency is known. With the flow choked at turbine inlet the non-dimensional mass flow at compressor inlet can be found in terms of the turbine inlet condition. The compressor working line is then defined.

Although this section was devoted to a simple jet engine the approach is immediately transferable to other gas turbines, even those with a power turbine replacing the propulsive nozzle: land based or marine prime movers. If the power turbine is choked the analysis is then virtually identical. As we shall see, it is also applicable to the turbofan and it could be applied to the turboprop.

12.3 A TWO-SHAFT TURBOJET ENGINE

The tendency of the working line to move the compressor operating point towards the surge line, evident in Figure 12.4a is a problem for all engines at reduced power. The problem gets more acute as the design overall pressure ratio at design point is increased. The most significant way of alleviating the effect is to use separate high-pressure (HP) and low-pressure (LP) compressors on concentric shafts which are able to rotate at different speeds. The compressors are able to select the speed at which they can best meet the requirement for the non-dimensional mass flow and pressure ratio. Section 12.2 considered the simplest form of gas turbine, the single-shaft turbojet, and that treatment provides the approach here to a slightly more complicated engine with two shafts. On each shaft there is a compressor and turbine but, being a straight turbojet (i.e. there is no bypass), all of the air flows through each component.

The engine is shown schematically in Figure 12.1b with the numbering system for the stations shown. The power from the LP turbine must be equal to that absorbed by the LP compressor so again, neglecting the mass flow rate of fuel and the cooling and bleed flows,

$$c_p \left(T_{023} - T_{02} \right) = c_{pe} \left(T_{045} - T_{05} \right) \tag{12.13a}$$

and likewise for the HP shaft

$$c_p \left(T_{03} - T_{023} \right) = c_{pe} \left(T_{04} - T_{045} \right). \tag{12.13b}$$

Just as for the single-shaft engine the sum of the pressure rises in the compression process and the pressure drops in the expansion process must be equal. Furthermore the mass flow must be equal throughout, a fact which is used in the turbine and the propulsive nozzle. It will be assumed that both the LP and HP turbines are effectively choked and that the final propulsive nozzle is also choked at all conditions of interest. The non-dimensional mass flow at HP turbine inlet, LP turbine inlet and propulsive nozzle inlet are all equal assuming constant c_p and γ:

$$\overline{m}_4 = \overline{m}_{45} = \overline{m}_9 = 1.389$$

i.e. $$\frac{\dot{m}\sqrt{c_{pe}T_{04}}}{A_4 p_{04}} = \frac{\dot{m}\sqrt{c_{pe}T_{045}}}{A_{45} p_{045}} = \frac{\dot{m}\sqrt{c_{pe}T_{05}}}{A_9 p_{05}} = 1.389. \qquad (12.14)$$

In the expression for $\overline{m}_9$ it has again been assumed that there is negligible change in stagnation pressure or temperature between LP turbine outlet and the propulsive nozzle exit, giving $p_{05} = p_{09}$ and $T_{05} = T_{09}$. Using Equation 12.14 across the HP turbine leads to

$$\left(\frac{p_{045}}{p_{04}}\right)^2 = \left(\frac{A_4}{A_{45}}\right)^2 \times \frac{T_{045}}{T_{04}}. \qquad (12.15)$$

As before, it is assumed that the turbine efficiencies are sufficiently insensitive to incidence that the polytropic efficiencies of both may be taken to be constant. Then across the HP turbine

$$\frac{p_{045}}{p_{04}} = \left(\frac{T_{045}}{T_{04}}\right)^{\frac{\gamma}{\eta(\gamma-1)}}. \qquad (12.16)$$

Equations 12.15 and 12.16 can be simultaneously satisfied only for particular values of stagnation pressure and stagnation temperature ratios and combining these two equations gives

$$\frac{p_{045}}{p_{04}} = \left(\frac{A_4}{A_{45}}\right)^{\frac{2\gamma}{2\gamma-\eta(\gamma-1)}} \quad \text{and} \quad \frac{T_{045}}{T_{04}} = \left(\frac{A_4}{A_{45}}\right)^{\frac{2\eta(\gamma-1)}{2\gamma-\eta(\gamma-1)}}. \qquad (12.17a)$$

For $\gamma = 1.30$ and assuming that $\eta_p = 0.9$ it follows that the numerical value of the index for pressure ratio is 1.12 and temperature ratio is 0.232.

Because for this engine the propulsive nozzle is choked the results for the LP turbine are entirely equivalent for the HP turbine above,

so $$\frac{p_{05}}{p_{045}} = \left(\frac{A_{45}}{A_9}\right)^{\frac{2\gamma}{2\gamma-\eta(\gamma-1)}} \quad \text{and} \quad \frac{T_{05}}{T_{045}} = \left(\frac{A_{45}}{A_9}\right)^{\frac{2\eta(\gamma-1)}{2\gamma-\eta(\gamma-1)}}. \qquad (12.17b)$$

The pressure and temperature ratios of both turbines are therefore uniquely fixed by the ratio of areas. As the numerical values show, the pressure ratio across each turbine is almost linearly proportional to the area ratio and area ratio is consequently the key determinant of engine matching and performance.

With T_{045}/T_{04} fixed across the HP turbine it follows that the HP turbine power per unit mass flow rate is given by

$$\dot{W}_{HP}/\dot{m} = c_{pe}(T_{04} - T_{045}) = c_{pe}T_{04}(1 - T_{045}/T_{04}) = c_{pe}k_{HP}T_{04}$$

where it is convenient to write

$$(1 - T_{045}/T_{04}) = k_{HP}. \qquad (12.18a)$$

and k_{HP} is a constant fixed by the relative areas of the two turbines. In an analogous way for the LP turbine T_{05}/T_{05} is fixed

$$\dot{W}_{LP}/\dot{m} = c_{pe}\,(T_{045} - T_{05}) = c_{pe}T_{045}\,(1 - T_{05}/T_{045})$$

$$\text{i.e} \quad = T_{04}\,T_{045}/T_{04}\,(1 - T_{05}/T_{045}).$$

Because, however, the ratio T_{045}/T_{04} is a constant given by Equation 12.17b, the LP turbine work is also proportional to the temperature entering the HP turbine, T_{04}. This is then conveniently written in terms of T_{04} so that

$$\dot{W}_{LP}/\dot{m} = k_{LP}\,T_{04}. \tag{12.18b}$$

Using Equations 12.18a and 12.18b the LP and HP compressor temperature rises can be shown to be equal to $k_{LP}T_{04}c_{pe}/c_p$ and $k_{HP}T_{04}c_{pe}/c_p$ respectively. The pressure ratio for the LP compressor is then given, as it was for the single-shaft turbojet, Equation 12.10, by

$$\frac{p_{023}}{p_{02}} = \left(1 + k_{LP}\frac{c_{pe}}{c_p}\frac{T_{04}}{T_{02}}\right)^{\gamma\eta/(\gamma-1)} \tag{12.19}$$

with a similar expression for the pressure ratio of the HP compressor.

An example engine – the olympus 593

There are not many examples of two-shaft turbojet engines. By the time the technology was sufficiently advanced to be able to build two-shaft engines it was also evident that for most applications a bypass stream was desirable.[1] For very high speed propulsion, however, the turbojet is still a viable alternative and the Olympus 593 used in the Concorde was an example of such an engine. It is illustrated in Figure 5.2. Some information about its operation at cruise is as follows:

Olympus 593 at cruise

Altitude	51,000 ft
Mach number	2.0
Overall pressure ratio in compressors p_{03}/p_{02}	11.3

The LP compressor and the HP compressor each have seven stages, the LP and HP turbines each have one stage. The engine was highly developed and polytropic efficiencies of 90% for all components may be pessimistic. At 51,000 ft for the International Standard Atmosphere (ISA) the ambient pressure is 11.0 kPa and the temperature is 216.65 K.

---

[1] It is sometimes supposed that the early jet engines were single-shaft turbojets because the designers knew no better. In fact, the delay in introducing multi-shaft bypass engines arose principally from the mechanical problems which had to be overcome with even the simplest turbojet. Dr. A. A. Griffiths, working for Rolls Royce in the 1940s, drew an engine scheme in 1941 with a bypass ratio of 8, a suggestion not far from what is now regarded as optimum for subsonic propulsion.

Exercise

12.5 a Find the stagnation pressure and temperature of the air entering the Olympus 593 engine at cruise. Neglect the stagnation pressure drop in the intake.   **(Ans:** $p_{02} = 87.0$ kPa; $T_{02} = 390$ K)

b Assuming equal pressure ratios across the LP and HP compressors, find the stagnation pressure and temperature out of each. What are the temperature drops in each of the turbines? Neglect the mass flow rate of fuel and the pressure drop in the combustor.

(Ans: $p_{023} = 291$ kPa, $T_{023} = 573.0$ K, $p_{03} = 976$ kPa, $T_{03} = 841.8$ K, $\Delta T_{0HP} = 217.2$ K, $\Delta T_{0LP} = 147.8$ K)

c If the temperature of the gas leaving the combustor is 1300 K, find k_{HP} and k_{LP} for the two turbines. Determine the temperature and pressure downstream of the LP turbine and find the jet velocity assuming that the nozzle is isentropic and fully expanded. Find, per unit mass flow through the engine, the HP and LP turbine inlet areas and the propulsive nozzle area.

(Ans: $k_{HP} = 0.167$, $k_{LP} = 0.114$, $T_{05} = 935$ K, $p_{05} = 199.7$ kPa, $V_j = 1065$ m/s, $A_{HP} = 0.943 \ 10^{-3}$ m^{2}kg^{-1}s, $A_{LP} = 2.0710^{-3}$ m^{2}kg^{-1}s, $A_{prop} = 3.91 \ 10^{-3}$ m^{2}kg^{-1}s)

Note: The afterburner is not referred to here because it was *not* used at the cruise condition.

Dynamic scaling and non-similar conditions

In Chapter 8 dynamic scaling and use of dimensional analysis was considered. It will be recalled that for this to be applied to an engine it was essential that the conditions inside the engine remained the same, that is to say all the pressure ratios, temperature ratios, non-dimensional speeds $N/\sqrt{c_p T_0}$ for all the shafts, etc. were held constant. It was perhaps remarkable how much could be learned in this way, but it was not possible to make deductions when the engine conditions were significantly different.

Such circumstances are most important when the flight speed is high. For the cruise conditions adopted for a subsonic airliner the stagnation temperature entering the engine is not ever likely to be much above about 259 K, not far removed from the sea-level temperature of the standard atmosphere 288 K. For a flight speed of Mach 2 the inlet stagnation temperature is very much higher, about 390 K. With such a high inlet temperature the allowable compression ratio in the compressor is limited on material strength grounds, which is why the *opr* for the Olympus was so low. The forward motion of the aircraft at $M = 2$, however gives a pressure ratio $p_{02}/p_a = 7.8$.

Exercises

12.6 a Taking the results from Exercise 12.5 for the Olympus, find the ratio of turbine inlet to compressor inlet temperature T_{04}/T_{02} at cruise. If this ratio were held constant, what would T_{04} be on a static sea-level test? If during take-off the turbine inlet temperature were increased to 1450 K, what was T_{04}/T_{02}?

(Ans: $T_{04}/T_{02} = 3.33$; $T_{04} = 960$ K; $T_{04}/T_{02} = 5.03$)

b For take-off with $T_{04} = 1450$ K find the overall pressure ratio, the pressure in the jet pipe and the jet velocity with no afterburner in use. (Assume that the flow in the nozzle was isentropic.) At this condition the engine passed 186 kg/s of air – what is the gross thrust?

(Ans: $p_{03}/p_{02} = 24.2$; $p_{05}/p_a = 4.95$; $V_j = 895$ m/s; $F_G = 167$ kN)

12.7 For the Olympus 593 at cruise use information in Exercise 12.6 to find the mass flow, and the gross and net thrust at the cruise condition with the nominal nozzle area.

(**Ans:** $\dot{m} = 78.0$ kg/s, $F_G = 82.9$ kN, $F_N = 37.0$ kN)

12.8 If the final propulsive nozzle area were increased by 10% on the Olympus 593 whilst at cruise what would be the effect on the engine if the inlet conditions and the turbine inlet temperature were held constant? Find the new k_{LP} and the jet velocity.

(**Ans:** $k_{LP} = 0.130$, $p_{023}/p_{02} = 3.87$, $p_{03}/p_{023} = 3.21$, $p_{05}/p_{045} = 0.444$, $p_{05} = 197$ kPa, $V_j = 1051$ m/s)

Note: on the Olympus 593 the variable nozzle was used to improve efficiency and reduce the noise during take-off and initial climb. By opening the nozzle the LP power was increased and the LP shaft rotated faster. This in turn increased the overall pressure ratio and the mass flow into the engine. The overall effect was to get the same amount of thrust with a higher mass flow and lower jet velocity than would have been the case with a constant propulsive nozzle area. Jet noise was the crucial issue.

12.9 Find the overall pressure ratio at take-off for $T_{04} = 1450$ K when the propulsive nozzle area was increased by 10%. What was the mass flow, jet velocity and gross thrust with this increased nozzle opening? Assume that losses in the engine were unchanged and the nozzle remains isentropic.

(**Ans:** $p_{03}/p_{02} = 27.2$; $\dot{m} = 203$ kg/s; $V_j = 885$ m/s; $F_G = 180$ kN)

The engine working lines

As explained above for the two-shaft engine, as for the single-shaft engine of Section 12.3, the compressor power is effectively fixed by the turbine and propulsive nozzle area ratios and the ratio of turbine inlet temperature to engine inlet temperature. From this the temperature ratios in the compressors and the pressure ratios can be found. To find the non-dimensional mass flow into each of the compressors we recognise that the HP turbine nozzle is choked and it is this area which determines engine mass flow rate. The area will have been chosen so that at design point, with design pressure and temperature, the desired mass flow is achieved. To find the mass flow at off-design conditions requires the pressure rise and temperature rise in the compressors in conjunction with the HP turbine nozzle guide vane area. From the conditions for the choked HP turbine nozzle row,

$$\overline{m}_4 = \frac{\dot{m}\sqrt{c_{pe}T_{04}}}{A_4 p_{04}} = 1.389,$$

the non-dimensional mass flow into the LP compressor can be written as

$$\overline{m}_2 = \frac{\dot{m}\sqrt{c_p T_{02}}}{A_2 p_{02}} = \overline{m}_4 \frac{\sqrt{c_p T_{02}}}{\sqrt{c_{pe}T_{04}}} \frac{A_4}{A_2} \frac{p_{03}}{p_{02}} = 1.389 \frac{\sqrt{c_p T_{02}}}{\sqrt{c_{pe}T_{04}}} \frac{A_4}{A_2} \frac{p_{03}}{p_{02}}. \qquad (12.20a)$$

The pressure drop in the combustor is ignored here so, $p_{03} = p_{04}$. The ratio of areas A_4/A_2 will have been chosen by the designer to give the required compressor mass flow at the design conditions and is fixed; the ratio of specific heats may also be regarded as constant but the pressure and temperature ratio vary with engine operating condition. Although the overall pressure ratio, p_{03}/p_{02}, is a function of T_{04}/T_{02}, the variation in the pressure ratio is much larger than the variation in $\sqrt{(T_{04}/T_{02})}$. The

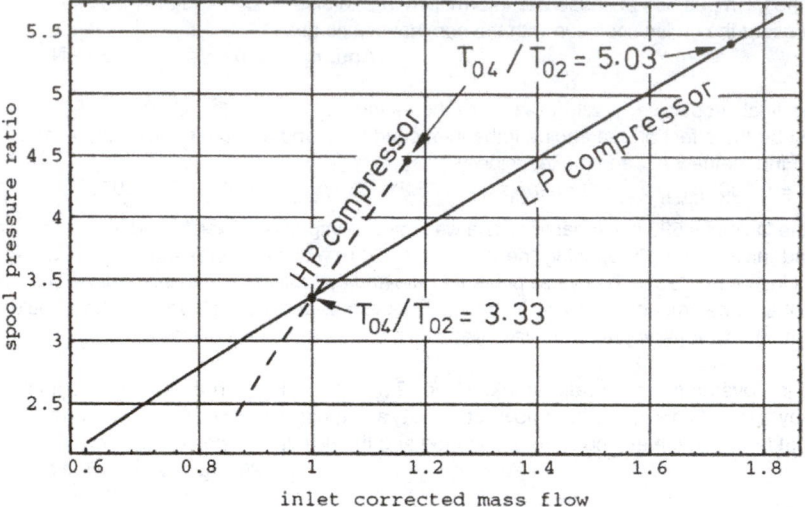

Figure 12.6. The working lines for the Olympus engine compressors. The LP and HP pressure ratios are assumed to be equal at cruise ($T_{04}/T_{02} = 3.33$), at which condition the non-dimensional mass flow into each is normalised to unity.

inlet non-dimensional mass flow, $\overline{m}_2$, is therefore mainly determined by the *overall* pressure ratio and is the same whether the engine is a single- or two-shaft engine. As a result $\overline{m}_2$ falls steeply as the pressure ratio is reduced (in other words, as the fuel supply to the engine is reduced). The variation of p_{023}/p_{02} with $\overline{m}_2$ gives the working line of the LP compressor. The engine rotational speeds would fall as $\overline{m}_2$ is reduced.

The non-dimensional mass flow into the HP compressor is likewise given by

$$\overline{m}_{23} = \overline{m}_4 \frac{\sqrt{c_p T_{023}}}{\sqrt{c_{pe} T_{04}}} \frac{A_4}{A_{23}} \frac{p_{03}}{p_{023}} = 1.389 \frac{\sqrt{c_p T_{023}}}{\sqrt{c_{pe} T_{04}}} \frac{A_4}{A_{23}} \frac{p_{03}}{p_{023}} \tag{12.20b}$$

and it is the pressure ratio across the HP compressor which primarily determines its variation. Since the pressure ratio across the HP is less than the overall pressure ratio, the excursion in non-dimensional mass flow at HP compressor inlet is correspondingly smaller than at LP compressor inlet. The variation of p_{03}/p_{023} with $\overline{m}_{23}$ defines the working line of the HP compressor.

Figure 12.6 shows the working lines calculated using Equations 12.19, 12.20a and 12.20b for the HP and LP compressors of the Olympus 593. The calculations have been carried out with T_{04}/T_{02} in the range 2 to 5.2, the latter being the value at take-off. In working out these the k_{HP} and k_{LP} have been taken from Exercise 12.5. (It was assumed in Exercise 12.5 that the pressure ratio across each compressor is equal at cruise, but this is not necessarily the optimum nor what was adopted.) For take-off the temperature ratio T_{04}/T_{02} is significantly higher than for cruise, partly because a higher turbine inlet temperature was allowed for the short take-off phase but primarily because T_{02} was much higher at cruise. Because of the higher T_{04}/T_{02} at take-off the overall pressure ratio at take-off is estimated to be 24.0, substantially greater than the cruise value of 11.3. It should be noted that the variation in non-dimensional mass flow into the engine, $\bar{m}_2$, is the same as it would be even if this were a single-shaft engine at the same overall pressure ratio. The variation of non-dimensional mass flow into the HP compressor is very much smaller because the range in $\bar{m}_{23}$ is essentially determined by the pressure ratio across only the HP compressor.

Compressor maps are not available for this engine, but a common property can be used to assess the changes in the shaft speeds needed to accommodate the changes in flow rate. For compressors with modest peak pressure ratios, such as that shown in Figure 11.5, the non-dimensional mass flow along a working line falls rapidly with rotational speed: about a 10% reduction in flow for a 10% reduction in speed is shown here. (For a machine giving a high-pressure ratio, such as that shown in Figure 11.6, the mass flow falls very much faster as the speed is reduced.) Because in Figure 12.6 the working line for the LP compressor shown has a much greater variation in mass flow than the corresponding HP compressor, the excursions in LP speed needed are very much greater. This is common to all multi-shaft engines, with the innermost shaft changing speed much less than the outer ones: for a three-shaft engine the non-dimensional rotational speed of the HP may vary by less than 5% over much of the engine operating range.[2]

Exercise

12.10 In Exercise 12.3 it was shown that for the single-shaft turbojet the compressor working line is defined by $\dot{m}\sqrt{(\Delta T_0)}/p_{03} = $ constant. Confirm that this applies to the HP compressor of a two-shaft engine and show that the working line of the LP compressor can be described by

$$\frac{\dot{m}\sqrt{(\Delta T_{0LP})}}{p_{023}} \frac{\sqrt{(\Delta T_{0HP})}}{\sqrt{(\Delta T_{0LP})}} \frac{p_{023}}{p_{03}} = \text{constant}$$

where ΔT_0 is the compressor temperature rise.

The need for multiple shafts

As already observed, the non-dimensional mass flow into an engine $\bar{m}_2$ falls as engine speed is reduced, with the tendency of the operating point to move towards the compressor surge line. Expressed differently showing the pressure ratio versus non-dimensional mass flow, the working line, is less steep than the surge line, so at low pressure ratios the working line can intersect the surge line, which is unacceptable. This is a problem for all engines and one which gets worse as the design pressure ratio is increased. The primary means of alleviating the effect is to adopt separate HP and LP compressors on separate shafts which are able to rotate at different speeds. The compressors are able to select the speed at which they can meet the requirement for the

---

[2] An analogous effect is found with turbocharged diesel engines. The reciprocating part of the engine is equivalent to the core of the jet engine, while the turbocharger is equivalent to the LP part of a jet engine. As the speed of the reciprocating part is altered the turbocharger speed alters proportionately much more.

non-dimensional mass flow and pressure ratio. The working line on the pressure ratio versus mass flow map for each compressor determines the rotational speed which is required. Having more than one shaft is, however, rarely sufficient mitigation and many compressors have variable stator blades (i.e. vanes that can be rotated to be more nearly tangential at low engine speeds) as well as bleeds part way along the compressor which can open to dump some of the compressed air at low rotational speed.

The use of multiple shafts has an additional benefit for the design point operation. The front stage of a compressor must not operate with a blade speed which is too high in relation to local sonic velocity lest the efficiency falls sharply because of the high relative Mach number. The temperature rises through the compressor, so that the blade speed (in m/s) for the later stages could be significantly higher without excessive relative Mach number and without loss in efficiency. With a compressor on a single shaft there is a severe limit to how much the blade speed can be increased (because this can only be done by increasing the mean radius) but by splitting the compressor into two or more parts it is possible to let the HP compressor rotate substantially faster than the HP.

Exercise
12.11 Sketch the working lines in Figure 12.6 on the compressor map, Figure 11.5. If the non-dimensional shaft speeds are 100% at the cruise condition, what are they at take-off?

$$\textbf{(Ans: } N_H\sqrt{T_{023}} \approx 110\%, \ N_L\sqrt{T_{02}} \text{ too large to be estimated on this figure)}$$

Note: The compressor used to produce the map in Figure 11.5 is very different from those in the Olympus engine. This comparison only serves to show how large the speed excursions need to be on the LP shaft compared with those on the HP.

SUMMARY OF SECTION 12.3

The behaviour of multi-shaft engines can be determined easily, if approximately, by using the particular flow features of the turbine and nozzle in a way introduced earlier for a single-shaft engine The turbine on each shaft behaves as if it were choked and the analysis is simple if the final propulsive nozzle can be taken to be choked. Furthermore the turbine is assumed to operate with constant efficiency over a wide range of speeds and pressure ratios. Combining this means that the ratios of stagnation temperature and pressure are constant across each turbine The ratio of turbine nozzle areas and the ratio of LP turbine nozzle area to propulsive nozzle area largely determines engine operating condition for a given T_{04}/T_{02}. The temperature ratio leads to the power per unit mass flow rate from *each* turbine being proportional to T_{04}, the temperature at entry to the HP turbine. Since the power from each turbine is the power into the compressor on the same shaft, the temperature rise in each compressor is also proportional to T_{04}. Assuming constant polytropic efficiency, the pressure ratios across the compressors can be found. The mass flow rate through a turbojet is fixed by the choking of the HP turbine.If efficiency is not adequately described as constant, or the assumption of choking is inadequate for one of the nozzles, the underlying principle of solution remains valid, but iterative solution is then necessary.

If the area of the propulsive nozzle is increased the pressure is lowered in the jet pipe. This has the effect of increasing the power output of the LP turbine, which raises the speed of the LP shaft, in turn raising the mass flow swallowed by the LP compressor. A similar approach can be used to predict the effect of changes in the area of the turbine nozzle guide vanes.

As the two-shaft turbojet is throttled back so the pressure ratio and non-dimensional mass flow fall, the front stages of a compressor move towards stall. Having two shafts allows the speed of each shaft to vary to meet the needs of the compressor, thereby reducing the mismatch of the front stages of each compressor. The changes experienced by the LP compressor are much greater than by the HP and the changes in LP rotational speed are correspondingly larger.

12.4 THE HIGH BYPASS TURBOFAN ENGINE

The groundwork has been laid for the consideration of a high bypass ratio engine of the type used in the design exercise in Part 1 of the book. The treatment is similar to the two-shaft turbojet for the core (HP compressor, combustor and HP turbine), but for the LP compressor and turbine it is somewhat different because only a small fraction of the flow through the fan enters the core compressor. The simplest possible layout of high bypass ratio engine would be that shown in Figure 12.7a. In this two-shaft engine the flow downstream of the fan is approximately uniform, and the pressure ratio of the core stream across the fan is similar in magnitude to that of the bypass stream. All the remainder of the core compression would be in the HP compressor. The choice here for the engine for the New Efficient Aircraft is a fan pressure ratio of 1.5 and an overall pressure ratio of 45 and would require the core compressor to produce a pressure ratio of 30 at cruise, which is currently unrealistic. Recalling Chapter 11, where the practical difficulties in using compressors with large pressure ratios on one shaft was addressed, the maximum for most engines has been below about 20.

There are two alternative solutions which are shown in Figures 12.7b and 12.7c. The simpler solution is shown in Figure 12.7b, where there is a so-called booster compressor on the LP shaft. Because it is simpler this was the style of engine used for the engine designed in Chapter 7 of Part 1. At cruise this engine had an overall pressure ratio of 45, with 2.5 produced in the fan root and booster and 18 in the HP compressor. The practical realisation of a design similar to this is the General Electric GEnx, Figure 5.4b. The alternative solution to avoid excessive pressure ratios in one compressor is to adopt the three-shaft design favoured by Rolls-Royce and represented by the Trent 1000 in Figure 5.4a. Because the two-shaft design is easier to describe and calculate (fewer stations) it is considered here first before looking at the three-shaft solution.

The core flow

Just as for the two-shaft turbojet considered in Section 12.3, the HP turbine in the high bypass engine is assumed to be choked. The restriction formed by the nozzle guide vanes (NGV) of the

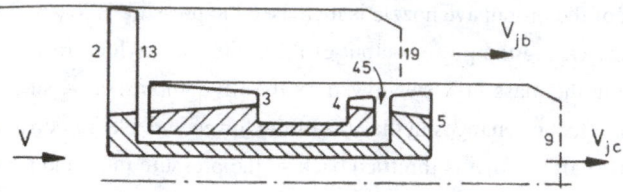

(a) Two-shaft engine, simplified configuration

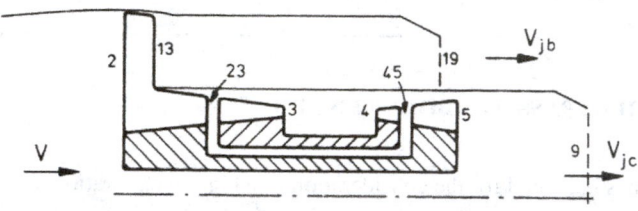

(b) Two-shaft engine, typical configuration

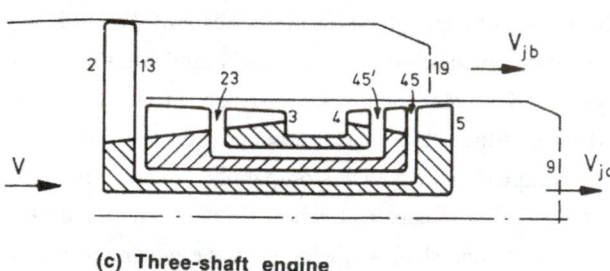

(c) Three-shaft engine

Figure 12.7. Schematic layouts and engine station numbering schemes for high bypass ratio engines.

HP turbine therefore fixes the mass flow rate through the core,[3] $\overline{m}_c$. As in the turbojet the choked HP turbine gives, with $\gamma = 1.30$, as in Equation 12.14

$$\overline{m}_{c4} = \overline{m}_4 = \frac{\dot{m}_c \sqrt{c_{pe} T_{04}}}{A_4 p_{04}} = 1.389.$$

The area A_4 can be found from an on-design calculation. The mass flow out of the HP turbine is the inlet flow to the LP turbine. The pressure ratio across the LP turbine will be large enough for it to be effectively choked at inlet for all conditions of interest so that

$$\overline{m}_{45} = \frac{\dot{m} \sqrt{c_{pe} T_{045}}}{A_{45} p_{045}} = 1.389,$$

---

[3] When it is necessary to denote core flow the subscript c is used, as in $\overline{m}_c$, but when there is no ambiguity the c will be dropped for sake of brevity, as in $\overline{m}_{c45} = \overline{m}_{45}$.

Because both the HP and LP turbine are assumed to be choked at inlet, the non-dimensional mass flow $\overline{m}$ is equal into both, $\overline{m}_{45} = \overline{m}_4$. Just as for the two-shaft turbojet the mass flow $\dot{m}$ into the HP and LP turbines is equal, neglecting for the present the cooling air, and this provides a relationship between pressure ratio, temperature ratio and area ratio, see Equation 12.5. In addition the pressure ratio and temperature ratio across the HP turbine are linked by polytropic relation, Equation 12.2

$$\frac{T_{045}}{T_{04}} = \left(\frac{p_{045}}{p_{04}}\right)^{\eta(\gamma-1)/\gamma}.$$

The equality of mass flow, the fixed values of non-dimensional mass flow at inlet and exit to the HP turbine, and the polytropic relation mean that the pressure ratio and temperature ratio across the HP turbine can be expressed in terms of turbine nozzle flow areas, A_4 and A_{45}. Then, just as for the turbojet in Equation 12.17a

$$\frac{p_{045}}{p_{04}} = \left(\frac{A_4}{A_{45}}\right)^{\frac{2\gamma}{2\gamma-\eta(\gamma-1)}} \quad \text{and} \quad \frac{T_{045}}{T_{04}} = \left(\frac{A_4}{A_{45}}\right)^{\frac{2\eta(\gamma-1)}{2\gamma-\eta(\gamma-1)}}.$$

With the approximation $\gamma = 1.30$ and assuming that $\eta_p = 0.9$ the pressure ratio depends on area ratios raised to the power of 1.12 and the temperature ratios to the power of 0.232. As noted before, pressure ratio is almost inversely proportional to area ratio.

Again, as for the turbojets, using the temperature ratio the power per unit mass flow rate for the HP turbine can be expressed as

$$c_{pe}(T_{04} - T_{045}) = k_{HP}c_{pe}T_{04}. \tag{12.21}$$

Because of the neglect here of fuel mass flow, bleed flows and cooling flows the mass flow here is equal in both the HP turbine and HP compressor and the power balance (equating turbine and compressor power) for the HP shaft is,

$$c_{pe}(T_{04} - T_{045}) = c_p(T_{03} - T_{023}) = k_{HP}c_{pe}T_{04}.$$

The mass flow rate through the core $\overline{m}_c$ is found in a way exactly analogous to that of the turbojet. The HP turbine, actually the first row of nozzle guide vanes, gives with $\gamma = 1.30$,

$$\overline{m}_4 = 1.389$$

from which the non-dimensional mass for the core flow at fan inlet, with area A_2 can be written

$$\overline{m}_2 = \frac{\dot{m}_c\sqrt{c_pT_{02}}}{A_2p_{02}} = \overline{m}_4\frac{\sqrt{c_pT_{02}}}{\sqrt{c_{pe}T_{04}}}\frac{A_4}{A_2}\frac{p_{03}}{p_{02}} = 1.389\frac{\sqrt{c_pT_{02}}}{\sqrt{c_{pe}T_{04}}}\frac{A_4}{A_2}\frac{p_{03}}{p_{02}}.$$

The LP system

In the case of the two-shaft turbojet the final propulsive nozzle was choked and and $\overline{m}$ was constant upstream and downstream of the HP and LP turbines. It was then possible to produce an equation for the temperature ratio across the LP turbine (Equation 12.17b) analogous to the equation for the HP turbine (Equation 12.17a). For the high bypass engine we can write a similar equation for the temperature ratio across the HP turbine but the core propulsive nozzle of recent large high-bypass engines is unchoked even at cruise, so that $\overline{m}$ is not constant downstream of the LP turbine. This makes a simple relation like 12.17b impossible for the LP turbine of the high-bypass engine. For earlier generations of large bypass engines a choked propulsive nozzle remained a reasonable assumption for the core flow and was adopted for editions 1 and 2 of Jet Propulsion. Even now, smaller engines, such as those on business jets, may still have pressure ratios for the core nozzle which are large enough to cause choking but for the current generation of large high bypass engines this is no longer appropriate or adequate. The explanation is that with a low fan pressure ratio, such as $fpr = 1.5$, the bypass jet velocity is low. For high propulsive efficiency the core jet velocity will be chosen to be of similar magnitude to the bypass jet velocity. The flow out of the LP turbine is at much higher temperature than the bypass air, so equal jet velocity is produced with a pressure ratio across the core nozzle, of about 1.4, well below the pressure ratio for choke which is about 1.832. As a result the non-dimensional mass flow into the propulsive nozzle, $\overline{m}_9$ is not constant but is a function of the pressure ratio across it,

$$\overline{m}_9 = \frac{\dot{m}\sqrt{c_{pe}T_{05}}}{A_9 p_{05}} = \overline{m}_9(p_{09}/p_a),$$

where the function is given in Equation 12.3. Here, as for the turbojet, it is assumed that there are no losses downstream of the LP turbine, so $p_{05} = p_{09}$ and $T_{05} = T_{09}$. At entry to the LP turbine the flow is choked with $\overline{m}_{45} = 1.389$.

The assumption that polytropic efficiency is constant across the LP turbine is again used leading to equations similar to Equations 12.6 which were obtained for the single-shaft turbojet. Hence

$$\frac{p_{05}}{p_{045}} = \left(\frac{\overline{m}_{45}}{\overline{m}_9\,(p_{09}/p_a)}\right)^{\frac{2\gamma}{2\gamma-\eta(\gamma-1)}} \left(\frac{A_{45}}{A_9}\right)^{\frac{2\gamma}{2\gamma-\eta(\gamma-1)}} \tag{12.22a}$$

and

$$\frac{T_{05}}{T_{045}} = \left(\frac{\overline{m}_{45}}{\overline{m}_9\,(p_{09}/p_a)}\right)^{\frac{2\eta(\gamma-1)}{2\gamma-\eta(\gamma-1)}} \left(\frac{A_{45}}{A_9}\right)^{\frac{2\eta(\gamma-1)}{2\gamma-\eta(\gamma-1)}}, \tag{12.22b}$$

As before, for $\gamma = 1.30$ and $\eta_p = 0.90$, the indices are 1.12 for pressure ratio and 0.232 for temperature ratio.

The pressure and temperature ratios across the LPT therefore depend upon the pressure ratio across the unchoked propulsive nozzle p_{09}/p_a, unlike the ratio across the HP turbine for which $p_{045}/p_{04} = $ constant and $T_{045}/T_{04} = $ constant. The power from the LPT per unit mass flow rate through the core is

$$\dot{W}_{LP}/\dot{m} = c_{pe}\,(T_{045} - T_{05}) = c_{pe}T_{045}\,(1 - T_{05}/T_{045})$$
$$= c_{pe}T_{04}\,[T_{045}/T_{04}\,(1 - T_{05}/T_{045})] \tag{12.23}$$

but in this case T_{05}/T_{045} a function of nozzle pressure ratio whilst T_{045}/T_{04} across the HP turbine, remains a constant depending on area ratio.

The bypass and flow through the fan

The bypass nozzle for the engine with a design fan pressure ratio of 1.5 will always be choked at the cruise conditions or whenever the aircraft is flying fast enough to be in steady level flight at cruising altitude. With no loss in stagnation pressure between the fan and the bypass nozzle exit we can write

$$p_{019}/p_a = p_{013}/p_{02} \times p_{02}/p_a.$$

Choking corresponds to $p_{019}/p_a = 1.893$ while for cruise at $M = 0.78$, $p_{02}/p_a = 1.485$. The bypass nozzle is, however, well away from choke at take-off conditions. (In earlier engines, with fan pressure ratio close to 1.8 the fan would be choked or close to choke for take-off.)

For choked operation of the bypass nozzle, with $\gamma = 1.40$ and A_{19} denoting bypass nozzle area the non-dimensional mass flow through the bypass nozzle is given by

$$\overline{m}_b = \frac{\dot{m}\sqrt{c_p T_{013}}}{A_{19} P_{013}} = 1.281, \tag{12.24}$$

where no losses or heat transfer is assumed in the bypass duct so the temperature and pressure into the bypass nozzle are equal to those leaving the fan, $p_{013} = p_{019}$. When the nozzle is unchoked, such as for take-off, Equation 12.3 must be used to obtain $\overline{m}_b$ as a function of $p_{019}/p_a = p_{013}/p_a$. For flight when the Mach number exceeds about 0.58 we are correct in assuming that the bypass nozzle is choked with a fan pressure ratio of 1.5.

A major difference between the bypass engine and the turbojet comes when the power in the LP shaft is considered. When the mass of fuel and the air removed in bleeds from the compressor are neglected the equation is,

$$\dot{W}_{LP} = \dot{m}_c c_{pe}(T_{045} - T_{05}) = c_p \left\{ \dot{m}_b (T_{013} - T_{02}) + \dot{m}_c (T_{023} - T_{02}) \right\}$$

where T_{013} is the temperature downstream of the fan in the bypass, T_{023} is the temperature downstream of the booster stages and $\dot{m}_b$ and $\dot{m}_c$ are the bypass and core mass flow rates respectively. Using the bypass ratio $\dot{m}_b/\dot{m}_c$ the power balance on the LP shaft can be written

$$c_{pe} T_{04}/T_{02} \left[T_{045}/T_{04} (1 - T_{05}/T_{045}) \right] = c_p \left[bpr (T_{013}/T_{02} - 1) + (T_{023}/T_{02} - 1) \right]. \tag{12.25}$$

Equation 12.25 contains four unknowns in addition to T_{04}/T_{02} which is regarded here as the independent input variable. (The ratio T_{045}/T_{04} across the HP turbine is, as discussed above, a function of area ratio only and remains at the design point value.) The four unknowns are: the bypass ratio; the temperature ratio for the bypass stream across the fan, T_{013}/T_{02}; the temperature ratio for the core flow across fan and booster, T_{023}/T_{02}; and the temperature ratio across the LP turbine, T_{05}/T_{045}.

Because the bypass ratio alters with the engine operating point the fraction of the air through the fan which enters the core also changes. The flow pattern inside the fan is therefore also changed and a precise determination of T_{013}/T_{02} and T_{023}/T_{02} would require detailed knowledge of the way the fan behaves as a function of speed, overall mass flow and bypass ratio. To make the problem tractable with the minimum of empirical input we make the simple but plausible assumption that the temperature rise for the core flow across the fan and booster stages $T_{023} - T_{02}$ is proportional to the temperature rise of the bypass flow across the fan $T_{013} - T_{02}$. In other words, comparing the temperature rises on-design and off-design the ratios are constant,

$$\frac{T_{023} - T_{02}}{T_{013} - T_{02}} = \frac{\{T_{023} - T_{02}\}_{design}}{\{T_{013} - T_{02}\}_{design}} = k_t. \tag{12.26}$$

The power balance for the LP shaft, Equation 12.25, then simplifies to

$$\frac{c_{pe}}{c_P} T_{04}/T_{02} \left[T_{045}/T_{04} \left(1 - T_{05}/T_{045} \right) \right] = \{bpr + k_t\} \{T_{013}/T_{02} - 1\}$$

$$= \{bpr + k_t\} \{fpr^{\gamma-1/\eta\gamma} - 1\}, \tag{12.27}$$

with three unkowns. For the current simple treatment it is assumed that the efficiency of the compressors will remain constant (which is plausible provided the excursion from the design point is not too large) and that the compressor will be able to adopt a speed at which the simultaneous requirements of the working line for pressure ratio and mass flow are simultaneously satisfied. The flow through the bypass nozzle and the power balance for the LP shaft also need iterative numerical solution, as outlined in the next section.

A practical method of solution

The calculation of off-design performance starts with an on-design calculation where the temperature ratio T_{04}/T_{02} and the flight Mach number are given. (Specifying flight Mach number is equivalent here to giving p_{02}/p_a.) From the design point calculation invariant parameters such as HPT temperature ratio, T_{045}/T_{04}; non-dimensional mass flow for the core propulsive nozzle, $\overline{m}_9$; constants k_{HP} and k_t; and nozzle areas A_4, A_{45}, A_9 and A_{19} may be found.

Assume that T_{04}/T_{02} is altered to take the engine off-design. An iterative process is needed to find the off-design solution but, reasons presented earlier, the temperature and pressure ratio of the HP turbine are unchanged. The temperature drop in the HPT is equal to $k_{HP}T_{04}$ and from this the temperature rise in the HP compressor is known. To turn this into temperature ratio and pressure ratio for the HPC it is necessary to know the temperature of the air entering the compressor, namely the temperature out of the booster T_{023}; in other words the pressure and temperature ratios in the core depend on the operation of the LP system.

With specified T_{04}/T_{02} the off-design solution steps may be summarised as follows, starting from values at design point:

i) guess the new fan pressure ratio. This determines the temperature rise in the bypass stream and, using Equation 12.26, the temperature rise across the booster. The booster temperature rise gives the booster pressure ratio. From the temperature rise across the HPC the overall pressure ratio, $opr = p_{03}/p_{02}$ and T_{03}/T_{02} are known;

ii) given the HPT nozzle area A_4 and p_{04} (assumed here to equal compressor outlet pressure) as well as T_{04}/T_{02}, the core mass flow rate $\dot{m}_c$ can be found from $\overline{m}_4$ which is known since the turbine is choked;

iii) knowing the pressure and temperature downstream of the fan, the bypass mass flow $\dot{m}_b$ can be found since the area A_{19} is known. If the nozzle is choked (when p_{019}/p_a is greater than 1.893) $\overline{m}_{19} = 1.281$ can be used, but if the nozzle is unchoked $\overline{m}_{19}$ must be calculated using Equation 12.3;

iv) from core mass flow in step ii) and bypass mass flow in step iii) the bypass ratio can be calculated;

v) the bypass ratio together with the fan temperature ratio may be used with Equation 12.27 to calculate the temperature ratio across the LP turbine, T_{05}/T_{045}. From the temperature ratio and an assumed value of efficiency, the pressure ratio across the LPT may be found;

vi) the opr was found in step i), the pressure ratio across the HPT is constant and the pressure ratio across the LPT was obtained in step v). The pressure downstream of the LPT (upstream of the core nozzle) is therefore known and Equation 12.3 can be used to obtain $\overline{m}_9$ (the propulsive core nozzle is always unchoked). Since both temperature and pressure are known downstream of the LPT, as well as the nozzle area A_9, the core mass flow $\dot{m}_c$ can be calculated from $\overline{m}_9$;

vii) the core mass flow from step vi) is to be compared with that from step ii). If they are equal the choice of fan pressure ratio is correct, otherwise it must be adjusted and the calculation repeated.

SUMMARY OF SECTION 12.4

The treatment here in Section 12.4 has been for the civil transport engine which has formed the basis of the design in Chapter 7. The engine was assumed to have two shafts and a simple assumption, which is not expected to compromise the predicted trends substantially, was introduced to accommodate the pressure ratio of the booster stages on the LP shaft. The same approach is readily adapted to the three-shaft engine or to military engines.

As for the turbojets the ratio of nozzle areas are crucial in setting the engine operating condition: the HP turbine inlet nozzle area, the LP turbine inlet nozzle area and the propulsive nozzle areas. For modern high bypass ratio engines the fan pressure ratio is low enough that the core propulsive nozzle is likely to be unchoked even at cruise. The LP turbine is therefore no longer operating at a constant pressure ratio and temperature ratio but p_{045}/p_{05} across the LP turbine also depends on the propulsive nozzle pressure ratio $p_{05}/p_a = p_{09}/p_a$. This requires iterative solution to obtain the power from the LP turbine.

Most of the power from the LP turbine is used in compressing the bypass stream and the power to the bypass stream depends on the bypass mass flow and bypass pressure ratio. For cruise the bypass nozzle is choked and this determines the combination of mass flow and pressure ratio that the fan produces, in other words the fan working line; an iterative solution is required to determine *fpr* and *bpr*. (For stationary conditions the bypass nozzle is not choked for a modern large engine.)

The ideas carry across into the consideration of modern military engines which are mainly turbofans of the two-shaft type with typical design values of bypass ratio in the range 0.3 up to 1.0. Whereas the non-dimensional operating point for civil engines does not normally alter very much, for military engines it varies greatly and an understanding of the matching 'off-design' is essential. For military engines the propulsive nozzle is normally choked, which greatly simplifies the treatment.

Exercise

12.12 Verify that in Equation 12.26 all the terms can be shown to depend on T_{04}/T_{02} and p_{02}/p_a. In other words that these are the independent variables which define the non-dimensional operating point of the engine.

12.5 APPLICATION TO A TWO-SHAFT HIGH BYPASS TURBOFAN

Cruise with off-design temperature ratio

The approach of Section 12.4 will be applied first to the engine outlined in Exercise 7.2 to look at engine operation at cruise when turbine inlet temperature ratio is varied, as it might be to adjust the thrust. At the design point (start of cruise at 35,000 ft and $M = 0.78$, giving $T_{02} = 245.4$ K) the turbine inlet temperature is taken to be $T_{04} = 1500$ K, which gives the ratio $T_{04}/T_{02} = 6.11$. The design fan pressure ratio was chosen to be 1.5. The design overall pressure ratio is 45 and the pressure ratio for the core flow across the fan and booster on the LP shaft was taken to be 2.5 giving an HP compressor design pressure ratio of 18.

In Exercise 7.2 the gas properties were taken to be those of low temperature air, $\gamma = 1.4$, even for the hot combustion products. Here in Chapter 12, however, we use the more realistic description of gas properties for combustion products, with $\gamma = 1.30$ and $c_{pe} = 1244 \, \text{J kg}^{-1}\text{K}^{-1}$ used for the flow through the turbine and core nozzle. For the fan and compressor $\gamma = 1.40$ and $c_p = 1005 \, \text{J kg}^{-1}\text{K}^{-1}$ are retained. In Chapter 12 we adopt *polytropic* efficiencies and for the bypass engine this has been set equal to 0.90 for all components – this is not strictly compatible with the

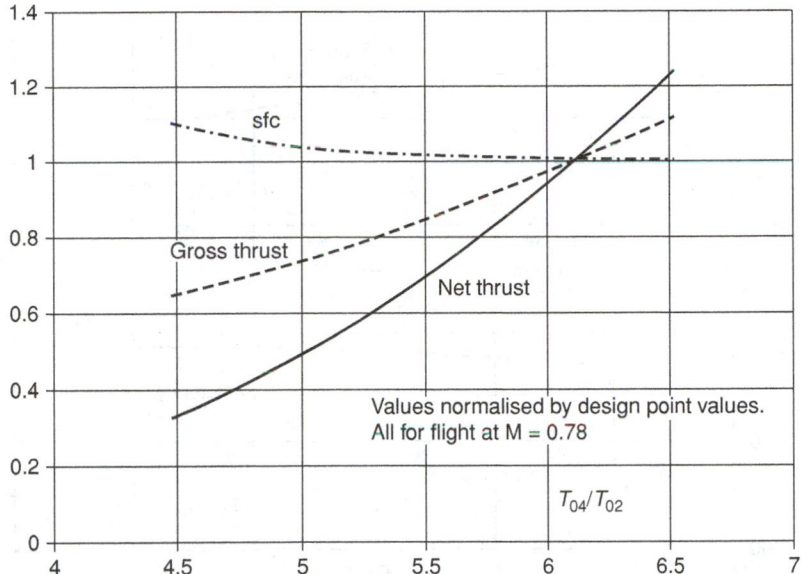

Figure 12.8. Computed thrust and specific fuel consumption as turbine inlet temperature is varied. Design point corresponds to $T_{04}/T_{02} = 6.11$. For cruise at $M = 0.78$.

assumptions in the design leading to Exercise 7.2 where *isentropic* efficiencies equal to 0.90 were used. However, the discrepancy will be relatively small and the convenience in using polytropic efficiency great.

The bypass ratio is predicted at design point using the efficiency definition and gas properties of Chapter 12 to be about 13.5, whereas it was equal to 11.2 in Chapter 7. The value found in Chapter 12 is therefore further than that for Chapter 7 from the *bpr* of real new engines, which with $fpr \approx 1.5$ is about 10. One explanation is that some of the inadequacies in Chapter 7 cancelled. By adopting $\gamma = 1.30$ for combustion products in Chapter 12 the specific heat is larger than in Chapter 7 and the turbines give substantially more work. Here in Chapter 12 we are still not including complicating factors like cooling air for the turbines and pressure drop in the combustor. These issues are addressed in Chapter 19. In summary, the bypass ratios predicted here in Chapter 12 are high for given fan pressure ratio. Because the performance of the engine depends on fan pressure ratio rather than bypass ratio, the high values of *bpr* do not detract from the usefulness of the calculations.

Figure 12.8 shows the computed thrust and specific fuel consumption as turbine inlet temperature is altered at cruise altitude and $M = 0.78$. Raising T_{04} from 1500 K to 1600 K, that is raising $T_{04}/T_{02} = 6.52$ the net thrust increases by about 24%. It should be recalled that the net thrust considered here is the so-called bare-engine thrust, which neglects the nacelle drag and bypass duct losses. In a real application the proportional changes in installed thrust would be larger because the drag on the nacelle would not alter significantly with T_{04}/T_{02} if the flight speed were unaltered. Gross thrust also varies proportionately much less than net thrust with T_{04}/T_{02}.

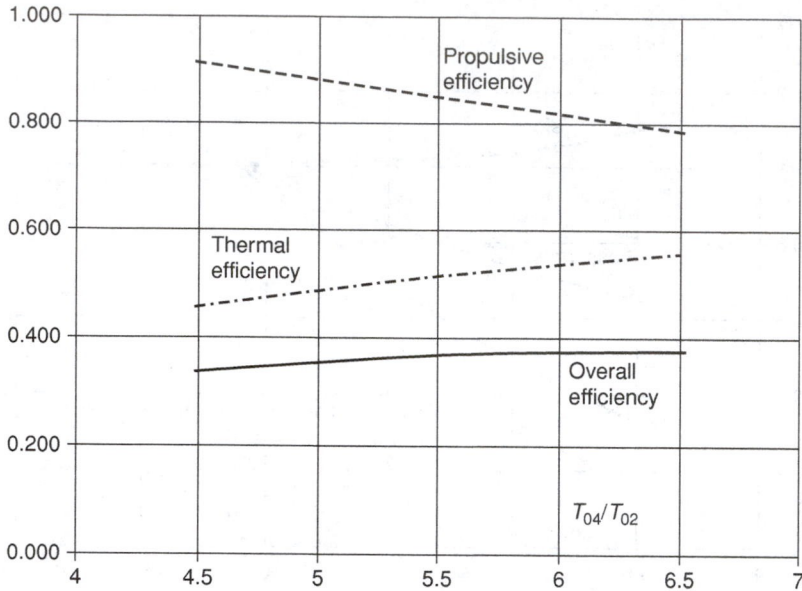

Figure 12.9. Computed efficiencies as turbine inlet temperature is varied. Design point corresponds to $T_{04}/T_{02} = 6.11$. For cruise at $M = 0.78$.

The rise in specific fuel consumption with reduction in temperature ratio evident in Figure 12.8 is explored in Figure 12.9, which shows engine performance in terms of efficiencies. The overall efficiency is inversely proportional to *sfc*, that is $\eta_0 = V/(LCV.sfc)$, where V is flight speed and LCV is calorific value of the fuel. The overall efficiency therefore falls slightly as turbine inlet temperature falls, consistent with *sfc* rising, but the slight fall is the result of two quite large effects operating in different directions. Propulsive efficiency rises when T_{04}/T_{02} is reduced because the jet velocity falls, whereas the thermal efficiency decreases. The reduction in thermal efficiency is both because of the direct dependence on temperature ratio and through the reduction in overall pressure ratio. The transfer efficiency is equal to about 0.90 and hardly varies with T_{04}/T_{02}.

The impact of varying T_{04}/T_{02} on overall pressure ratio and fan pressure ratio is shown in Figure 12.10. The overall pressure ratio is approximately halved from design at $T_{04}/T_{02} = 6.11$ to $T_{04}/T_{02} = 4.5$, whilst fan pressure ratio falls to about 1.2. With the fall in fan pressure ratio bypass ratio is increased to almost 20. The rise in bypass ratio with reduction in thrust from an engine is common experience in all engines. Likewise, if an existing engine is "up-thrusted" to create a higher thrust version by increasing the allowable temperature ratio T_{04}/T_{02}, it is found that the bypass ratio decreases whilst all the pressure ratios increase; this too can be seen in Figure 12.10.

For the operation of the fan and compressors it is the working lines, sometimes called the operating lines, which are most relevant. Figure 12.11a shows a working line for the fan where the pressure ratio is for the bypass stream of the fan but the corrected mass flow is for the whole fan, bypass and core stream. Figure 12.11b shows the working line for the HP compressor.

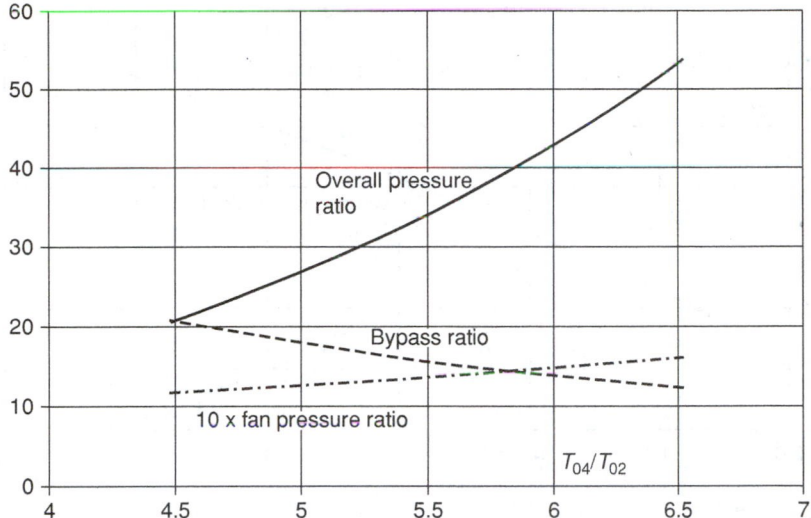

Figure 12.10. Pressure ratio and bypass ratio as turbine inlet temperature is varied.
Design point corresponds to $T_{04}/T_{02} = 6.11$. For cruise at $M = 0.78$.

The corrected mass flow for the HP compressor is based on core mass flow and its entry pressure and temperature, p_{023} and T_{023}. In both parts of Figure 12.11 the upper end of the working lines correspond to $T_{04}/T_{02} = 6.52$ ($T_{04} = 1600$ K) and the lower end to $T_{04}/T_{02} = 4.48$ ($T_{04} = 1100$ K). The corrected mass flow is normalised by the value at design, $T_{04} = T_{02} = 6.11$.

The variation in pressure ratio and in corrected mass flow is always greater for the lower-pressure device, in this case the fan. In an engine the shaft speeds would adjust to achieve the conditions along the working line and the proportional change in speed would have to be larger for the LP shaft. If constant speed lines, showing the pressure rise versus corrected mass flow for the fan and compressor, were available they could be superimposed on graphs of the working lines and variation in rotational speed found. Where the working line for a real engine enters regions of altered component efficiency the numerical values in the relevant equations would be changed and the working line would move. With varying efficiency a further level of iteration has to be introduced into the method of solution.

A key quantity involved in the iterative solution of conditions for this engine is the pressure ratio across the LP turbine, p_{045}/p_{05}, since this pressure ratio determines the pressure and the temperature into the propulsive nozzle. As noted above the mass flow through this nozzle must be equal to that through the HP turbine and equating these mass flows is essential to the method of solution. The LP turbine pressure ratio is shown in Figure 12.12. The curve is not straight and at the design point, $T_{04}/T_{02} = 6.11$, the curve is levelling off, in other words it is approaching the choking condition for the propulsive nozzle beyond which the pressure ratio would be constant. (If the static pressure at exit from the propulsive nozzle is lowered relative to inlet stagnation pressure the nozzle chokes when the pressure ratio across the LP turbine becomes $p_{045}/p_{05} \simeq 12.3$. Once

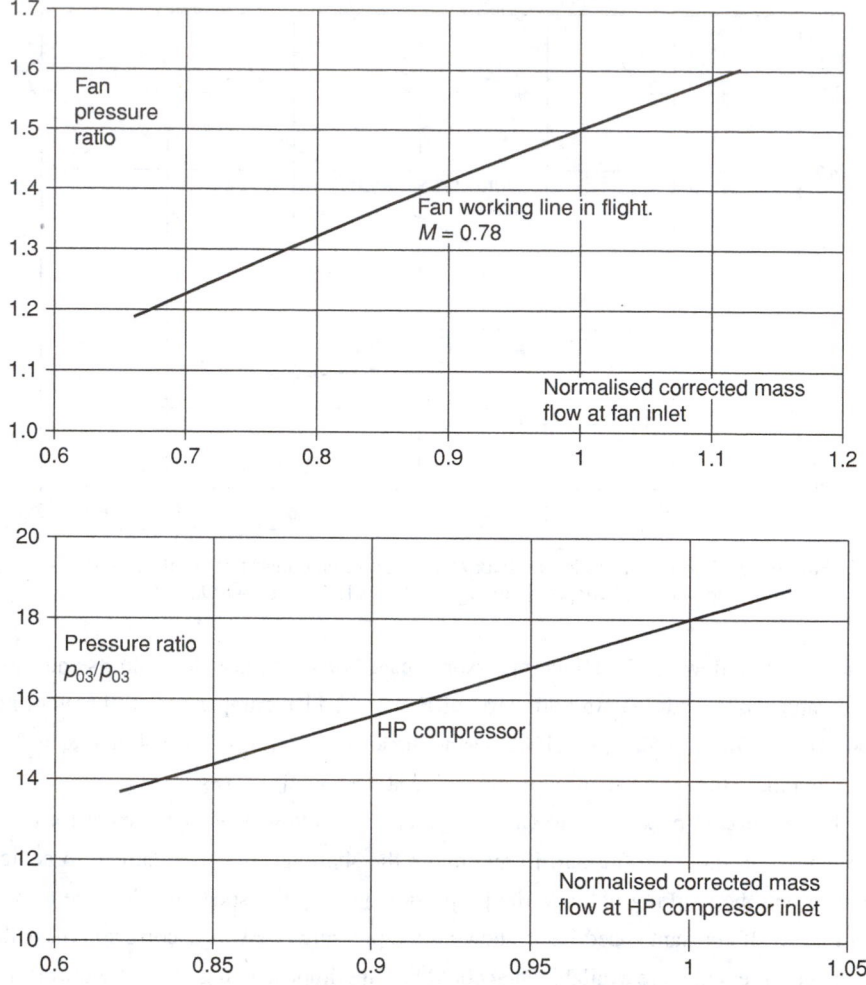

Figure 12.11. Computed working line of the fan (bypass stream) and the HP compressor for cruise at $M = 0.78$ as turbine inlet temperature is varied. Design point corresponds to $T_{04}/T_{02} = 6.11$, left end of lines at $T_{04}/T_{02} = 4.48$, right end $T_{04}/T_{02} = 6.52$.

choked this pressure ratio becomes independent of T_{04}/T_{02}). The further the propulsive nozzle is from being choked, the stronger is the dependence of p_{045}/p_{05} on T_{04}/T_{02}.

Variation in inlet Mach number – sea-level static operation

It is a feature of the engine being considered that the core propulsive nozzle is unchoked even at cruise. The pressure inside the engine, from the fan inlet to the inlet to the LP turbine, is proportional to inlet stagnation pressure p_{02} and almost independent of ambient pressure p_a. It is not totally independent of p_a because some of the core pressure ratio is from the booster driven by the LP turbine. Because the propulsive nozzle is not choked the ambient pressure does have an effect

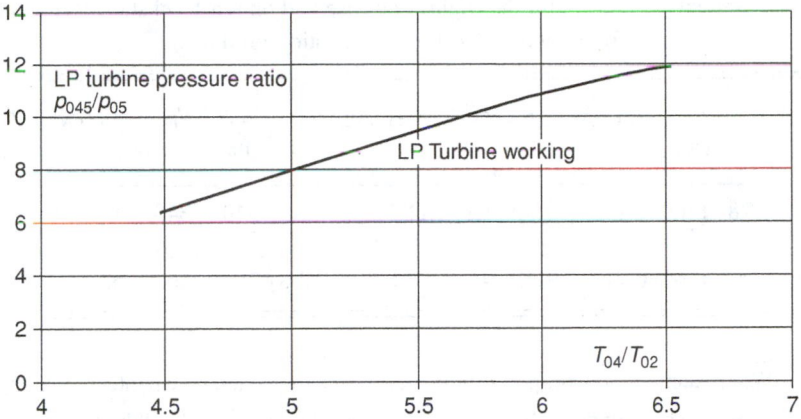

Figure 12.12. Computed working line of LP turbine for cruise at $M = 0.78$ as turbine inlet temperature is varied. Design point corresponds to $T_{04}/T_{02} = 6.11$, left end of line at $T_{04}/T_{02} = 4.48$, right end $T_{04}/T_{02} = 6.52$.

downstream of the LP turbine and in consequence the LP turbine pressure ratio depends on p_{02}/p_a as well as T_{04}/T_{02}. If the propulsive nozzle were choked this dependence on p_{02}/p_a would be removed for all components inside the engine. The most striking change in p_{02}/p_a is achieved by going from design conditions at cruise, $p_{02}/p_a = 1.49$ at $M = 0.78$, to the static test bed, when $p_{02}/p_a = 1$.

For conditions at cruise the product of p_{02}/p_a and the fan pressure ratio was sufficient to choke the bypass nozzle. For a static test the design fan pressure ratio of the engine considered here, $fpr = 1.5$, is no longer sufficient to choke the bypass and $\overline{m}_{19}$, the non-dimensional mass flow rate through the bypass nozzle, is a function of the bypass pressure ratio. It is assumed there are no losses in the bypass duct so $p_{019} = p_{013}$ and form of $\overline{m}_{19}(p_{019}/p_a)$ is given by Equation 12.3. Conditions for the engine have been computed when operating on a sea-level static test bed at the temperature ratio for design, $T_{04}/T_{02} = 6.11$ using the approach of Section 12.4. An iteration is required to get LP turbine pressure ratio and fan pressure ratio. Some salient values are shown in Table 12.2 comparing cruise, the design point, with operation on the static test bed. As might be anticipated, the largest perturbation to the internal ratios is the pressure ratio across the LP turbine, since the reduction in p_{02}/p_a is felt at propulsive nozzle exit. Because the pressure ratio across the core propulsive nozzle is so low, $p_{05}/p_a \approx 1.4$, the LP turbine pressure ratio responds strongly to the change in p_{02}/p_a.

The fan pressure ratio in Table 12.2 is slightly lower at sea-level static than design at cruise, which is counter-intuitive. The impact of approximations is considered in Chapter 19, where the calculations have been repeated using gas properties (γ and c_p) derived from the accurate approximations and where cooling air and combustor pressure loss are included. Anticipating Chapter 19 it can be stated here that these effects must be included in order to get the *fpr* and *bpr* correct at off-design conditions, since the effect of the inaccuracies tends to collect in the

Table 12.2   Salient parameters for the engine at cruise and on sea-level static test bed ($\gamma = 1.4$ for air and $\gamma = 1.3$ for combustion products)

	T_{04} (K)	T_{04}/T_{02}	fpr	opr	p_{045}/p_{05}	bpr	V_{jc} (m/s)	V_{jb} (m/s)	F_N/F_{ND}	sfc (g s^{-1}kN^{-1})
35kft, $M = 0.78$ (design)	1500	6.11	1.50	45	11.1	13.5	340	340	1	14.4
Sea-level static	1761	6.11	1.48	44.4	8.6	12.8	266	263	5.97	7.46

LP system. When accurate gas properties are used in the approach of Chapter 12 the appropriate increase in *fpr* would be found for the static test bed but the *bpr* would still be too great unless the impact of cooling air is included.

The rise in thrust (actually gross thrust for static tests) can be compared with the requirements of the aircraft. For take-off thrust it is plausible to assume a required thrust equal to 0.3 times maximum take-off weight. For initial cruise the aircraft weight will be only a little less than take-off weight, perhaps 2% less, so the initial thrust requirement will be approximately equal to this weight divided by lift-drag ratio, which is 21.6. The ratio of take-off and initial cruise thrust is therefore approximately 6.6. As Table 12.2 also shows this is close to the value for $F_N/F_{ND} = 5.97$ computed when the temperature ratio T_{04}/T_{02} at sea-level static was equal to the design value at cruise, $T_{04}/T_{02} = 6.11$. With modest increases in temperature ratio the take-off thrust can be achieved. This confirms the assertion in early chapters of the book that the engine with low fan pressure ratio is well balanced, since take-off thrust and cruise thrust are achievable with almost identical non-dimensional parameters in the engine. This implies that components can be near peak efficiency at both conditions and that excess thrust capacity (i.e. a bigger heavier engine) is not needed for one or other condition.

The results in the table bring home a key issue for the testing and development of engines with low fan pressure ratio: tests on a sea-level test bed cannot fully reproduce the conditions in flight. The engine can be tested with a larger bypass nozzle than that for cruise to bring the fan operating conditions to those it is designed for. The core, however, cannot be satisfactorily modified in this way. All the large engine companies now have flying test beds (large aircraft adapted to allow new engines to be tested at realistic conditions) and some tests are also carried out in special ground based facilities. One such facility is in the Arnold Engineering Development Center where large engines can be tested on the ground in what is a huge wind tunnel at representative cruise conditions, specifically the correct Mach number, pressure and temperature.

SUMMARY OF SECTION 12.5

This section has looked at the variation in engine parameters, like thrust, *sfc*, pressure ratios and bypass ratio, when turbine inlet temperature has been varied at constant cruise Mach number and when the engine has been operated on a static test bed. In Chapter 8 dynamic scaling of the engine was considered when the non-dimensional operating point was held constant; one way of specifying this constancy was that T_{04}/T_{02} was constant. In the present chapter T_{04}/T_{02} was found to be a key independent variable determining the operating condition. For the bypass engine with low fan pressure ratio the unchoked core propulsive nozzle means that the conditions inside the engine depend on p_{02}/p_a as well as on T_{04}/T_{02}. In line with Chapter 8, keeping T_{04}/T_{02} constant does not fix the engine operating point for current generation large engines for which the core nozzle is unchoked even at cruise.

The variation in specific fuel consumption has been considered by looking at the overall, propulsive and thermal efficiency variations. Reducing T_{04}/T_{02} has the effect of increasing bypass ratio and reducing jet velocity in both streams, thereby increasing propulsive efficiency. However, reducing T_{04}/T_{02} decreases thermal efficiency so the net effect is a small reduction in overall efficiency, giving a rise in *sfc*, together with a rapid fall in net thrust.

Operation of an engine with low fan pressure ratio (*fpr* $\approx$ 1.5) on a sea-level test bed at the same T_{04}/T_{02} as that for design at cruise will not achieve the same internal pressure and temperature ratios as those at cruise. Both the core and bypass nozzles are unchoked on a static test bed for such an engine. The largest discrepancy is in the pressure ratio for the LP turbine.

The predicted values of engine performance are highly dependent on the properties used for the gas (γ and c_p). The errors accumulate in the LP turbine and therefore become errors in the fan pressure ratio and the bypass ratio. Accurate estimates for bypass ratio require the effect of cooling air and combustor pressure loss to be included.

Exercises

Note: Exercise 12.13 essentially repeats calculations of Chapter 7 but using *polytropic* efficiency and using more realistic description of the gas properties.

12.13 a A 2-shaft high bypass engine is designed to have an overall pressure ratio of 45 at cruise ($M = 0.78$, 35,000 ft) with a fan bypass pressure ratio of 1.5 and a pressure ratio in the fan root and booster of 2.5 (the HP compressor ratio is 18). The turbine inlet temperature is 1500 K. Assume *polytropic* efficiency of 90% for the fan, compressors and turbines. Take $\gamma = 1.4$ for the unburned air (so $c_p = 1005\,\mathrm{J\,kg^{-1}K^{-1}}$) and $\gamma = 1.3$ for the combustion products (so $c_{pe} = 1244\,\mathrm{J\,kg^{-1}K^{-1}}$). Calculate T_{013}, T_{023}, T_{03} and T_{045}. Also find the pressure at HP turbine exit, p_{045}.
(**Ans:** $T_{013} = 279.0$ K, $T_{023} = 328.3$ K, $T_{03} = 821.6$ K, $T_{045} = 1101.4$ K, $p_{045} = 362.4$ kPa)

b If the pressure ratio across the LP turbine is 11.13, find the stagnation pressure and temperature entering the core propulsive nozzle and hence the core jet velocity. Confirm that this is equal to the bypass stream velocity.
(**Ans:** $p_{05} = 32.6$ kPa, $T_{05} = 667.7$ K, $V_j = 340$ m/s)

c Find k_{HP} (as in Equation 12.21) and k_t (as in Equation 12.26). For unit mass flow rate through the core also find the HP and LP turbine nozzle areas, the core propulsive nozzle area and bypass nozzle area.
(**Ans:** $k_{HP} = 0.266$, $k_t = 2.46$, $A_{HPT} = 0.613$, $A_{LPT} = 2.32$, $A_9 = 22.0$, $A_{19} = 104.1\ \mathrm{m^2\,kg^{-1}s}$)

12.14 a For the engine of Ex 12.13 at $M = 0.78$ and 35,000 ft the turbine entry temperature is increased to 1600 K at same altitude and Mach number. Use k_{HP} to find the temperature out of the HP turbine and hence the pressure ratio across the HP turbine and the HP compressor. Show that if the fan pressure ratio is 1.60 and the pressure ratio across the LP turbine is 11.9 the constraints for mass flow rate across the

LP turbine nozzle and propulsive nozzle are satisfied. (First use the LP turbine pressure ratio to find the turbine work.

b Use the pressure ratio in the fan and booster, which is found with k_t, to get the overall pressure ratio. Hence find the pressure and temperature into the core propulsive nozzle and then the core jet velocity. Also calculate the core mass flow as a ratio of the mass flow rate at design point, m_{rc}.

(**Ans:** $T_{045} = 1174.8$, $pr_{HPT} = 4.42$, $pr_{HPc} = 18.77$, $p_{05} = 36.38$ kPa, $T_{05} = 702.6$ K, $V_{jc} = 403$ m/s, $m_{rc} = 1.154$)

c Calculate the bypass ratio and the bypass jet velocity and the ratio of bypass mass flow to that at design point, m_{rb}.   (**Ans:** $bpr = 12.3$, $V_{jb} = 355.4$ m/s, $m_{rb} = 1.056$)

d Calculate the net thrust for the whole engine as a ratio of the net thrust for the design condition in Exercise 12.13.   (**Ans:** $F_N/F_{Nd} = 1.247$)

12.15 For the two-shaft high bypass engine discussed in the previous section, show that the design point pressure ratio of the core compressor, 18.0 would intersect a constant speed line at about 96% speed for the E^3 compressor shown in Figure 11.6, close to the marked working line. Take the 96% speedline to be the design speed for the HP compressor of the new engine, occurring when $T_{04}/T_{02} = 6.11$. Take combinations of pressure ratio and normalised mass flow at the two ends of the HP compressor working line in Figure 12.11 (for $T_{04}/T_{02} = 4.5$ and 6.5) and plot them on Figure 11.6. Draw a straight line between these points to indicate the approximate working line of the new engine – does the simple model give a working line of an engine close to that drawn by GE?

Use the speed lines to estimate the HP speed when $T_{04}/T_{02} = 4.5$ as a percentage of that speed at the engine design point.   (**Ans:** For $T_{04}/T_{02} = 4.5$, $N_{HP}/\sqrt{T_{02}} \approx 96\%$ of value at design point)

12.6 A THREE-SHAFT HIGH BYPASS TURBOFAN ENGINE

The treatment of the three-shaft engine is similar to that of the two-shaft engine, though there is some increase in complexity, mainly in nomenclature. The way two- and three-shaft engines behave off-design is rather different, and this is discussed towards the end of the section.

With the configuration illustrated schematically in Figure 12.7c, the LP spool consists of only the fan on one end and the LP turbine on the other. The overall pressure ratio is the same as that considered for the two-shaft engine, namely 45. The combined intermediate pressure (IP) and high pressure (HP) compressor therefore together produce a pressure ratio of 30. This pressure ratio is split to be 7.5 in the IP and 4 in the HP compressor, which gives similar temperature rise in the two compressors and therefore similar power per unit mass flow from the IP and HP turbines. (This is the subject of Exercise 12.16.) We will assume that the stagnation pressure and temperature are uniform in the radial direction downstream of the fan and are equal for the core and bypass streams; conditions at this station will be denoted by 13 from hub to casing with $p_{013}/p_{02} = 1.5$ at design point. The conditions for the core flow out of the IP compressor and into the HP are denoted by 23. For the turbine the numbering system is, as before for the two-shaft engine: 4 into the HP turbine, 45 into the LP turbine and 5 out of the LP turbine. The conditions leaving the HP turbine and entering the IP turbine are denoted by 45′.

The LP turbine is choked at inlet so both the HP and IP turbines each operate between choked nozzles. The pressure ratio and temperature ratios across the HP and IP turbines therefore depend only on area ratios, as described for the two-shaft turbojet and the power balance on these shafts will be analogous to the power balance on the HP shaft of the 2-shaft engine, Equation 12.21. Thus for the HP shaft

$$\text{for the HP shaft} \quad c_p (T_{03} - T_{023}) = c_{pe} (T_{04} T_{045'}) = k_{HP} T_{04}$$
$$\text{and for the IP shaft} \quad c_p (T_{023} - T_{013}) = c_{pe} (T_{045'} - T_{045}) = k_{IP} T_{04},$$

(12.28)

where k_{HP} and k_{IP} are found at design conditions.

The LP turbine, like the LP turbine of the 2-shaft engine, has a choked nozzle at inlet and the unchoked propulsive nozzle downstream of it, so the treatment is similar. Similarly Equation 12.24 for the mass flow through the bypass is the same, with the bypass nozzle choked at cruise. As for the two-shaft engine, Equation 12.27 for the power balance in the LP shaft must be solved numerically to find fpr and T_{013}/T_{02}, the temperature ratio across the fan; with the approximation adopted here for the fan in the 3-shaft engine, k_t in Equation 12.26 is equal to unity.

In the off-design treatment of the three-shaft engine the overall performance variables, such as the variation in thrust, sfc, bypass ratio and overall pressure ratio are essentially identical to those for the two-shaft engines. This is because both engines were set up with the same design pressure ratios and equal and *constant* polytropic efficiencies. (If the variation in component efficiency with off-design operation is included differences do show up between two- and three-shaft engines.) These overall results, e.g. thrust and sfc, are therefore not repeated for the three-shaft engine. Also the fan working line in the bypass stream is essentially identical for the two- and three-shaft engines and is not therefore shown again here. The difference occurs in the working lines (i.e. pressure ratio–mass flow variations) for the core compressors. In place of the curve for the HP compressor in Figure 12.11 for the two-shaft engine is Figure 12.13 for the three-shaft engine at cruise Mach number, with the same range in temperature ratio T_{04}/T_{02} from 4.48 to 6.52. Both the IP and HP working lines are shown together in Figure 12.13, with the corrected mass flow normalised by the value at the design point. It is immediately apparent that the range of operating conditions is very much greater for the IP than the HP. It may also be noticed that the excursion in corrected mass flow for the IP is very similar to that for the HP compressor of the two-shaft engine in Figure 12.11 and would be identical if the booster stages were part of the HP compressor for the two-shaft engine but the overall pressure ratio maintained. This is a direct consequence of the mass flow being determined at HP turbine inlet and the overall core pressure ratio being involved in the specification of the core non-dimensional mass flow at inlet.

Although the reduced non-dimensional mass flow at core compressor entry for low values of T_{04}/T_{02} is a consequence of the overall pressure ratio and is similar for two- and three-shaft engines, the way the compressors are forced to behave is rather different. The working line for the IP compressor tends to be rather shallow, so that at low non-dimensional mass flow rates the pressure ratio is comparatively high. The rotational speed is largely determined by the compressor responding to the magnitude of the mass flow so, as the non-dimensional mass flow rate is reduced, the IP compressor slows down to accommodate this. The power output from the IP turbine remains

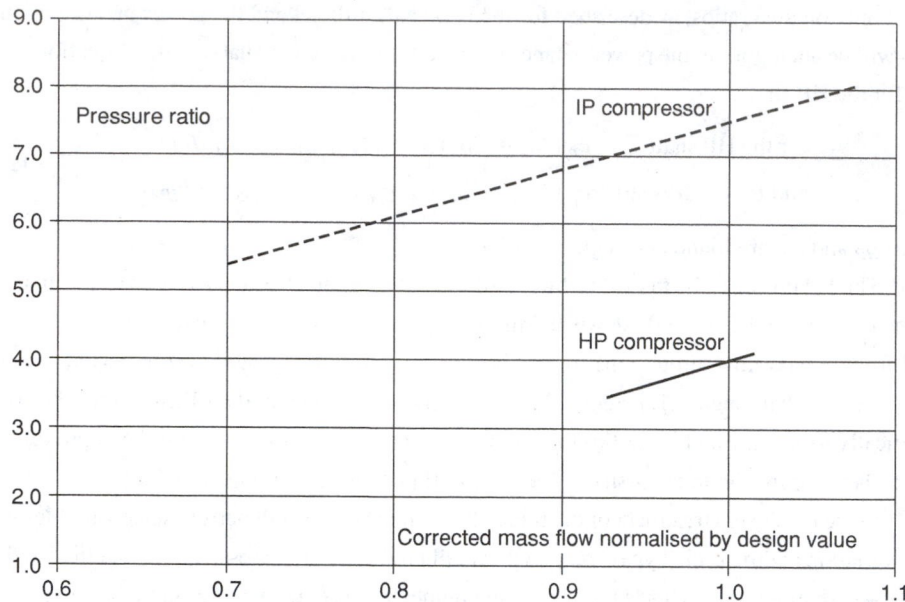

Figure 12.13. The working lines for the IP compressor and HP compressor for a three-shaft engine with a design point $opr = 45$, $fpr = 1.5$, cruise Mach number 0.78. Left end $T_{04}/T_{02} = 4.48$, right end $T_{04}/T_{02} = 6.52$.

relatively large, which causes the pressure rise in the IP compressor to be relatively high, and the working line 'moves up' the constant speed line to higher pressure ratio to achieve this. The problem for the HP compressor of the two-shaft engine is the high incidence in the front stages at low values of T_{04}/T_{02} consequent on the large reduction in non-dimensional mass flow at inlet and relatively small reduction in rotational speed. The corresponding issue in the 3-shaft engine is high incidence and a tendency to stall in the front of the IP compressors. For 2- and 3-shaft engines the issues are similar; although the whole compressor may be well removed from the stall line the front stages may be approaching stall and the mitigation is variable stagger stator blades for the front stages.

Exercises

12.16 Suppose each spool of a three-shaft engine has a single stage of turbine. It is decided that the temperature drop in each turbine should be equal; in consequence the temperature rises in the IP and HP compressors are equal. If r_i is the temperature rise in the IP compressor and r_h the temperature rise in the HP compressor show that the condition for equal temperature rise in each compressors, is given by $r_h - 1 = (r_i - 1)/r_i$. If the combined core compression ratio in the IP and HP compressors is to be 30, find the two pressure ratios. Take the polytropic efficiency to be 0.90. Calculate the isentropic efficiency for each compressor, see Exercise 11.4.   **(Ans:** $r_i = 8.49$, $\eta_i = 0.867$; $r_h = 3.53$, $\eta_h = 0.881$)

12.17 From the HP working lines in Figure 12.13 determine the pressure ratios and normalised mass flows for $T_{04}/T_{02} = 4.48$ and 6.52. Plot the HP values on the compressor map in Figure 11.5 and connect them with straight lines as an approximate working lines.

Estimate the reduction in speed for the HP spool compressor when T_{04}/T_{02} is reduced from the design point value of 6.11 down to 4.5. Do any potential difficulties arise?   (**Ans:** $N_{HP}/\sqrt{T_{023}} = 93.5\%$)

Note: the design pressure ratio of the IP compressor, 7.5, is significantly higher than the values shown on Figure 11.5, so there is no possibility of an exercise for the IP similar to what can be done for the HP.

SUMMARY OF CHAPTER 12

The behaviour of multi-shaft engines can be determined easily, if approximately, by using the particular flow features of the turbine and nozzle. Each of the turbines behaves at the inlet to it as if it were choked. In addition the efficiency of each turbine is well approximated as constant over a wide range of speeds and pressure ratios. When a turbine is between two choked conditions, such as the HP turbine in the high bypass engines or the turbines of a turbojet, the non-dimensional mass flow through each choked section is equal and then the ratio of flow areas into and downstream of the turbine determine its pressure ratio and temperature ratio. With the ratios fixed the power output per unit mass flow rate is known and is proportional to T_{04}, the temperature at entry to the HP turbine. Since the power from each turbine is the power into the compressor on the same shaft, the temperature rise in each compressor is also proportional to T_{04}. Assuming constant polytropic efficiency, the pressure ratios in the compressors can then be found.

For most operating conditions the mass flow through the core in an engine is fixed by the choking of the HP turbine nozzle guide vanes and the non-dimensional mass flow is therefore constant at this location. The condition at turbine entry determines the non-dimensional mass flow at the front of the HP compressor (or, in the case of three-shaft engine, the IP compressor). It is not essential for nozzles to be choked for a solution to be obtained off-design, but with choked nozzles approximate solutions are possible without the need for an iterative method of solution. For some engines, such as bypass engines with low fan pressure ratio, the core propulsive nozzle is not choked even at cruise. The area of the core propulsive nozzle again plays a key role in determining the pressure and temperature ratio across the LP turbine, with an iterative solution needed to adjust LP turbine pressure ratio to find the mass flow through the propulsive nozzle that is equal to the mass flow rate through the HP turbine.

Neglecting pressure drop in the combustor, the non-dimensional core mass flow into the engine is proportional to $(p_{03}/p_{02})\sqrt{(T_{02}/T_{04})}$. Since at compressor inlet the pressure ratio varies much more rapidly than the square root of temperature ratio the non-dimensional mass flow falls rapidly as the temperature into the turbine falls. This leads to high incidence and possible stall of the front stages of the compressor and it is principally to reduce the extent of this that the multi-shaft configuration is used.

The present analysis defines the working lines for the compressor and engine. It does not in itself fix the rotational speed; nor does it determine whether the compressors would allow the engine to operate at the condition chosen. When the working line is superimposed on the compressor map with constant speed lines of pressure ratio versus non-dimensional mass flow the shaft rotational speed at any value of T_{04}/T_{02} can be obtained. If the working line takes the compressor into a region where the efficiency is low, a recalculation is necessary. If the working line extends above and to the left of the surge line, operation is not possible without a redesign of the compressor.

The approach described in this chapter is designed to allow *simple* calculations which will illustrate the way in which engines operate off-design. The most important simplifications are holding the component efficiencies constant and assuming all the turbines remain effectively choked at inlet. Simplifications which can be removed with increase in complexity are the neglect of the turbine cooling air and the pressure drop across the combustor. Improved gas properties can also be introduced. We have also ignored power off-take (electrical and mechanical) and air bleed in addition to that used for turbine cooling. It is not conceptually difficult to include these and cooling air is included in the treatment of combat engines in Chapters 16 and 17 of Part 3 of this book as well. In Chapter 19 all of these effects are introduced for the civil engine so that the errors can be assessed. To repeat, the purpose of the treatment in Chapter 12 is *not* to provide the most accurate estimate possible but to demonstrate principles and constraints.

Part 3

Design of Engines for a

New Fighter Aircraft

CHAPTER 13   A NEW FIGHTER AIRCRAFT

13.0 INTRODUCTION

This part of the book begins the consideration of the engine requirements of a new combat aircraft. In parallel with the treatment in earlier chapters for the engines of the New Efficient Aircraft, the approach chosen is to address the design of engines for a possible new aircraft so that the text and exercises can be numerically based with realistic values. The specifications for the New Fighter Aircraft used here have a marked similarity to those available for the new Eurofighter or Typhoon.

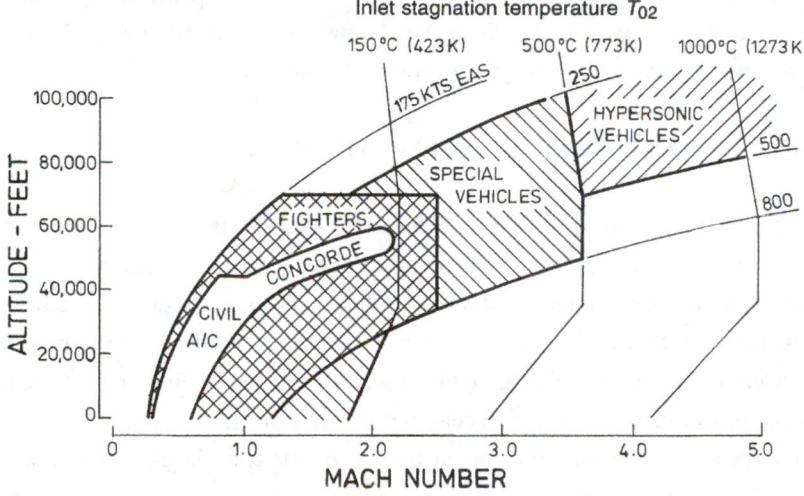

Figure 13.1. Aircraft operating envelopes, altitude versus Mach number, with contours of inlet stagnation temperature. (From Denning and Mitchell, 1989.)

The topic of the present chapter is the nature of the combat missions and the type of aircraft involved. Figure 13.1 shows the different regions in which aircraft operate in terms of altitude versus Mach number, with the lines of constant inlet stagnation temperature overlaid. We are concerned here with what are referred to in the figure as fighters, a major class of combat aircraft. Figure 13.1 shows the various boundaries for normal operation. Even high-speed planes do not normally fly at more than $M = 1.2$ at sea level because in the high density air the structural loads on the aircraft and the physiological effects on the crew become too large. At high altitude high-speed aircraft do not normally exceed $M \approx 2.3$, largely because the very high stagnation temperatures preclude the use of aluminium alloys without cooling. The boundary to the left in Figure 13.1 corresponds to the aircraft having insufficient speed to create the necessary lift; this is the stalling boundary and can be expressed in terms of the equivalent airspeed, that is the airspeed at sea level having the same value of $\frac{1}{2}\rho V^2$. The stalling boundary flattens at very high altitude to give the operational ceiling. Some special aircraft operate outside these boundaries but they will not be considered here.

For recent civil transport aircraft at subsonic conditions the lift-drag ratio is around 21. For high-speed aircraft the lift-drag ratio is lower and a value of between 6 and 7 seems realistic. This shifts the optimum type of engine from a relatively heavy one with low fuel consumption for civil transport to one which is lighter but with higher fuel consumption. In any case the requirements in a fighter for acceleration, turning and high speed demand a much higher thrust-weight ratio engine than is to be expected for transport aircraft. Engine weight is very important. An increase in the weight of one component, for example the engine, requires changes in other parts of the aircraft, most obviously wing area and structural strength of the airframe, and this adds additional aircraft weight. A rule of thumb is that growth in the aircraft will take place such that a 1 kg increase in the weight of any component leads to between 4 and 5 kg increase in overall aircraft weight for a subsonic aircraft and between 6 and 10 kg increase for an aircraft capable of supersonic flight. In other words, 1 kg of extra weight in each engine of a two-engine aircraft able to fly at supersonic speeds would increase the aircraft weight by between 12 and 20 kg.

13.1 THE TYPES AND NEEDS OF COMBAT AIRCRAFT

The civil transport has a comparatively easily defined task – to carry the largest payload between airports at the lowest cost. Although the precise optimum depends on the distance between the airports and the fraction of the maximum payload which is likely to be used, the duty is relatively straightforward. In contrast the task of a combat aircraft is normally complicated with the performance classified under three headings: field requirements (for take-off and landing), mission requirements (where the performance, in particular fuel consumption, over the whole mission must be adequate) and point requirements (to enable acceleration, turning, etc., to take place). Indeed, if an aircraft is recognisably deficient in one area, an opponent can immediately seek out this vulnerability.

Different aircraft are designed for different duties, but pressures to reduce cost often lead to large overlaps. Sometimes there is a decision to design an aircraft combining several disparate missions from the start. A clear example of this was the European Multi-Role-Combat-Aircraft, later called the Tornado, which had quite different requirements for the different countries taking part in the programme. Although there were variations between the different classes of Tornado, the same configuration (and, in particular, the same engines) had to be able to operate as a high-altitude interceptor or as a low-altitude ground attack aircraft. The most interesting aircraft with multiple roles at the moment is the American F-35 Lightning II, being built by Lockheed-Martin. This aircraft first flew in 2006 and is scheduled to be introduced to service initially with the US Air Force in a conventional form in 2016. Other versions are suitable for carrier operation or for short and vertical take-off and landing. This aircraft is therefore designed to cover a wide range of different tasks. A key feature of the F-35 is the attention given to reduced observability, which involves removing features giving strong radar reflections, coating the surface with non-reflecting material and minimising the infra-red signal.

Some of the classifications of combat aircraft and their attributes are summarised below; they can be divided into two broad classes. In the first (air superiority and interception) the aircraft is involved in combat with other aircraft, in the second (battlefield interdiction and close air support) the idea is to use the aircraft to attack ground targets. The situation becomes more confused when aircraft designed with one clear primary mission are adapted for a totally different one. An example of this might be the F-15, which was designed primarily as a high-speed, air-superiority fighter, but is also equipped to carry bombs in a secondary attack role.

Air superiority

This is the role of the traditional fighter aircraft. Manoeuvrability is the key performance issue and speed now seems to be held to be less crucial – it is believed that most fighting would take place in the subsonic range between $M \approx 0.7$ and 0.9, though there is no recent experience of real combat to base this on. At the highest speeds that many aircraft are capable of, around $M = 2$, the fuel consumption is so large that the duration of combat would necessarily be very short; in addition the turning circle is so high that conventional fighting may be impossible. Examples of air-superiority aircraft are the US Air Force F-15 and F-16, the US Navy F-14 and F-18, the Russian Mg-29 and Su-27, the French Mirage 2000. The new[1] F-22, the French Rafaele and the European Typhoon are of this class. The F-22 is different in having a much greater degree of low-observable technology.

Manoeuvrability is the ability of the aircraft to be able to accelerate and turn very quickly. To be able to turn quickly means that there must be a relatively large wing area for the mass of aircraft. The area of wing is usually expressed in terms of the wing loading, the load per unit area of wing when the lift is equal to the normal maximum weight of the aircraft. The normal weight of the aircraft corresponds to 1 g loading, in other words, weight $= g \times$ mass. In many manoeuvres the load on the wings is much greater than this, perhaps going as high as 9g, (referred to as a *load factor* of 9) which is about the upper level which the human body can withstand. A low wing loading corresponds to large wings for the normal weight of the aircraft; large wings allow a very large lift force to be created before the wings stall, which allows rapid turns with a comparatively small radius of curvature. Large wings, however, create more drag and can, by their inertia, make the aircraft less responsive in roll. They can also make the ride exceedingly bumpy in low-altitude, high-speed flight.

---

[1] The term 'new' takes on a rather special meaning here. For example, the need for what has become the F-22 was formally identified in 1981. In 1986 two companies were selected by the USAF to build prototypes and in 1991 the Air Force selected the Lockheed F-22 with the P&W F119 engine, awarding a $9.55 Billion contract for engineering and manufacturing development. The aircraft entered service with USAF in 2006. Eurofighter Typhoon came from a proposal in 1979, first flew in 1994 and entered service in 2003.

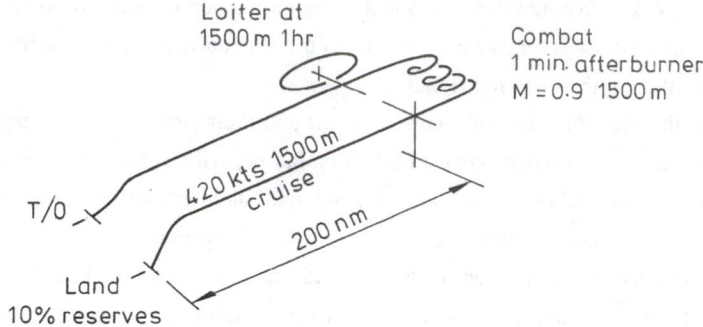

Figure 13.2. A simple, idealised mission for an air-superiority aircraft.
(Derived from Garwood, Round and Hodges, 1995.)

In specifying the optimum aircraft it is normal to calculate the performance over idealised typical missions. Such a mission is shown in Figure 13.2 for an air-superiority aircraft where a significant period of loitering is involved; loitering involves remaining in an area for relatively long periods using as little fuel as possible. It should be noted that whereas the whole mission in Figure 13.2 lasts about two hours, the combat portion is only one minute. Fuel consumption during the cruise and loiter period can therefore be far more influential to the total aircraft weight at take-off, and therefore its size and cost, than the fuel used during the brief period of combat.

Both rapid turns and high acceleration in the direction of travel require a high ratio of engine thrust to aircraft weight. During cruise and loiter phases of a mission the thrust will be much less than the maximum used during combat and the principal requirement then is low fuel consumption. Some of this variation in thrust is achieved by the use of the afterburner, sometimes also called reheat, which is common to most fighter aircraft. In afterburning operation fuel is burned in the jet pipe, downstream of the turbine but upstream of the propulsive nozzle, to give a boost in thrust of the order of 50%. When the afterburner is not being used the operation is referred to as 'dry'. The use of the afterburner increases fuel consumption several-fold and gives a huge infra-red signature which makes the aircraft vulnerable to heat-seeking missiles. One of the key areas of interest for engines is now the variable-cycle engine which can adapt to the different thrust requirements without the use of the afterburner.

The balance which determines the design of the aircraft and its engines depends on geo-political factors as well as technical issues. An important issue relates specifically to fuel consumption whilst not engaged in combat. During the cold war, for example, the United States needed to be able to ferry its combat aircraft over the Atlantic to meet a potential threat in Europe and, even with in-flight refuelling, the fuel consumption at the optimum condition for range was very critical. The Soviet Union, on the other hand, did not have the requirement to ferry its planes long distances over water and could give less attention to the range and duration of flight.

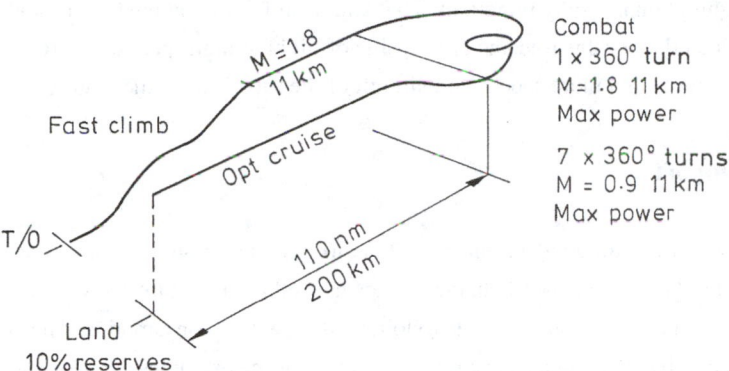

Figure 13.3. A simple, idealised mission for an interceptor aircraft.
(Derived from Garwood, Round and Hodges, 1995.)

Interception

This type of aircraft is part of an air defence system, with the aircraft despatched to intercept intruders detected by radar. When attack by bombers was perceived as a major strategic threat special interceptor aircraft were built, but the need for these special aircraft has receded somewhat and modern air-superiority fighters also usually serve as interceptors.

For the specialised interceptor long range is not normally required, but what is needed is high climb rate and speed, with manoeuvrability and fuel consumption being less important. Many of the aircraft specifically of this type are old, for example the British Lightning. The Russian MiG-25, with a maximum speed of Mach 3.3 at high altitude, is a good, if extreme, example of this type. Aircraft such as the American F-15, with its high maximum operational speed, and the F.3 version of the European Tornado are also able to operate as interceptors. The new American F-22 also has interception as one of its design missions.

Figure 13.3 shows a typified intercept mission with a supersonic dash at $M = 1.8$ over a considerable distance to the combat zone. Here the interception would be carried out beyond visual range (BVR) using missiles. There are advantages, including reduced fuel consumption, if this supersonic cruise can be carried out without use of the afterburner, a condition referred to as 'supercruise'. Some tight turns are envisaged as part of avoidance manoeuvres after the release of air-to-air missiles, and operation without the afterburner is highly desirable here too since heat-seeking missiles can more easily lock into the afterburner plume.

Interdiction

The proposed role here is to operate 50 km or more behind the battle front and attack roads, bridges, railways, radar installations, etc. The requirement for the aircraft and engine is transonic operation at low altitude, medium-to-long range and good manoeuvrability. In terms of the engine this means that fuel consumption is vitally important. The American F-111 and the F/A-18 and the European Tornado are intended for this role and are capable of flight at high speeds. The Russian MiG-27 is similar. An example of a more specialised aircraft of the type is the American A7.

Close air support

This category can be sub-divided, but the role is to operate within 50 km of the battle front to attack ground targets. The general characteristics needed are the ability to cruise at low level, to carry a substantial payload and to be able to loiter with low fuel consumption. The aircraft must be highly manoeuvrable, but range can be relatively short. Some examples are the American A10, the British/American Harrier AV8 and the Russian Yak-38. The US Marines intend using their F35s with vertical take-off and landing capability in this capacity.

A broader categorisation: versatility

It will be apparent that the range of combat missions which *might* be envisaged is wide and that an aircraft procured with one type of mission may be used quite differently. Clearly, when procurement may take something like 20 years between proposal through to an aircraft in proper service (as opposed to initial training duties), the use of the deployed aircraft may be significantly different from those originally envisaged. Two features which build in versatility are the ability to fly long ranges (not necessarily very fast, subsonic flight may suffice) and the ability to loiter (to fly around for a long time in a chosen vicinity ready to respond at short notice).

The threats which a combat aircraft need to counter have varied over the years. In particular, combat between roughly comparable fighter aircraft is now virtually unknown and, despite the specification of aircraft recently entering service, is not foreseen as likely in the future. The principle threat to an aircraft comes from missiles. As missiles have become more effective, a low radar signature and reduction of infra-red emissions have therefore become ever more important. The intake and exhaust of engines such as those of the previous generation generate strong radar reflections and the trend now is to replace straight ducts with serpentine ducts lined with material able to absorb radar signals. To reduce the infra-red radiation the nozzle is generally shielded from the ground and cooled using bypass air.

The uncertainty about the future, as well as the huge cost of most new aircraft,[2] encourages the versatility to carry out many sorts of task. Engines, as well as the aircraft, which are specifically optimised for the different tasks might be significantly dissimilar and effective compromise is essential in design. As noted above, a recent trend has been to design what are called variable-cycle engines to accommodate the large differences in requirements for different modes of operation

---

[2] For example the Eurofighter Typhoon has a unit price of about 90 million Euro and almost double this if development and production costs are included.

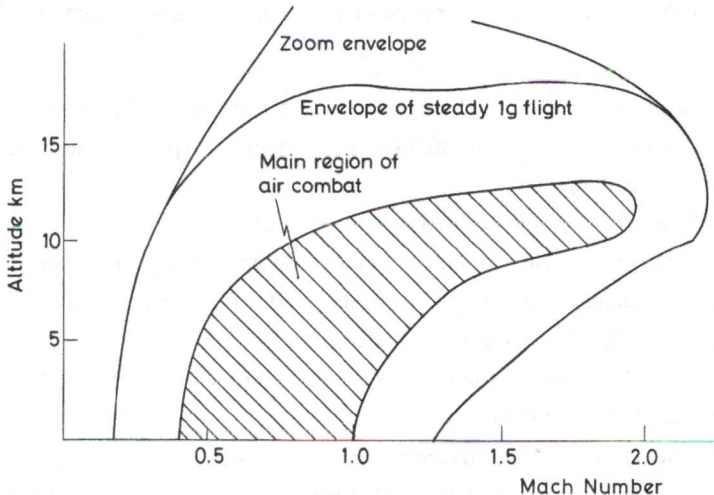

Figure 13.4. The flight envelope for a typical combat aircraft showing the main region envisaged for combat. (The zoom envelope represents the extreme altitude which can be attained taking full advantage of kinetic energy at lower altitude.)

corresponding to different missions or phases of mission. Versatility, supported by variable-cycle engines, with low observables is the way design is currently heading.

13.2 THE REQUIREMENTS FOR A NEW FIGHTER AIRCRAFT

To make the treatment more concrete, a hypothetical aircraft is considered for which the design decisions can be made. So far as the engines are concerned the most demanding role is for the air-superiority aircraft and the subject of the design is therefore a New Fighter Aircraft focussed on this. In fact it is because there are so many different performance requirements for an air-superiority aircraft that considerable adaptability is needed.

The entire flight envelope for an air-superiority and interception aircraft are sketched in Figure 13.4 with the main region envisaged for air combat shown shaded. Of particular note is that most combat is expected to take place at comparatively low speed, mainly below about $M = 1.5$, and mostly below an altitude of about 11 km (36,089 feet). This altitude is the tropopause in the International Standard Atmosphere, above which the temperature is assumed to remain constant at 216.7 K. (It may be noted that whereas for civil aviation feet are the normal units of altitude, in military aviation metres are widely used.) The temperature at altitude is approximately defined by the standard atmosphere, but at sea level the range can be enormous: perhaps 230 K (ISA–58°C) in the Arctic and 320 K (ISA+32°C) in some hot regions of the world. For the present, as in earlier sections of the book, the temperature at sea level will be taken to be the standard value of 288 K. The stagnation temperature into the engine T_{02} has a powerful effect on the behaviour and performance. Temperature T_{02} is not only determined by the ambient static, assumed here to be given by the standard atmosphere, but also by the flight Mach number.

Some of the possible important operational points for a new combat aircraft are set out below:

1. Low level combat and escape manoeuvres at Mach numbers between about 0.5 and 0.9 and altitudes below about 4 km. (This combination of speed and altitude requires very tight and rapid turns.)
2. Acceleration near sea level from around $M \approx 0.3$ to 1.2.
3. Subsonic cruise and loiter (with low fuel consumption) at both low and high altitude.
4. Medium-altitude combat around 6 km at $M \approx 0.7$ to 1.2, with acceleration being important in this altitude range from $M \approx 0.5$ to 0.9.
5. High-altitude combat in the range 9 km to 11 km at $M \approx 0.9$ to 1.6. Acceleration is important from $M \approx 0.9$ to 1.6.
6. Subsonic cruise over long distances and loiter at both low and high altitude.
7. Supercruise (without afterburner), M at least 1.5 and perhaps up to 1.8, altitude 11 km or above.
8. Supersonic steady level flight at $M \approx 2.0$ or higher, altitude 11 km or above.

Not every design of aircraft would be able to meet all the requirements, and judgement would be required to determine which were most important. Small changes to the same basic design might allow different versions to excel in some, but not all, of the possible missions. Only by trying a large number of different combinations of airframe and engine could anything like an optimum be said to have been chosen. The exercises in this book for a New Fighter Aircraft will concentrate on the two extremes for temperature of the standard atmosphere, which occur at sea level and at the tropopause – both are vitally important to a fighter aircraft and encompass many of the critical operating points. At the tropopause, operation will be considered at three Mach numbers (0.9, 1.5 and 2.0) whilst at sea level the static condition and $M = 0.9$ will be used.

A paramount additional requirement for most new combat aircraft is some measure of stealth and this has been referred to earlier. To minimise infra-red visibility it is desirable to avoid needing an afterburner and to have the hot final nozzle shielded from the ground by the wing or tail plane. Moreover, bypass air is used to cool the nozzle, which discourages the use of turbojet engines. The worst sources of radar reflections, common for older aircraft, are the engine inlet and exhaust. Frequently now the exhaust nozzle is rectangular, which makes it easier to shield it from below by the wing or tail. The intake is often curved (serpentine) and vanes may be used to absorb the incident radar wave. Surfaces of the intake, and the aircraft itself, may be covered in radar absorbing material. The intake is to some degree compromised aerodynamically and the fan has to be designed to cope with the non-uniform flow which it receives.

13.3 THE PROPOSED NEW FIGHTER AIRCRAFT

The hypothetical New Fighter Aircraft (NFA) provides an example, analogous to the New Efficient Aircraft (NEA) for civil application considered in the Chapters 1 to 10. Just as the New Efficient

Table 13.1   Major parameters for five existing fighter aircraft and a possible New Fighter Aircraft (NFA)

	F-15	F-16	Dassault Rafale	Eurofighter Typhoon	F-35	NFA
Number of engines	2	1	2	2	1	2
Date of first flight	1972	1976	1986	1994	2006	
Operating weight empty, W_e	12.7	8.6	10.3	11.0	13.2	13.3
Loaded weight, W_L	20.2	12.0	15.0	16.0	22.5	18.0
Nominal max. weapon weight	10.7	5.8			8.2	
Nominal max. internal fuel weight	6.1	3.2	4.7	4.5	8.3	
Engine weight, W_L	0.135	0.127	0.128	0.125	0.076	0.10
Max. dry thrust, W_L	0.66	0.62	0.72	0.76	0.57	0.66
Max. thrust with ab, W_L	1.07	1.00	1.10	1.15	0.87	1.00
Wing loading (for W_L) kN/m^2	3.6	4.2	3.0	3.06	5.16	3.5
Maximum Mach number, sea level	1.2	1.2	1.1	1.2		1.2
Maximum Mach number (at 40 kft)	2.5	2.05	1.8	2.0	1.6	2.0
Service ceiling (1000 ft)	65	60	55	55	60	55
Service radius, no external fuel tanks	1060	340	1000	750	584	500
Ferry range, with external fuel tanks	2880	2415	2000	2350	1200	2000
Maximum acceleration by wing lift	9g	+9g	−3g	9g	+9g	−3g

Notes. The afterburner, described in Chapter 14, is referred to by 'ab' and 'dry' refers to operation without the afterburner. The loaded take-off weight corresponds to the maximum load of fuel and weapons for a typical mission. All weights in metric tonne and ranges in nautical miles. Gaps in Table 13.1 correspond to information being unavailable for the existing aircraft or not yet decided for the New Fighter Aircraft.

Aircraft was presented as a step forward from existing large civil aircraft, any new combat aircraft would also take advantage of previous experience. In the present case it is convenient to use as precursors the American F-15 and F-16, both very successful aircraft designed and developed in the early 1970s. The F-15 has two engines, as does the Rafale and the Typhoon, whilst the F-16 and the F-35 have only one. Two engines are selected here for the NFA, but the design could be modified quite easily to be for one.

Table 13.1 gives the loaded take-off weight of the New Fighter Aircraft to be 18 tonne but the more extreme manoeuvres would not be carried out at this weight. For the sake of simplicity a single value of weight, 15 tonne, will be used for all the calculations (other than take-off which will be at 18 tonne) allowing for some fuel to have been burned in take-off, climb and loiter.

It is also helpful to relate some of the statistics to those of a current civil aircraft and as an example Table 13.2 compares the F-16 and the Boeing 787. The fraction of the weight which is fuel is similar. For the fighter the maximum thrust–weight ratio at take-off is about 0.66 without the afterburner in use and about 1.00 when the afterburner is used. The ratio is much larger than for civil aircraft; for the New Efficient Aircraft the take-off thrust was 0.3 times the maximum take-off weight. The fraction of the total weight due to the engines is

Table 13.2   Comparison of salient parameters for a current
large civil aircraft and current fighter aircraft. Maximum and
loaded take-off weights for civil and military aircraft.

	787–8	F-16
Take-off thrust/take-off weight	0.29	0.66 'dry'
		1.00 a/b
Total engine weight/take-off weight	0.047	0.13
Max. fuel weight/take-off weight	0.45	0.40
Maximum wing loading (kN/m^2)	6.9	4.2

about three times greater for the fighter, which puts particular pressure to increase the engine thrust-to-weight ratio for the fighter. The engine for the Eurofighter Typhoon has a thrust–weight ratio of about 10 with the afterburner whilst the engine for the F-35 Lightning has a ratio of about 11.5.

The fighter clearly has much greater thrust in relation to weight at off, but at high altitude the difference would be even more marked. The net thrust for the fighter engine decreases proportionately less with forward speed because of the high jet velocity from the low bypass ratio engine.

Exercises

13.1 For the New Fighter Aircraft use the data in Table 13.1 to find the wing area.   (**Ans:** 50.5 m^2)

During combat the aircraft mass may be assumed to be 15 tonne (3 tonne less than the maximum loaded take-off mass). Find the wing loading for this condition.   (**Ans:** 2917 N/m^2)

13.2 The Pratt & Whitney F100-PW-100 engine (powering the F-15 and F-16 aircraft) has an approximate maximum thrust–weight ratio on a sea-level test bed of 7.9 with the afterburner lit and 4.7 'dry'. The specific fuel consumption is about 2.6 kg h^{-1}kg^{-1} with the afterburner and about 0.69 when dry. Find the time required in each operating condition to burn a mass of fuel equal to the engine mass.
(**Ans:** with afterburner, 2.9 minutes; 'dry' 18.5 minutes)

It is perhaps relevant here to refer to one additional aspect of combat aircraft, the unmanned (sometimes called uninhabited) aircraft. At one extreme is the cruise missile, with a small gas turbine, and at the other an aircraft which might be expected to engage conventional fighters in aerial combat. In between there is a wide range of reconnaissance vehicles, from the very small up to the size of small manned aircraft. In terms of the aerodynamics and thermodynamics of the propulsion the task is essentially the same as that for manned aircraft. In terms of other aspects of the engine there may be major differences. Naturally, if the aircraft is to be destroyed in the mission (as for a cruise missile) a short life (of a few hours) suffices and low cost is important. If the aircraft is loaded with expensive reconnaissance equipment the aircraft and engine would expect to be used many times and long life would be important. Some of the other requirements may be more surprising. For manned combat aircraft in times of peace practice flying is the major use of the aircraft, but this would not be the case for an unmanned aircraft. For example an unmanned fighter would not need to be flown for practice by the pilot sitting at remote control station – the practice could presumably be provided by a simulator. The unmanned aircraft engine must therefore be capable of being stored for several years without deterioration and then to start reliably and quickly.

SUMMARY OF CHAPTER 13

The fighter aircraft is the topic of Chapters 13–18 and this chapter has described some features of this class of aircraft, comparing it with some other types of combat aircraft. The fighter is required to be manoeuvrable and it will be seen in Chapter 14 that to be able to make rapid turns requires a high level of thrust if altitude or speed are not to be lost. Fighter aircraft therefore must have a high thrust-weight ratio and the same is true of the engines themselves.

Although fighter aircraft must have a higher thrust-weight ratio than civil aircraft, which indicates a lower bypass ratio, this is not the feature which makes the aerothermal design of the engines so interesting. Because fighter aircraft are normally expected to be able to fly fast, the inlet temperature can rise well above the temperature experienced for a sea-level take-off. This, it will be shown, has a radical effect on the behaviour of the engine and therefore on its design.

CHAPTER 14 — LIFT, DRAG AND THE EFFECTS OF MANOEUVRING

14.0 INTRODUCTION

A fighter aircraft is required to be agile, which requires it to turn sharply, to accelerate rapidly and usually to travel fast. It is no surprise that accelerating rapidly or travelling fast require large amounts of thrust from the engine. What may be more of a surprise is that rapid changes in direction require high levels of engine thrust and achieving high turn rates is probably the most important reason for needing high thrust-weight ratios. The reason is that the drag of the aircraft rises approximately with the square of the lift coefficient and making rapid turns demands high lift from the wings. An aircraft normally banks in order to turn so that the resultant of the gravitation acceleration and the centripetal acceleration is normal to the plane of the wings (Figure 14.1), and the force they produce is exactly balanced by the wing lift. It is normal to express the increase in acceleration in terms of the *load factor*, denoted by n. A load factor of unity corresponds to an acceleration g perpendicular to the wing, when the lift is the normal weight of the aircraft, whereas a load factor of 5 corresponds to an acceleration of $5g$ and the lift is five times the weight. For a modern fighter aircraft the structure is designed to withstand the approximate limit on acceleration set by the human pilot and load factors can be as high as 9.

For the civil airliner the turns are normally so gentle that the lift on the wings is little more than the weight of the aircraft, and the size of the engine is normally fixed by requirements at the top of the climb. As will be shown below, for the fighter the size of the engine is more likely to be set by the maximum rate of turn required during some parts of the flight envelope.

The treatment of aircraft aerodynamics is necessarily very superficial here, but a much more complete treatment is given by Anderson (2011).

14.1 LIFT AND ACCELERATION

The lift force L is related to the dynamic pressure $\frac{1}{2}\rho V^2$ by the lift coefficient

$$C_L = \frac{L}{\frac{1}{2}\rho A V^2},\qquad(14.1)$$

where A is the wing area. For steady level flight the lift force is equal to the weight of the aircraft mg and at this condition mg/A defines the *wing loading*. Although the shape of the wings affects the maximum lift force which can be produced for a given flight speed and air density, the most

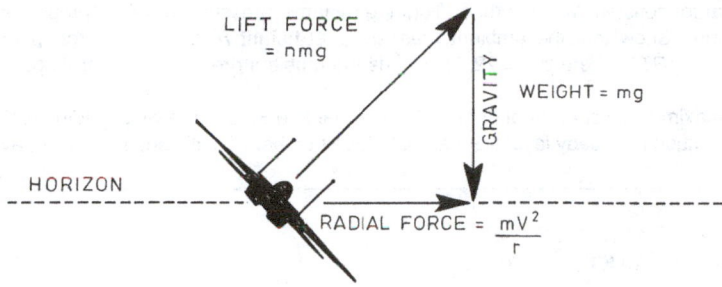

Figure 14.1. An aircraft making a horizontal banked turn.

important parameter in determining maximum lift is the area of the wings. The wing area is set by the aircraft weight and the selected wing loading – reducing the wing loading therefore increases the maximum lift which a wing can produce. Note from Table 13.2 how the wing loading for the F16 is substantially less than the value for the Boeing 787. The low wing loading, meaning large wings in relation to the aircraft weight, is needed to give high manoeuvrability. For the New Fighter Aircraft the wing loading is taken to be 3.5 kN/m².

For take-off and landing the lift coefficient is often raised by the use of flaps, slats or blowing, but for high speed flight this is less common. The upper limit on lift coefficient depends on Mach number, but is likely to vary little between different designs of fighter aircraft. For the present purpose a maximum value of lift coefficient of about 1.00 is realistic at $M = 0.7$, falling to about 0.4 at $M = 1.5$; for simplicity a linear dependence on Mach number will be assumed between these values.

The maximum acceleration normal to the flight direction, or load factor, that can be achieved is determined by the peak value of lift and this means that the maximum turning rate also depends on lift coefficient. The maximum turn rate that the wings can produce is known as the maximum *attained* turning rate and will normally cause the flight speed and/or altitude to decrease, for reasons described below. The engine performance has only a small effect on maximum attained turn rate, which is mainly fixed by the wing area and the dynamic pressure.

Exercises

14.1 The New Fighter Aircraft flies with a mass of 15 tonne. The wing loading is given in Exercise 13.1.

a For a constant-altitude 9g banked turn at sea level (ambient temperature 288 K) and a Mach number of 0.9, find the angle of bank and the radius of curvature. What is the lift coefficient?
(**Ans:** bank = 83.6°; radius = 1.07 km; $C_L = 0.457$)

b* Find the radius of curvature and lift coefficient for 3g turns at Mach numbers of 0.9, 1.5 and 2.0 at the tropopause (11 km altitude; $T = 216.65$ K, $\rho = 0.365$ kg/m³).
(**Ans:** radius = 2.54, 7.06, 12.55 km; $C_L = 0.680, 0.245, 0.134$)

c If the lift coefficient is not to exceed 0.85, find the highest turning acceleration possible flying at $M = 0.9$ at the tropopause. If the maximum lift coefficient is 1.00, find the lowest Mach number at which a 9g turn is possible close to sea level.
(**Ans:** 3.76g; $M_{min} = 0.609$)

14.2 Show that for constant Mach number flight the dynamic pressure $\frac{1}{2}\rho V^2$ is proportional to the local ambient pressure. Show that the ambient pressure p at height h_T above the tropopause is given by $p/p_T = \exp\{-gh_T/RT\}$, where $p_T = 22.7$ kN/m^2 is the ambient pressure at the tropopause and h_T is in metres.

If the maximum lift coefficient at $M = 0.9$ is equal to 0.85 and the wing loading is 2917 N/m^2, find the maximum altitude for steady level flight at that Mach number (if sufficient thrust were available).

(Ans: 19.4 km)

14.2 DRAG AND LIFT

For the civil transport aircraft treated in Chapter 2 consideration was only given to the case when the lift was equal to the weight of the aircraft. In other words the case treated was when the load factor was taken to be equal to unity and any effect of turning was neglected. At this condition the drag was obtained from a fixed lift-drag ratio. For combat aircraft, however, the load factor is frequently sufficiently high that the effect of lift on drag must be included.

The drag coefficient C_D is defined by

$$C_D = \frac{D}{\frac{1}{2}\rho A V^2},\tag{14.2}$$

which can be written, with some simplification, as

$$C_D = C_{D0}(M) + k(M)C_L^2,\tag{14.3}$$

where C_L is the lift coefficient and C_{D0} is the drag coefficient at zero lift. Both C_{D0} and the parameter k are dependent on flight Mach number as well as the aircraft geometry. (For combat aircraft the minimum drag corresponds fairly closely with the drag at zero lift, which is not the case for transport aircraft.) C_{D0} is altered by the configuration of the aircraft, and is much increased by weapons or extra fuel tanks attached to the outside of the aircraft. For combat the extra fuel tanks would normally be dropped, even if partly full. The exact values for C_{D0} and k for specific aircraft are normally secret, but Figure 14.2 shows approximate[1] values which serve to demonstrate how drag is affected by aircraft Mach number and lift coefficient. These results are displayed as *drag polars* in Figure 14.3a and for comparison some measured results for the F-16 are shown in Figure 14.3b.

At subsonic flight speeds the lift-dependent drag, $k(M)C_L^2$, is usually referred to as the induced drag and is produced by trailing vortices behind the wing. These are revealed in Figure 14.4 which shows a Tornado making a steeply banked turn, producing high lift on the wings and creating strong trailing vortices. At the core of the vortices the pressure and temperature are low enough for water vapour present in the air to condense and to reveal their presence. The strength of the vortices, and therefore the velocity associated them, is linearly proportional to the lift coefficient. The induced velocity requires the wing incidence relative to the direction of motion to be increased

---

[1] Figure 14.2 was originally compiled by Fred Jonas of the US Air Force Academy and incorporated in the book by Mattingley, Heiser and Daley (1987). These plots are used in the solution of exercises in Chapter 14. It is hard when reading numbers from such diagrams to obtain better accuracy than two significant figures. The answers to the exercises are nevertheless given to three significant figures to assist consistency, to show changes and to help in checking.

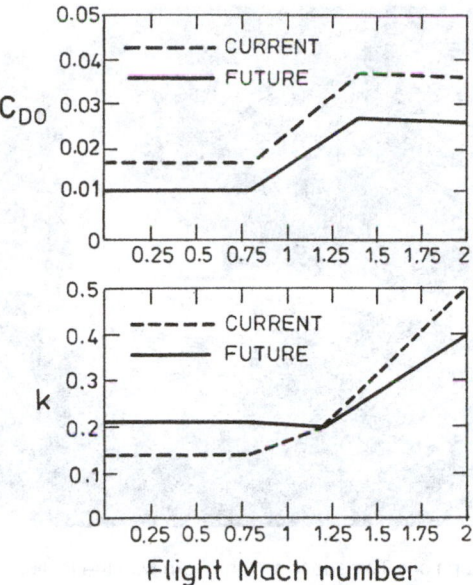

Figure 14.2. Drag coefficient for zero lift and lift-dependent drag factor versus Mach number for fighter aircraft. (Derived from Mattingley, Heiser and Daley, 1987.)

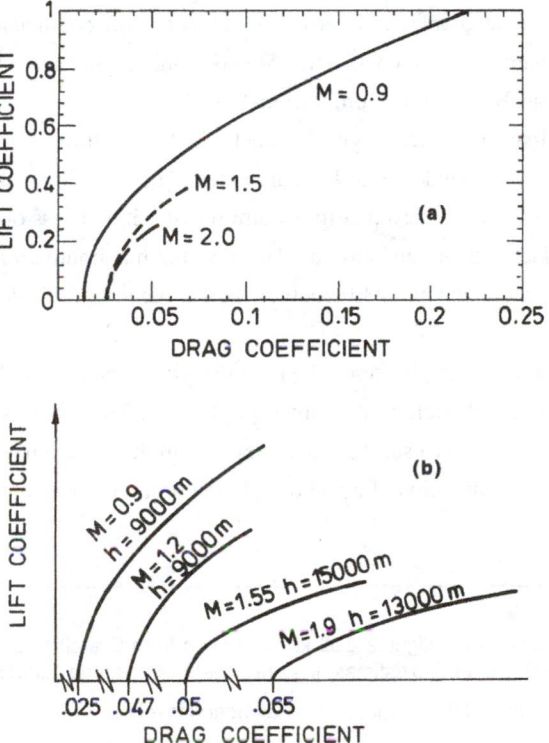

Figure 14.3. Drag polars, showing lift plotted versus drag for different Mach numbers (a) derived from the curves in Figure 14.2; (b) measured results for F-16 (from Buckner and Webb, 1974).

Figure 14.4. A photograph of a Tornado in a high-g turn with the trailing vortices made visible by condensation of water vapour. (Reproduced by permission of BAE Systems Heritage Dept – Warton.)

in order to maintain the lift force, inclining the wing backwards. The lift vector therefore has a component acting backwards parallel to the direction of travel. Since the backwards inclination of the lift due to the induced velocity is proportional to the lift coefficient, the induced drag is proportional to the square of the lift coefficient. At supersonic flight speeds the lift-dependent drag is principally wave drag, but again it is proportional to C_L^2.

Because the drag rises steeply with lift coefficient, high turn rates require high levels of engine thrust if speed and altitude are to be maintained. The maximum *sustained* turning rate is the highest rate which can be achieved at maximum thrust without loss of speed or altitude; it is a function of flight Mach number and altitude. Whereas the maximum *attained* turn rate depends mainly on the wing area and maximum lift coefficient, the maximum *sustained* rate is engine thrust dependent.

We are neglecting a complication when an aircraft operates at high incidence. The jet is then directed at a substantial angle to the direction of flight. The thrust is then no longer in the direction of the drag and there is a substantial component in the direction of the lift. Although this would be important for specification of a real design, the omission does not affect the conclusions reached here.

Exercises

14.3 If the wing loading in steady flight is 2.92 kN/m² (taken from Exercise 13.1) use the future-aircraft curves for C_{D0} and k in Figure 14.2 to estimate the drag coefficients at the following conditions:

a 11 km altitude and $M = 0.9$, 1.5 and 2.0 in straight and level (i.e. 1g or $n = 1$) steady flight;

b sea level at $M = 0.9$ straight and level (i.e. 1g) and also for 5g and 9g turns;

c 11 km at $M = 0.9$ and 1.5 for 3g turns.

(**Ans:** a) 0.0246, 0.0289, 0.0268; b) 0.0145, 0.0275, 0.0577; c) 0.110, 0.0435)

14.4 Use the results of Exercise 14.3 to estimate the thrust needed from *each* of the two engines of the New Fighter Aircraft (with a mass of 15 tonne) for the following conditions:

a* 11 km altitude and $M = 0.9$, 1.5 and 2.0 in straight and level (i.e. 1g) steady flight;

b* sea level at $M = 0.9$ straight and level (i.e. 1g) and also *sustained* 5g and 9g turns;

c* 11 km at $M = 0.9$ and 1.5 for *sustained* 3g turns;

(**Ans:** a) 8.0 kN, 26.0 kN, 44.3 kN; b) 21.0 kN, 39.8 kN, 83.5 kN; c) 35.8 kN, 39.3 kN)

14.5 a For the New Fighter Aircraft flying steady and level ($n = 1$) at the tropopause at $M = 1.5$ find the lift-drag ratio, L/D using Exercise 14.3a. (**Ans:** $L/D = 2.83$)

b Show that in steady level flight the required thrust from the engines is a minimum if the net thrust is directed downwards at an angle θ to the horizontal, where $\tan\theta = (L/D)^{-1}$, where L/D is assumed constant.

c Find the value of θ for minimum net thrust and the magnitude of net thrust from each engine. Then, if the jet velocity relative to the aircraft is 1100 m/s, determine the angle β at which the jet should be directed downward for minimum engine thrust. (**Ans:** $\theta = 19.4°$, $F_N = 24.5$ kN, $\beta = 11.4°$)

14.3 ENERGY AND SPECIFIC EXCESS POWER

The *energy state* is the combined potential and kinetic energy of the aircraft, mass m, given by

$$E = m\left(gh + V^2/2\right),$$

with the specific energy

$$E_S = gh + V^2/2. \tag{14.4}$$

At constant energy state it is possible to trade kinetic energy for potential energy, so possessing a high energy state gives a combat aircraft great initiative. Neglecting any difference between engine thrust and aircraft drag (i.e. maintaining constant energy state) an aircraft travelling at $M = 0.9$ at sea level ($V = 306$ m/s) can 'zoom' to an altitude of 4777 m, at which point it would have no forward speed. More significantly, an aircraft at an altitude of, say, 15 km at $M = 0.9$, for which $V = 266$ m/s, could zoom to 18.6 km, beyond the altitude at which the engines produce enough thrust for it to fly continuously. Manoeuvres in which the aircraft drag exceeds the engine thrust, that is manoeuvres which exceed the sustainable turn rate, lead to a reduction in energy state. Conversely, if the thrust exceeds the drag the energy state is increasing.

The ability to accelerate or climb depends on the excess thrust over and above that required to balance the drag in steady level flight. The measure of this is the excess thrust divided by the aircraft weight mg, which can be written

$$(F_N - D)/mg,$$

where F_N is the maximum net engine thrust and D is the aircraft drag. (Both engine thrust and drag are functions of altitude and speed.) Conventionally the above quantity is multiplied by the

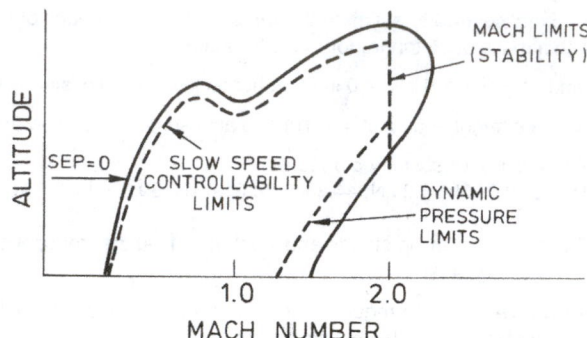

Figure 14.5. A schematic plot of altitude versus Mach number showing the line $SEP = 0$ for a combat aircraft. Possible operating limits for steady flight are superimposed. (Derived from Shaw, 1985.)

flight speed V to give *specific excess power, SEP*

$$SEP = V(F_N - D)/mg, \qquad (14.5)$$

with units of velocity. During manoeuvres, such as turning, the drag rises, as discussed in Section 14.3. If prior to the commencement of the turn the thrust equalled drag and during the turn the thrust stays constant the SEP will fall. The available maximum rate of climb at constant speed is equal to SEP, whereas the available rate of forward acceleration in level flight (i.e. increase in forward speed) is equal to the specific excess thrust $(SEP/V)\, g$. The specific excess power is related to the specific energy by

$$SEP = \frac{d\,(E_S/g)}{dt}, \qquad (14.6)$$

and $SEP = 0$ corresponds to flight at constant speed and constant altitude with the engine producing maximum thrust. Figure 14.5 shows a schematic plot of the $SEP = 0$ curve for a representative aircraft at the condition when the lift on the wings is equal to the aircraft weight, the $1g$ case. The boundary of the $SEP = 0$ curve is set by the aircraft aerodynamics and weight and by the maximum thrust of the engine.

For Figure 14.5 the engine thrust would be at its maximum available at every altitude and Mach number. Because the aircraft drag rises around $M = 1.0$ the $SEP = 0$ curve dips in this region, implying a lower steady ceiling around sonic velocity. The aircraft may be forced to operate somewhat inside the boundary set by $SEP = 0$ for reasons indicated on the diagram.

Curves of $SEP = $ constant are shown in Figure 14.6 for a fighter aircraft at three levels of load factor. The figures show contours plotted on axes of Mach number and altitude; for the plot at a load factor of 1, which corresponds to straight and level flight, curves of energy state, expressed as altitude, are superimposed. The curve $SEP = 0$ corresponds in each case to the limit of steady operation at maximum engine thrust and moving inside this contour the value of SEP rises.

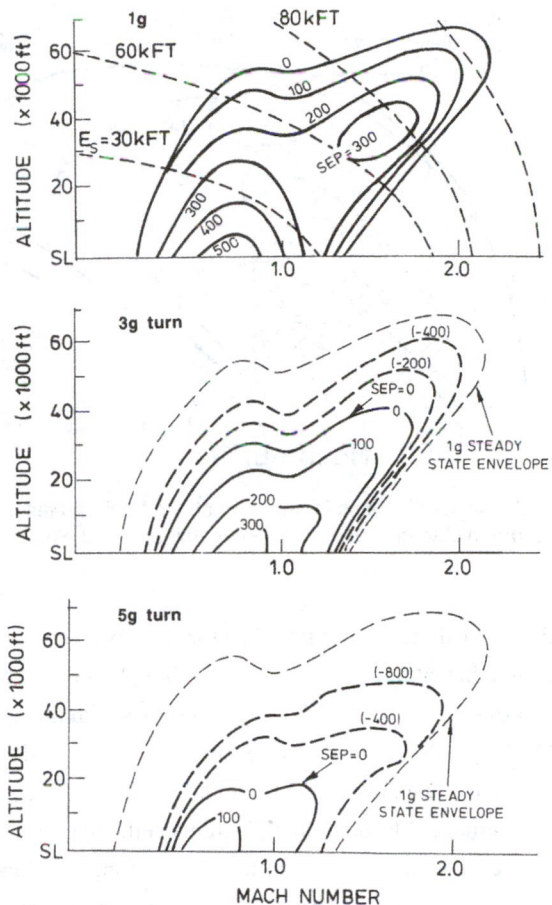

Figure 14.6. Schematic plots of altitude versus Mach number for a combat aircraft
at load factors of 1,3 and 5. Shown are lines of *SEP* = constant (units ft/s), drawn solid when
positive, broken when negative. The *n* = 1 case also shows lines of constant energy state (units ft).
(Derived from Shaw, 1988.)

As the load factor is increased the size of the region for which $SEP \geq 0$ rapidly decreases.
For turning, when the load factor is larger than unity, the $SEP = 0$ boundary gives the limit of
sustained turn rate. The highest values of *SEP* occur in all cases at low altitude and at Mach
numbers in the high subsonic range. At this condition the high density air gives low values of lift
coefficient (and therefore low induced drag) and also high net thrust from the engine.

To be more concrete, the curves in Figure 14.6 show that for a 3*g* turn at 30,000 ft altitude
(9.1 km) and $M = 0.9$ the value of *SEP* is about zero. In other words 3*g* gives the maximum
sustained turn rate at this combination of altitude and Mach number so that at this condition
the turn could be accomplished at constant energy state. The corresponding plot for 1*g* flight
at this combination of altitude and M shows that $SEP \approx 150$ ft/s, so with the engines giving
their maximum thrust the aircraft would be increasing its energy state, either climbing at 150 ft/s

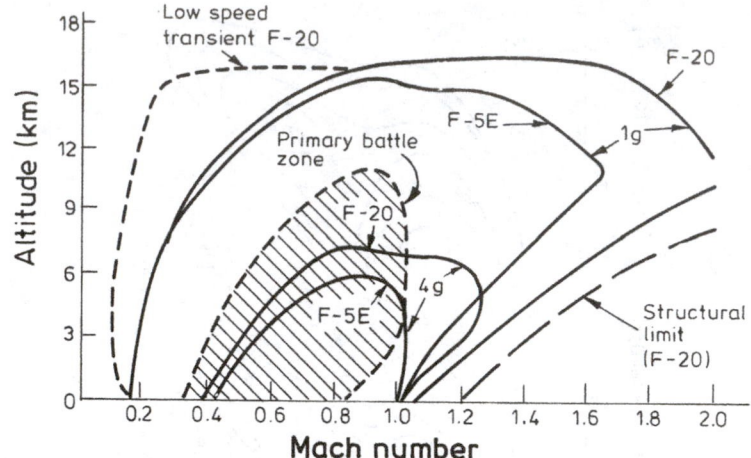

Figure 14.7. Altitude–Mach number diagrams for two aircraft: the F-5E and the later derivative the F-20. (From L'Aeronautique et L'Astronautique, 1983–85.)

(47.5 m/s) or accelerating in the direction of flight. The forward speed for $M = 0.9$ at this altitude is 273 m/s and the acceleration in that direction is given by $(SEP/V)g = (47.5/273)g$, or $0.17g$. For a $5g$ turn, however, the corresponding plot in Figure 14.6 shows with this combination of altitude and Mach number that $SEP \approx -400$ ft/s $(-122$ m/s$)$ and the aircraft would either be losing altitude at this rate or decelerating at the rate of $(122/273)g = 0.45g$.

Figure 14.7 shows altitude–Mach number diagrams pertaining to two aircraft, the F-5 and a later derivative, the F-20. The wider steady, sustained operating envelope of the F-20 is very clear, but still more obvious is the smaller operating range in terms of speed or altitude at a load factor of $4g$ compared with $1g$. As already noted, the primary battle zone is presumed to occur at altitudes from sea level to the tropopause and for Mach numbers below unity. The region for combat is restricted because of the low density at high altitude, turning becomes difficult, whilst at high supersonic Mach numbers the turning radius becomes so large that combat is difficult.

14.4 OPERATION AT LOW THRUST AND DRAG

Up to now most of the discussion has centred on the manoeuvres which require large lift and create large drag, therefore requiring high levels of engine thrust if they are to be sustained. High thrust almost invariably involves high fuel consumption and this can shorten the mission. For many combat missions there are phases of cruise to and from the combat zone and periods of loitering or patrolling. For both the cruise and the loiter it may be possible to fly at a condition approaching that for minimum fuel consumption. In this simplified treatment we will assume that the specific fuel consumption (mass flow rate of fuel/net thrust) is constant in the subsonic range of conditions likely. Under this assumption fuel consumption is therefore proportional to drag.

Condition of minimum fuel consumption at constant speed and altitude

This condition corresponds to that for maximum flight duration and occurs for minimum drag. The drag is given by

$$D = qAC_D = qA\left(C_{D0} + kC_L^2\right),\tag{14.7}$$

where $q = \frac{1}{2}\rho V^2$ is the dynamic pressure whilst C_{D0} and k are defined in Section 14.2. The weight of the aircraft mg is fixed, neglecting here the variation as fuel is burned or weapons released, so the lift coefficient C_L can be replaced by mg/qA, where A is the wing area, and mg/A is the wing loading. The drag is therefore given by

$$D = qAC_{D0} + \frac{k}{q}A(mg/A)^2\tag{14.8}$$

and the minimum drag will occur when

$$q = (mg/A)\sqrt{\frac{k}{C_{D0}}}.\tag{14.9}$$

Minimum fuel consumption during cruise

The condition for this is identical to that required for maximum range, considered in Chapter 2 for the civil transport. The requirement is that VL/D should be a maximum. This can be found by differentiating VL/D after again replacing the lift coefficient by the wing loading divided by the dynamic pressure, see Exercise 14.7.

Exercises

14.6 The New Fighter Aircraft may be required to loiter at a Mach number of 0.6. Find the dynamic pressure and thence the altitude which will give the minimum drag. Hence determine the thrust required from each engine. Take the aircraft mass to be 15 tonne.   (**Ans:** 12.75 kPa, 5.6 km, 7.1 kN)

14.7 Show that the minimum fuel consumption to cover a given distance at given altitude will occur when the dynamic pressure is given by

$$q = (mg/A)\sqrt{\frac{3k}{C_{D0}}}.\tag{14.10}$$

Hence find the optimum altitude and thrust necessary for cruise of the New Fighter Aircraft with a mass of 15 tonne at $M = 0.8$.   (**Ans:** 5.7 km, 8.1 kN)

14.5 VECTORED THRUST

There are some benefits in being able to vary the pitch of the propulsive nozzle and thereby alter the direction of the jet and the gross thrust. (The net thrust also varies in direction, but the direction of ram drag is fixed by the forward motion of the aircraft, of course.) It is well known that the F-22 and all versions of the F-35 will have vectored thrust. Some Russian combat aircraft also have vectored thrust and the Harrier has used it for many years.

Various benefits are claimed for vectored thrust. It can be used to lower landing speed, shorten take-off distance and reduce cruise drag. Following the theme of this chapter, it is immediately apparent that deflecting the jet is a way of producing large thrust normal to the direction of flight when the flight speed is low but the air density high. In other words it can be used to give a large normal force when the wings are not capable of giving large lift, and can therefore be used to increase manoeuvrability at low speed. The normal force is equal to the *gross* thrust times the sine of the angle of deflection and at high speed quite small deflections of the jet away from the direction of travel produce a large normal force. This is because at high speed the ram drag is a large fraction of the gross thrust so the gross thrust is generally large in relation to the net thrust.

The disadvantages of vectored thrust are the weight, complexity and cost of the variable nozzle. On top of this is the issue of reliability, especially if the vectored thrust is used in place of control surfaces on the tail.

SUMMARY OF CHAPTER 14

Turning manoeuvres are an essential part of combat and load factor is used to express the increase in lift needed for the turn. The drag increases in proportion to the square of the lift coefficient.

The maximum *attained* turn rate is fixed by the wing area and the dynamic pressure $\frac{1}{2}\rho V^2$ but the drag at this condition is likely to exceed the net thrust from the engines. In this case there is a reduction in flight speed or altitude during the turn; equivalently, the specific excess power is negative and there is a reduction in energy state. The maximum turning rate which can be maintained without loss in speed or height (when the engine thrust equals the drag) is the *sustained* turning rate and at this condition the specific excess power is zero ($SEP = 0$) so energy state remains constant.

Most combat is expected to take place between sea level and the tropopause and at Mach numbers below unity. A requirement also exists for cruising at higher Mach numbers. The specification of the New Fighter Aircraft demands that at take-off there is a thrust–weight ratio of 1.0 with the afterburner and 0.66 'dry', but the maximum thrusts required of the engines at a sample of other conditions were found in Exercise 14.4. The maximum required thrusts are summarised in Table 14.1 for use in later chapters.

Any one of the conditions which require high thrust could be imagined to prescribe *the* design point for the engine, but any engine will be expected to operate satisfactorily at many of these points. The steady level flight condition for $M = 1.5$ does not call for very high thrust, but for this case the practical interest is whether the engine can produce sufficient thrust 'dry' (i.e. without using the afterburner) to achieve 'supercruise'.

Table 14.1 Required net thrust (kN) from each engine for the New Fighter Aircraft for total aircraft mass of 15 tonne

			Mach number		
			0.9	1.5	2.0
Tropopause	Steady level flight	$1g$	8.0	26.0	44.3
	Banked turn	$3g$	35.8	39.3	
Sea level	Steady level flight	$1g$	21.0		
	Banked turn	$5g$	39.8		
		$9g$	83.5		

Take-off thrust required at sea level with total aircraft mass 18 tonne:

| 'dry' | thrust–weight ratio $= 0.66$ | $F_N = 58.3$ kN |
| afterburning | thrust–weight ratio $= 1.00$ | $F_N = 88.3$ kN |

Note. At tropopause $T_a = 216.7$ K, $p_a = 22.7$ kPa; at sea level $T_a = 288.2$ K, $p_a = 101$ kPa.

In Chapter 15 characteristics of the engines for combat aircraft will be explored. In Chapter 16 different conditions will be taken as design points and the performance of the corresponding engines calculated and compared.

In Exercise 14.4 it is shown that the highest thrust for a $9g$ turn at sea level was over 84 kN per engine, whilst from Exercise 14.5 the thrust required for loiter could be as low as about 7 kN; the engine needs to be able to accommodate these large variations. In Chapter 17 the performance at off-design conditions will be examined, both for high thrust conditions critical for the mission and also for low thrust, such as cruise or loiter, when low fuel consumption becomes a primary concern.

CHAPTER 15   ENGINES FOR
COMBAT AIRCRAFT

15.0 INTRODUCTION

Figure 15.1 shows a cross-section through a modern engine for a fighter aircraft and the large differences between it and the modern engines used to propel subsonic transport aircraft, Figure 5.4, are immediately apparent. Above all the large fan which dominates the civil engine, needed to provide a high bypass ratio, is missing. Engines used for combat aircraft typically have bypass ratios between zero (when the engine is known as a turbojet) and about unity; most are now in the range from 0.3 up to about 0.7 at the design point, though the bypass ratio does change substantially at off-design conditions.

This chapter seeks to explain why fighter engines are the way they are. It begins with some discussion of specific thrust, equal to net thrust per unit mass flow, since this is a better way of categorising engines than bypass ratio; fighter engines have higher specific thrust than civil transport engines. Then the components of the combat engine are described, pointing out features common to the civil engine and drawing attention to their differences. Features peculiar to the combat engine, such as the mixing of the core and bypass stream, the high-speed intake, the afterburner and the variable area propulsive nozzle are also considered. A brief treatment of the thermodynamic aspects of high-speed propulsion leads into the constraints on the performance of engines for combat aircraft and the rating of engines.

In earlier chapters the cooling air supplied to the turbine was neglected in calculating the cycles, so too was the mass flow rate of fuel in the gas through the turbine. These are now included, together with more realistic properties for the gas downstream of the combustor, where it is assumed that $\gamma = 1.30$ and $c_p = 1244 \, \mathrm{J \, kg^{-1} K^{-1}}$.

15.1 SPECIFIC THRUST

The net thrust from an engine is given by

$$F_N = V_j(\dot{m}_a + \dot{m}_f) - V\dot{m}_a, \tag{15.1}$$

where $\dot{m}_a$ is the mass flow of air entering the engine, $\dot{m}_f$ is the mass flow of fuel, V_j is the jet velocity and V is the flight speed. The specific thrust is the net thrust per unit mass of air flow, and is given by

$$F_N/\dot{m}_a = V_j(\dot{m}_a + \dot{m}_f)/\dot{m}_a - V \tag{15.2}$$

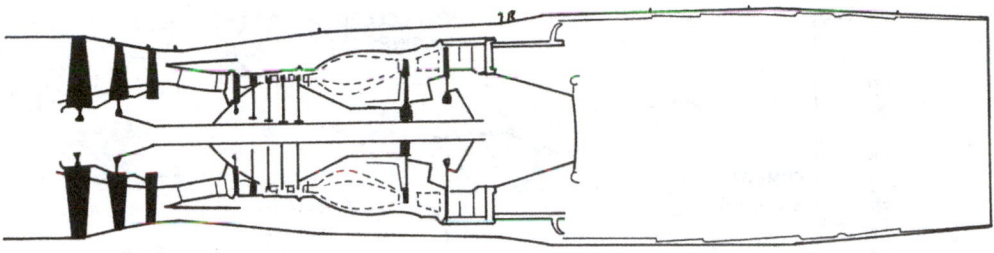

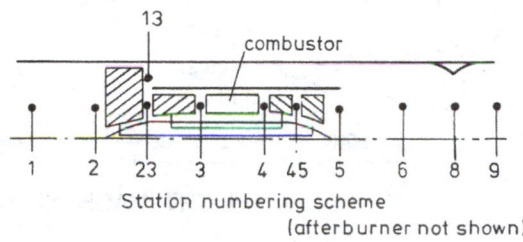

Figure 15.1. A cross-section through a modern combat engine with a schematic indicating the numbering system adopted.

with units of velocity (m/s). When the mass flow of fuel is neglected the specific thrust reduces to

$$F_N/\dot{m}_a = V_j - V. \tag{15.3}$$

Although the correct units of specific thrust are those of velocity, m/s, the industry often expresses it in units of pounds thrust per unit mass flow in pounds per second; the numerical value is the same in pounds or in kilograms. (The magnitude in m/s is 9.81 times the value in $\mathrm{kg\,kg^{-1}s^{-1}}$.)

For sound reasons it is more normal to characterise combat aircraft engines by their specific thrust than by their bypass ratio. Specific thrust for a given engine varies with altitude and Mach number, so the value quoted always refers to a particular flight condition. The size (i.e. diameter) of an engine is primarily determined by the mass flow rate of air it swallows, so the combination of aircraft thrust requirement and engine specific thrust essentially determines the engine size. The high specific thrust combat engine generally has a low bypass ratio and its diameter/length ratio is much less that the low specific thrust high bypass ratio civil engine. The energy released in combustion at high pressure is converted into mechanical energy relatively efficiently and this has to appear as kinetic energy of the jet streams. The high jet velocity, which leads to high specific thrust, also leads to low propulsive efficiency at subsonic flight speeds and to a large amount of noise. (The type of engine required for supersonic transport aircraft would also need to be of the low bypass type and the problem of noise is particularly serious.)

Figure 15.2 shows the regions where aircraft operate and the different range of specific thrust involved. Whereas civil turbofan engines operate with values of specific thrust below 200 m/s, combat aircraft have values between about 500 and 1000 m/s. For propulsion at Mach numbers above about 2.5 a conventional engine becomes an encumbrance and the normal solution is to dispense

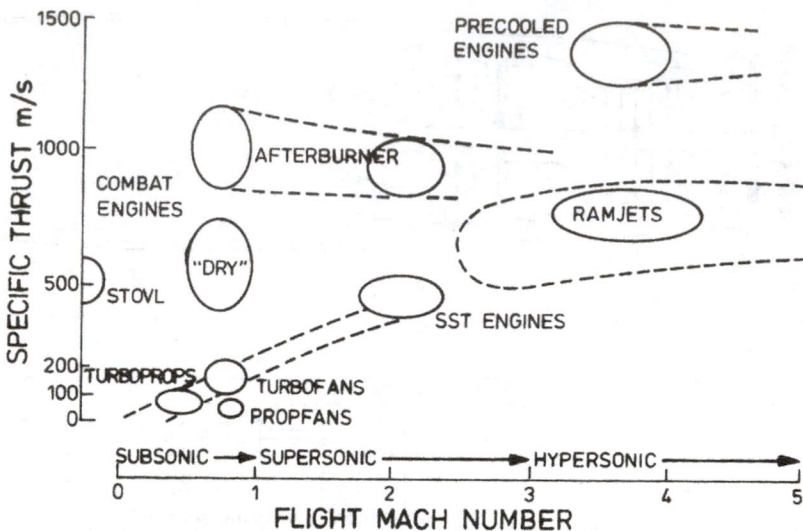

Figure 15.2. The variation in specific thrust with design flight Mach number. (SST refers to supersonic transport. From Denning and Mitchell, 1989.)

with the compressor and turbine to become a ram-jet; a more innovative possibility is to use the fuel to pre-cool the air before it enters the compressor, but this has not been proven to be feasible.

The reason for selecting high specific thrust will be explored further in this chapter and the next, but certain general reasons can be given. There is intense pressure to reduce weight, mainly to increase acceleration and manoeuvring ability. Low specific thrust engines have a large diameter to pass a large mass flow and high specific thrust engines, which give a large increase in velocity to a small mass flow of air, normally weigh less for the same thrust.

The net thrust is less than the gross thrust because of the ram drag, $\dot{m}_a V$. The proportional reduction in thrust with forward speed is therefore less when the jet velocity is very high, in other words when the specific thrust is high. The reduction in thrust with forward speed is normally referred to as thrust lapse and thrust lapse is less when the specific thrust is high.

High-speed propulsion gives special reasons for wanting compact engines which pass relatively small amounts of air. At supersonic speeds it is important to shape the aircraft to produce low drag, and this would be impossible with low specific thrust engine because of the large frontal area. In addition there is a loss in stagnation pressure in the inlet to an engine at supersonic flight speeds. For a low specific thrust engine, which passes a large mass flow but gives it only a small increment in kinetic energy, the stagnation pressure of the jet is much nearer to the inlet stagnation pressure than for a high specific thrust engine. This means that for the low specific thrust engine the loss in stagnation pressure across the shocks of the inlet is proportionately larger and low specific thrust engines are undesirable at supersonic speeds.

The bypass air is used to cool the outside of the engine and it makes certain variations in the nozzle easier, so current practice favours a bypass ratio of at least 0.3. As the bypass ratio increases the boost in thrust with afterburning is increased. This means that the difference in thrust between the engine operating 'dry' to achieve low fuel consumption at cruise, for example, and afterburning, to achieve maximum thrust, is also increased. The choice of bypass ratio and specific thrust at the design point will be discussed in the next chapter.

15.2 THE LAYOUT OF HIGH SPECIFIC THRUST ENGINES

The diagram of an engine in Figure 15.1 reveals many of the important features of a combat engine. It can be seen that the LP compressor (colloquially referred to as the fan) is driven by the single-stage LP turbine, whilst the core compressor is driven by the single-stage HP turbine. One of the most distinctive features of the majority of combat engines is the mixing of the core and the bypass streams in the jet pipe upstream of the final propulsive nozzle. In this chapter the gas properties of the mixed stream are calculated by an approximate method based on the relative mass of core gas (with $c_{pe} = 1244 \, \text{J} \, \text{kg}^{-1} \text{K}^{-1}$ for the gas which has gone through the combustor) and bypass air (with $c_p = 1005 \, \text{J} \, \text{kg}^{-1} \text{K}^{-1}$). The specific heat of the mixed-out gas c_{pm} is given by

$$(1 + bpr) \, c_{pm} = c_{pe} + bpr \, c_p. \tag{15.4}$$

The value of γ_m for this mixed stream can be obtained from $\gamma_m = c_{pm}/(c_{pm} - R)$ where $R = 287 \, \text{J} \, \text{kg}^{-1} \text{K}^{-1}$. When the afterburner is lit it is assumed that $\gamma = 1.30$ and $c_p = 1244 \, \text{J} \, \text{kg}^{-1} \text{K}^{-1}$ for the exhaust gas. Because the mixing of the core and bypass is so important for the performance of the engine, this is addressed first.

Mixing of core and bypass streams

All combat aircraft, with the exception of the Harrier, have engines which mix the core and bypass streams before they enter the final nozzle. There is a relatively long jet pipe, comparable in length to the rest of the engine, to allow mixing to take place and it is normally assumed in simple treatments that the flow is fully mixed out to uniform before the final nozzle.

The condition where mixing takes place is most accurately modelled by assuming equal *static* pressure for each stream. The remainder of the cycle analysis, however, uses *stagnation* pressure, so it is comparatively complicated to use the static value. (In addition, to use static pressures would require the flow areas to be specified so that the Mach numbers, and thence the ratio of stagnation to static pressures, could be found.) In fact the Mach numbers out of the core and in the bypass duct are normally relatively low, so equating static and stagnation pressures for this purpose does not introduce a major error. Furthermore, if the Mach numbers are well below unity the loss in stagnation pressure in the mixing process itself will be much less than the absolute value of the stagnation pressure and we will neglect this loss here.

Because the core and the bypass are assumed to have equal stagnation pressure when they mix the pressure rise across the fan must be just sufficient for its outlet flow to have equal stagnation pressure to the flow leaving the LP turbine. This puts an extra constraint on the matching of the fan and core streams.

The two streams are mixed because there is an increase in thrust by doing so. A simple explanation for the gain in thrust by mixing can be made by considering two streams of equal stagnation pressure, each having a constant mass flow. For a given stagnation pressure upstream of the final nozzle the jet velocity is proportional to $\sqrt{c_p T_0}$ and the gross thrust is proportional to the jet velocity. Imagine that in place of a constant pressure mixing the two streams are kept separate but there is a constant pressure heat transfer from the hot core stream to the relatively cool bypass stream until both have the same temperature. If the bypass ratio were unity, the temperature change of each stream would be equal (assuming for simplicity that c_p is equal for both). The *proportional* increase in stagnation temperature, however, for the cooler bypass stream would inevitably be greater than the *proportional* drop in temperature of the warmer core stream. This gives a larger increase in V_j for the bypass than the corresponding fall in V_j for the core and consequently a gain in thrust. The true situation is more complicated than this and mixing inevitably leads to a loss in stagnation pressure. Heating a moving gas also lowers stagnation pressure, but for low Mach numbers this loss in stagnation pressure is small.

The LP compressor or fan

For the engine of Figure 15.1 the fan has three stages and the pressure ratio in the jet pipe is therefore much higher than would be normal in a civil engine; a figure in excess of 4 on a sea-level test bed might be regarded as typical. The flow from the fan is divided at the splitter with about half going to the core and half to the bypass duct. The bypass flow mixes with the core flow downstream of the LP turbine and the passage of the mixed stream through the final nozzle therefore determines the back pressure behind the fan.

All the stages of the combat aircraft engine are heavily loaded in the aerodynamic sense, in other words $\Delta h_0/U^2$ is high, typically near the upper level at which satisfactory performance is possible. The high loading generally leads to reduced efficiency at maximum speed conditions and there is often a substantial rise in efficiency when the engine is 'throttled back' to a lower thrust setting and the Mach numbers in the blades are decreased.

The pressure and temperature leaving the fan will not in general be equal for the streams entering the bypass and the core but as a simplification which will suffice for the present purpose we will take $T_{023} = T_{013}$ and $p_{023} = p_{013}$, see the numbering scheme in Figure 15.1.

The core compressor

The engine in Figure 15.1 has a five stage HP compressor, suggesting a pressure ratio at design across this component of about 7, giving an overall pressure ratio for the engine on a test bed of about 30. The design will normally lead to efficiencies which are lower than those associated with a civil engine because of the higher value of $\Delta h_0/U^2$. It will be assumed that the back of the HP compressor has titanium alloy discs and blades, which limits the exit temperature T_{03} to about 875 K. This choice of temperature is absolutely critical in determining the on-design and, even more, the off-design performance of the engine.

The combustor

The combustor is similar in its duty to those in a civil engine though the exit temperatures are higher than service temperatures in a civil engine. At some operating points the mass flow of fuel is higher in relation to the mass flow of air (i.e. the equivalence ratio is higher) than for the civil engine because the compression ratio is lower so the compressor delivery temperature is lower too. As discussed in Chapter 11 it is plausible to assume a pressure loss of up to 4% of the stagnation pressure at compressor outlet in the combustor.

The turbines

The turbines are likely to be more heavily loaded than would be normal in a civil engine because of the incentive to keep the engine weight low. Efficiencies may be expected to be a little lower than those for a civil engine. The higher loading also means that the assumption that they are choked is more generally valid than was the case for a civil engine.

Combat engines normally have higher turbine entry temperatures than civil engines. High thrust-weight ratio is more important than long service life and the time between overhauls for combat engines is a small fraction of the time between overhauls for a commercial engine. The maximum entry temperature to the HP turbine will be taken here to be 1850 K; this temperature, T_{04}, is the mixed-out temperature at outlet from the HP nozzles and includes the effect of the cooling air provided to the nozzles. Some of the newest engines, and some of the proposals for the future, have temperatures in the region of 2000 K.

The afterburner

It has already been noted that the maximum thrust required may be more than ten times the minimum thrust required whilst the aircraft is loitering. One of the ways by which this large variation is accommodated is the use of the afterburner, sometimes known as reheat or thrust augmentation. The ability to switch the afterburner on and off gives a special flexibility to this type of engine and most combat engines have them.

An afterburner allows the exhaust to be raised in temperature. The jet pipe contains a system of fuel injectors and gutters to allow fuel to be burned downstream of the turbine. When the afterburner is not being used (so the engine is operating 'dry') the exhaust temperature is typically around 1000 K. When lit the afterburner gives an increase in temperature to about 2200 K with an

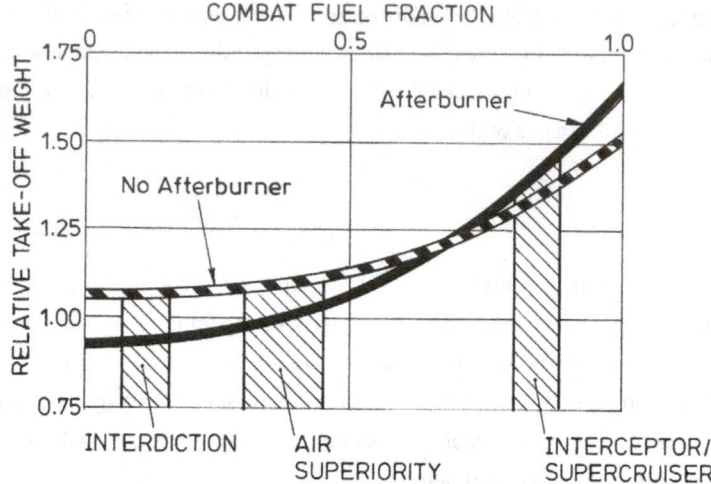

Figure 15.3. The effect of aircraft design role on the decision whether to have an afterburner. Relative take-off weight is the measure of quality (lighter is better) and combat fuel consumption is the fraction of fuel burned during the high thrust part of a typical mission. (Denning and Mitchell, 1989.)

increase in thrust up to about 50% and perhaps a three-fold increase in specific fuel consumption. The fuel flow with the afterburner lit may therefore be five or more times that with the engine 'dry', see Exercise 13.2. If an engine is called upon to produce high thrust for only short periods of time the afterburner can be the most efficient overall solution because the specific thrust can be high and the engine consequently small and light. If the high thrust is needed for a substantial fraction of the flight the afterburner may be inappropriate since the weight of fuel consumed would be so large.

This is illustrated in Figure 15.3, produced by Rolls-Royce, where two possible styles of engine are compared for a range of duties: one with an afterburner and the other without. The duty is characterised by the fraction of the total fuel which is used with the engine at combat rating (i.e. giving the maximum thrust). Three possible missions are marked on the figure, interdiction, air superiority and interception. From Chapter 13 it will be recalled that an air-superiority aircraft (the typical fighter) spends a significant amount of time getting to the combat zone or loitering, whereas an interceptor spends a large part of its mission travelling as fast as possible to intercept the enemy. The relative figure of merit in the figure is the relative take-off weight: basically the lighter the aircraft capable of carrying out the mission the lower the cost is likely to be. It can be seen from Figure 15.3 that the interceptor is likely to be slightly lighter (and therefore better) if it achieves its necessary thrust without use of an afterburner, whereas the other two aircraft will be substantially lighter if they are designed with an afterburner.

For military engines enough fuel may be added to the afterburner to use virtually all the oxygen and the temperature approaches the stoichiometric value. A value of 2200 K is a realistic value to use here. The amount of fuel can be less than this to give a smaller increase in thrust and fuel consumption. (In the case of the Concorde, for example, the afterburner was used for take-off and to pass through $M = 1.0$ and the temperature rise was more modest, the stagnation temperature in the jet being raised by the afterburner to about 1450 K for take-off and to about 1300 K passing through the sonic speed respectively.)

The propulsive nozzle

The conventional operation of the afterburner attempts to keep the jet pipe pressure unchanged when the afterburner is lit; with the jet pipe pressure unaltered the operating points of the compressor and turbine are also unchanged and the engine is unaware of whether the afterburner is on or off. To cope with the rise in temperature in the jet pipe when the afterburner is in use the propulsive nozzle must be made of variable area. The required change in propulsive nozzle throat area can then be found, to maintain $\overline{m}_8 = \dfrac{\dot{m}\sqrt{c_p T_{08}}}{A_8 p_{08}}$ constant at the throat of the nozzle, which is choked, and thereby to maintain the pressure in the jet pipe constant. (Note that here station 8 is the propulsive nozzle throat and station 9 is the exit plane. If the nozzle is only convergent, not convergent-divergent, then stations 8 and 9 coincide.) If the gas properties (c_p and γ) were unaffected by the afterburning the nozzle area would need to rise in proportion to the square root of the temperature. (In fact the temperatures from the afterburner are sufficiently high that the specific heat varies strongly with temperature. In the industry it is therefore found to be better to work with enthalpy rather than c_p and T.)

For a reversible nozzle the jet velocity is given by

$$V_j = \sqrt{2c_{pm}T_{08}\left\{1 - (p_a/p_{08})^{(\gamma-1)/\gamma}\right\}}. \tag{15.5}$$

The gas properties of the exhaust, the specific heat c_{pm} and the ratio of specific heats γ, depend on the amount of bypass air (and on the temperature T_{08} but we neglect this aspect). The other important term is the pressure ratio across the nozzle p_{08}/p_a. The pressure across the fan is in the range 2 to 5, but for high-speed flight the inlet stagnation pressure p_{01} is much higher than the ambient pressure p_a so the ratio p_{08}/p_a may get as high as 16. For the nozzle to be reasonably efficient at $p_{08}/p_a \approx 16$ it is important that it be of the convergent-divergent form, see Figure 11.3.

The behaviour of a convergent-divergent nozzle is given in Figure 6.2 (ignore for the present that this figure is for $\gamma = 1.40$ whereas the exhaust gases have a lower value). For a pressure ratio of 16 the exit Mach number would be about 2.46 and $\overline{m}_9$ approximately 0.4 times the throat value. In other words the area at nozzle outlet would be 2.5 times the throat area, which might be larger than the cross-sectional area of the rest of the engine. Such a large nozzle exit might be impractical and the area increase downstream of the throat may be limited to a smaller value than this. The expansion would then not be fully reversible, but the loss of gross thrust can be kept reasonably small. Figure 15.4 shows the variation in gross thrust with pressure ratio for three different nozzles as a function of pressure ratio p_{08}/p_a calculated for $\gamma = 1.30$, a value representative of exhaust gases. The nozzles are a convergent nozzle, a fully reversible convergent-divergent nozzle and a

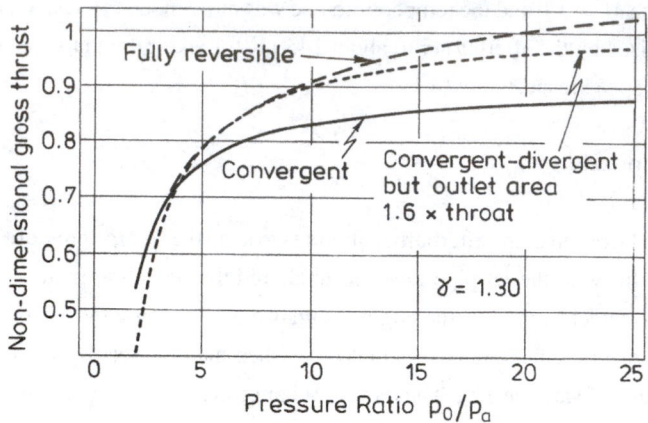

Figure 15.4. Non-dimensional gross thrust from nozzles, $F_G/(\dot{m}\sqrt{c_p T_0})$, versus the
ratio of upstream stagnation pressure to downstream static pressure.

nozzle which is truncated where the exit area is 1.6 times the value at the throat. (An exit area 1.6 times the throat area was selected here because it has been used for the F-16 fighter.) At a pressure ratio of 16 a loss in gross thrust of about 1.5% would be incurred as a result of the truncated divergent section, but the thrust is about 10% higher than that from a simple convergent nozzle. It can also be seen from Figure 15.4 that up to a pressure ratio of about 5 across the nozzle the penalty in thrust in having the simpler and lighter convergent nozzle is about 6%.

An ideal nozzle would vary the throat area A_8 and the exit area A_9 independently to maintain the correct area ratio corresponding to the pressure ratio imposed. The required throat area is determined by the mass flow, stagnation pressure and stagnation temperature from the engine so as to to achieve the required engine mass flow and pressure in the jet pipe. In some engines the two areas are made to vary independently, but in others there is a fixed schedule of A_9/A_8 as a function of A_8 so that a single set of actuators varies both areas at once. Having only a single set of actuators saves cost and weight, but with some loss in thrust and increase in *sfc* over parts of the operating envelope.

The requirement for stealth means that new aircraft frequently have nozzles of special shape. These are often rectangular and can be above the wing or tail plane to provide shielding from the ground. Aerodynamically the rectangular shape is likely to be less good than a simple round nozzle, but the principle of operation is unchanged.

The inlet for high-speed flight

By convention the inlet is the responsibility of the airframe manufacturer, but in terms of understanding the engine behaviour it needs to be included in any assessment of engine configuration. For flight at subsonic conditions a simple 'pitot' intake, similar in principle to those fitted to subsonic

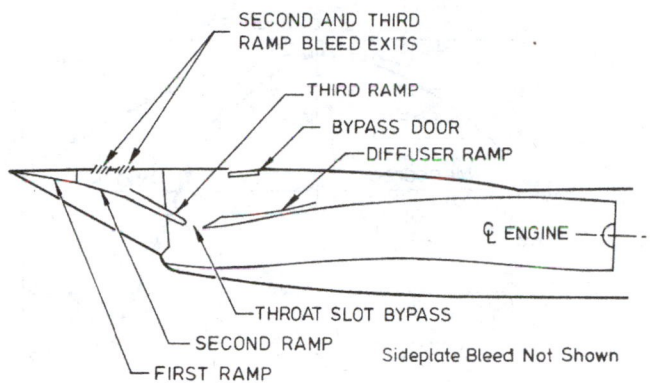

Figure 15.5. The engine intake for the F-15 fighter.

civil aircraft, with suitably rounded and shaped leading edges to accommodate incidence effects can produce very small loss. Such an intake can also be satisfactory at low supersonic conditions, but at supersonic flight speeds some loss is unavoidable. With the simple pitot intake at supersonic speeds the loss is created by a normal shock wave ahead of the intake. For Mach numbers approaching 2.0 the loss from a normal shock is relatively large and aircraft for which a critical role involves such high Mach numbers tend to have intakes which decelerate the inlet flow in a series of oblique shocks. The stagnation pressure loss decreases as the deceleration is spread through a greater number of shocks.

The American F-15 and F-16 aircraft were both designed at about the same time, but the F-15 is intended to be capable of operation up to $M = 2.3$, whereas the F-16 is not expected to have much role beyond $M = 1.6$, though it can reach $M = 2.0$. Because of their different roles the two aircraft have very different intakes. The F-16 is a pitot type, whereas the F-15 has variable ramps to produce deceleration in three oblique shocks followed by a weak normal shock, as shown in Figure 15.5.

An empirically based US military standard, MIL-E-5007/8, is used in the industry for design studies for intakes designed to minimise shock losses in supersonic flight. With p_{01} and p_{02} denoting the stagnation pressure at entry to and exit from the inlet, the empirical expressions for MIL-E-5007/8 for loss associated with shocks are given by

$$
\begin{aligned}
p_{02}/p_{01} &= 1.0 & \text{for } M \leq 1.0 \\
\text{and} \quad p_{02}/p_{01} &= 1.0 - 0.075(M-1)^{1.35} & \text{for } M > 1.0.
\end{aligned}
\tag{15.6}
$$

The pressure ratios in the F-15 and F-16 intakes are shown in Figure 15.6 as functions of flight Mach number together with the ratio for a normal shock and the ratio from the empirical relationship MIL-E-5007/8. As expected, the loss for the F-16 pitot intake is very close to the normal shock loss in the region where shock loss dominates. MIL-E-5007/8 gives losses in the supersonic region very close to those measured for the F-15. In the subsonic region the losses of the F-16 intake are generally low but the F-15 intake has comparatively sharp leading edges, so its loss at low

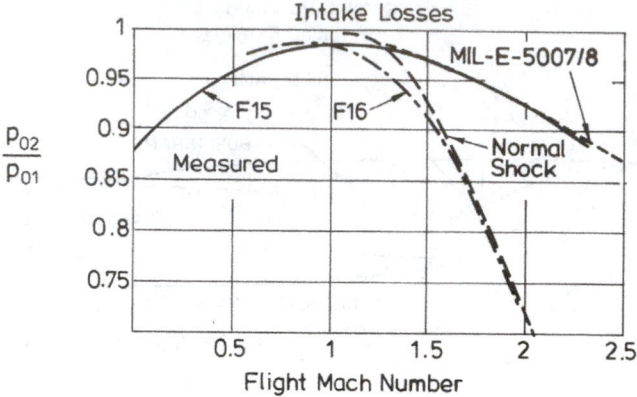

Figure 15.6. Intake pressure ratio showing measured results for the F-15 and F-16, the normal shock loss and the empirical relation from MIL-E-5007/8.

subsonic flight speeds is markedly higher. The three ramps in the F-15 intake have to be adjusted to accommodate variations in Mach number and engine condition and to operate these and to open the bypass door it is necessary to have three actuators, with associated control inputs, in each intake. It becomes easy to appreciate that such an intake entails substantial extra weight and cost compared with a simple fixed geometry pitot type, which can only be justified when high flight speeds are sufficiently important to the main design mission. The position is further complicated by the wish to include aspects of stealth technology, providing a curved path along the intake (to stop line of sight to the engine inlet face) and to include vanes coated in radar-absorbent material in the intake.

Exercises

15.1 a An aircraft flies at $M = 0.9$ at the tropopause ($T_a = 216.65$ K) and the fan pressure ratio is 4.5. Find the temperature downstream of the fan T_{013} if the compressor polytropic efficiency is 0.85.

If the bypass ratio is 0.67, find the value of c_{pm} for the mixed flow through the nozzle using a simple mass weighting of the two streams and taking $\gamma = 1.30$ for the flow leaving the turbine. Thence find γ_m for the mixed flow. What is the pressure ratio p_{08}/p_a across the propulsive nozzle?

(**Ans:** $T_{023} = 417.4$ K; 1148 J kg^{-1}K^{-1}, 1.333, 7.61)

b Consider two alternative configurations of the above turbofan engines. In one case the core and bypass mix with negligible loss in stagnation pressure before entering the Propulsive nozzle; in the other case each flow passes through a separate nozzle. The engines are identical upstream of where the streams can mix and in both cases the stagnation pressure *at nozzle entry* is equal to that out of the fan. In both cases the nozzles are reversible. If the temperature downstream of the LP turbine is $T_{05} = 1200$ K, find the mixed-out temperature T_{06}. Find the gross thrust per unit mass flow of air for the mixed and unmixed flows. (**Ans:** $T_{06} = 925$ K, unmixed $F_G/\dot{m} = 877$ m/s; mixed $F_G/\dot{m} = 919$ m/s)

15.2 For the mixed engine of Exercise 15.1, find the proportional increase in throat area needed when the afterburner raises the temperature to 2200 K. The jet pipe pressure (and all the properties of the engine upstream) remains unchanged by the afterburner. Assume that γ is equal to 1.30 for the jet with

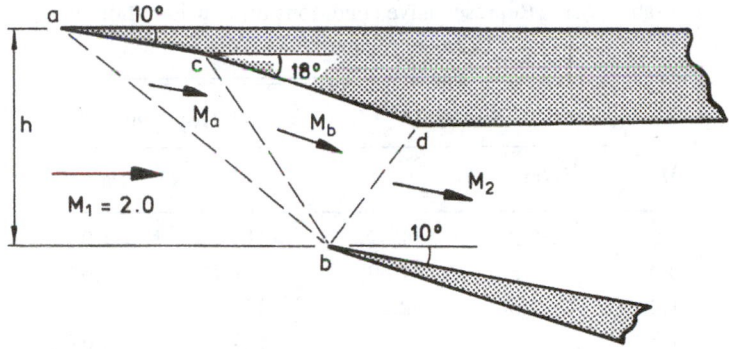

Figure 15.7. A simplified intake for an aircraft designed to operate at $M = 2.0$.

afterburning. Assume the nozzle is isentropic and neglect the mass flow rate of fuel. From Chapter 6 note that at the throat of a choked nozzle

$$\overline{m}_8 = \frac{\dot{m}\sqrt{c_p T_{08}}}{A_8 p_{08}} = \frac{\gamma}{\sqrt{\gamma - 1}}\left(\frac{2}{\gamma + 1}\right)^{(\gamma+1)/\{2(\gamma-1)\}}.$$

What is the increase in gross thrust F_G produced by the afterburner if the nozzle remains isentropic? For a flight Mach number of 0.9 at the tropopause, what is the increase in *net* thrust?

(**Ans:** $\Delta A_8 = 56\%$, $\Delta F_G = 56\%$, $\Delta F_N = 78.2\%$)

15.3 Figure 15.7 shows a possible intake for supersonic flight with inlet Mach number 2.0 and exit Mach number 0.89. The ramp angles are marked and the shocks are shown with broken lines.

Find the stagnation pressure ratio p_{02}/p_{01} and compare this with the pressure ratio for MIL-E-5007/8. Note that the pressure loss across a normal shock at $M = 2.0$ is given by $p_{02}/p_{01} = 0.721$.

(**Ans:** three oblique shocks $p_{02}/p_{01} = 0.959$; MIL-E-5007/8 $p_{02}/p_{01} = 0.925$)

If the height of the intake between points marked a and b (measured normal to the inlet flow direction) is h, what are the lengths of the ramps ac and cd? (**Ans:** ac $= 0.634h$; cd $= 1.060h$)

Note: To solve this exercise easily requires tables or charts for oblique shock behaviour. To facilitate this some relevant data is given below, where subscripts u and d refer to upstream and downstream of the shock respectively. The shock angle is measured from the upstream flow direction.

M_u	Flow deflection	M_d	Shock wave angle	p_d/p_u	T_d/T_u	
2.0	10°	1.64	39.3°	1.71	1.17	
1.64	8°	1.36	46.6°	1.49	1.12	
1.36	8°	0.89	71.0°	1.76	1.18	(Strong branch)

15.3 THE THERMODYNAMIC CYCLE OF COMBAT AIRCRAFT ENGINES

As will be discussed in Chapter 16, the operating conditions of a combat engine vary greatly between the different operating points. The most significant change comes from the variation in inlet stagnation temperature and pressure as the flight Mach number increases. This is summarised for some key operating conditions in Table 15.1.

Table 15.1   Representative conditions for consideration of combat requirements.

	Sea level		Tropopause	
M	T_{01}	p_{01}/p_a	T_{01}	p_{01}/p_a
0.0	288.2	1.0	216.7	1.0
0.9	344.8	1.69	251.7	1.69
1.2	371.3	2.24	279.0	2.42
1.5			314.1	3.67
2.0			390.0	7.82

Note: The tropopause is at 11 km altitude, $T_a = 216.7$ K, $p_a = 22.7$ kPa.

It will be recalled that for a gas turbine the ratio T_{04}/T_{02}, the ratio of turbine inlet temperature to compressor inlet temperature, is very significant. Since the turbine inlet temperature cannot be increased beyond a limit fixed by the material and the cooling technology, the maximum permissible value of T_{04}/T_{02} falls as T_{02} goes up as the flight speed increases. The effect of this is to reduce the engine operating point as speed increases, with consequent reductions in non-dimensional rotational speeds, non-dimensional mass flows and pressure ratios. For the present studies the maximum turbine inlet temperature, $T_{04} = 1850$ K, which is in line with current advanced performance, will be used for the maximum power conditions when T_{04} is what limits the upper performance.

The ratio of inlet stagnation pressure to ambient pressure p_{01}/p_a rises with Mach number even more rapidly than the temperature ratio since $p_{01}/p_a = (T_{01}/T_a)^{3.5}$ (For supersonic flight there is in general a loss in stagnation pressure in the inlet, so $p_{02} < p_{01}$, but the loss is likely to be less than about 10% within the normal operating range.) Because the pressure ratio produced in the inlet is so substantial at high flight Mach numbers, the pressure ratio required from the turbomachinery inside the engine is correspondingly lower; for $M = 2.0$ a fan pressure ratio of only 2 gives an overall pressure ratio of almost 16 across the propulsive nozzle, and as Figure 15.4 shows, there is little gain in thrust from increases in pressure ratio beyond this value. As noted earlier, this leads to the ramjet being attractive for design Mach numbers in excess of 2.5.

Exercise

15.4* For flight Mach numbers of 0.9, 1.5 and 2.0 at the tropopause find for the air entering the inlet the stagnation temperature T_{01} and stagnation pressure p_{01}.

(**Ans:** $T_{01} = 251.7$ K, 314.1 K, 390.0 K; $p_{01} = 38.3$ kPa, 83.3 kPa, 177.6 kPa)

For the same conditions find the stagnation pressure entering the engine p_{02} using the inlet loss given by MIL-E-5007/8. (Note that $T_{01} = T_{02}$.)   (**Ans:** $p_{02} = 38.3$ kPa, 80.8 kPa, 164.3 kPa)

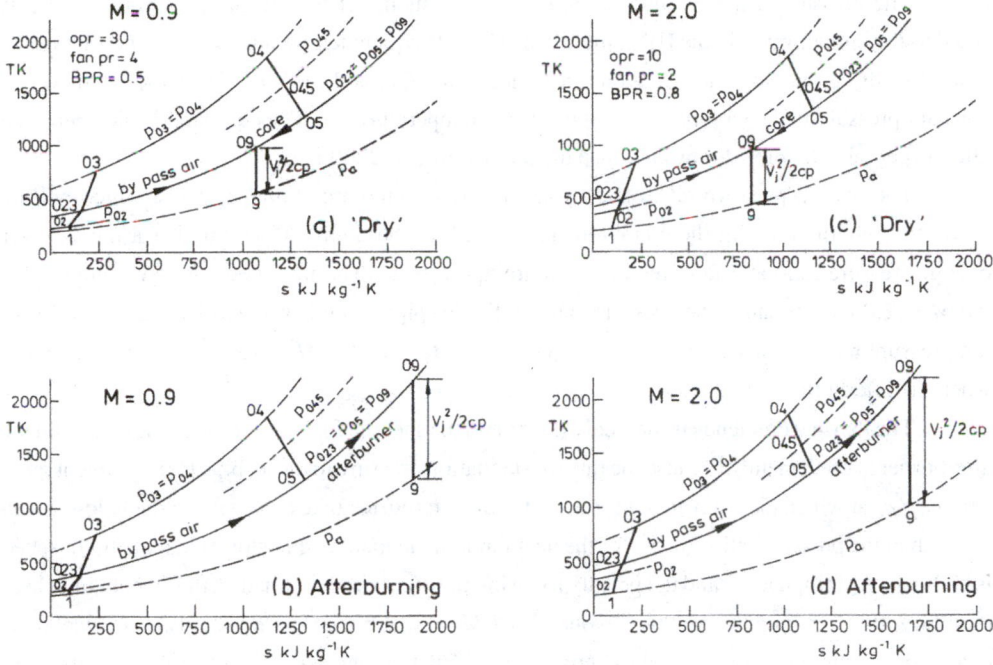

Figure 15.8. Temperature–entropy diagrams for combat engines at $M = 0.9$ and 2.0 with and without afterburning. (For these diagrams c_p taken as equal for burned and unburned gas.)

The thermodynamic behaviour of the engines may be better understood by looking at temperature entropy diagrams. These are shown in Figure 15.8 for two flight speeds at the tropopause, $M = 0.9$ and 2.0, with the engine operating 'dry' and with the afterburner in each case. The turbine inlet temperature is equal to 1850 K in all the cases and the afterburner temperature raises the jet stagnation temperature to 2200 K. For simplicity, in Figure 15.8 (but not elsewhere in the consideration here of combat engines) the gas properties are those associated with $\gamma = 1.40$ for the burned and unburned air. Losses in the intake are ignored. The chosen fan pressure ratio and overall pressure ratio in Figure 15.8 is lower for the engine at $M = 2.0$, in line with what will be found to be necessary.

Considering the 'dry' cases first, the pressure first rises in the intake, from Ambient pressure p_a to engine inlet stagnation pressure p_{02}. The pressure is then raised in the fan to $p_{013} = p_{023}$ (it being assumed here that the pressure is uniform in the radial direction at outlet from the fan) and this sets the stagnation pressure for the flow in the jet pipe, $p_{08} = p_{013}$. In words, the stagnation pressure of the flow entering the propulsive nozzle is therefore set by the fan exit pressure because the core and bypass flows are mixed at constant pressure. The pressure of the core flow is further raised in the HP compressor to p_{03} and then raised in temperature, at constant pressure, to the turbine inlet temperature, T_{04}.

The pressure at the HP turbine outlet, p_{045}, is such that the HP turbine enthalpy drop is equal to the enthalpy rise in the HP compressor. The subsequent temperature drop in the LP turbine is used to drive the fan. The bypass air at temperature T_{013} and the core air at T_{05} are mixed at constant pressure in the jet pipe to give the exhaust temperature T_{09}. The comparable plots with the afterburner take the mixed flow and raise the temperature to 2200 K.

The overall pressure ratios and bypass ratios selected are similar to those which will be found to be optimum with the conditions prescribed; at the higher Mach number a much lower overall pressure ratio and fan pressure ratio are appropriate. The jet kinetic energy is higher for the $M = 2.0$ case because the pressure ratio in the jet pipe is higher, notwithstanding the lower fan pressure ratio for this case. The higher jet velocity for flight at $M = 2.0$ is even more apparent when the afterburner is in use.

The thermal efficiency is markedly lower, and the specific fuel consumption higher, when the afterburner is used, mainly because the ratio of stagnation pressure in the jet pipe to the ambient pressure p_{08}/p_a at which the equivalent heat input for the afterburner takes place is relatively low, much lower than the pressure ratio p_{03}/p_a for the heat input in the main combustor. When the flight speed is high p_{08}/p_a is higher too and the penalty for using the afterburner is smaller; this benefit increases rapidly with Mach number so that beyond about $M = 2.5$ the optimum propulsion device is, as noted before, the ram-jet which relies entirely on the inlet compression to obtain the pressure rise.

Exercises

15.5 For a mixed-flow turbofan engine with an afterburner producing a constant exit temperature T_{0ab}, show that the jet velocity is a function of the fan pressure ratio and flight Mach number only. (Neglect any losses and assume the bypass and cores flows are fully mixed.)

If the fan pressure ratio is 4.5, the flight Mach number is 0.9 at the tropopause and $T_{0ab} = 2200$ K, find the jet velocity, the fuel flow per kg of air, the specific thrust and the *sfc*. Take $\gamma = 1.30$ for combustion products.   (**Ans:** $V_j = 1431$ m/s, $\dot{m}_f/\dot{m}_a = 0.0594$, $F_N/\dot{m}_a = 1250$ m/s, *sfc* $= 1.68$ kg h^{-1}kg^{-1})

15.6 Neglecting loss in the intake find the flight Mach number at which the inlet stagnation pressure is equal to 16 times the ambient pressure. For this Mach number at the tropopause what is the inlet stagnation temperature?   (**Ans:** $M = 2.46$, 478.4 K)

A ram-jet travels at the tropopause at the Mach number determined above. Fuel is burned to raise the stagnation temperature to 2200 K. If the combustion produces no loss in stagnation pressure and the expansion in the nozzle is reversible, find the jet velocity, the mass of fuel burned per kg of air, the specific thrust (net thrust per unit mass flow of air) and the specific fuel consumption. Take $\gamma = 1.40$ for the air and $\gamma = 1.30$ for the products of combustion, and for the fuel $LCV = 43.10^6$ J/kg.

(**Ans:** $V_j = 1608$ m/s, $\dot{m}_f/\dot{m}_a = 0.0538$, $F_N/\dot{m}_a = 969$ m/s, *sfc* $= 1.96$ kg h^{-1}kg^{-1})

Now consider the ram-jet with an intake loss at the above Mach number according to MIL-E-5007/8, and a further loss in stagnation pressure in the combustion process of 5%. Recalculate the jet velocity, the specific thrust and the *sfc*. The stagnation temperature of the jet is unchanged at 2200 K, so the fuel flow rate per unit mass flow of air is also unchanged.

(**Ans:** $V_j = 1569$ m/s, $F_N/\dot{m}_a = 928$ m/s, *sfc* $= 2.05$ kg h^{-1}kg^{-1})

Note: At this Mach number the ram-jet is an attractive engine, with high specific thrust and specific fuel consumption which is similar to an afterburning gas turbine engine. The complexity, weight and cost are low. The attractiveness of the ram-jet increases as the Mach number goes up, with the compressor and turbine of a gas turbine becoming encumbrances. One of the problems is accelerating the vehicle to a high Mach number to get the ram-jet working; this can be achieved by using rockets or by launching the vehicle from an aircraft at high speed.

15.4 SOME CONSTRAINTS ON COMBAT AIRCRAFT ENGINES

The most familiar limit on engine operation is the turbine inlet temperature T_{04}, which is fixed by the material properties of the turbine blades, the amount of cooling air used, how effective the cooling is and the expected life of the blades. For the current work an upper level of 1850 K will be assumed which would be permitted during take-off and for combat manoeuvres.

The non-dimensional operating point of the engine is set by the ratio of turbine inlet temperature to compressor inlet temperature T_{04}/T_{02}, where $T_{02} = T_{01}$ is the engine inlet stagnation temperature given by

$$T_{02} = T_a \left[1 + \frac{1}{2} (\gamma - 1) M^2 \right],$$

T_a being the ambient air temperature and M the flight Mach number.

Another limit on engine performance is the compressor delivery temperature T_{03}. The limit here comes from the material properties of the compressor disc and blades at the rear of the compressor, and a value of 875 K is taken to represent the current maximum allowed. (This temperature is an appropriate upper limit for titanium alloys, but with nickel-based alloy about 100 K higher temperature could be tolerated. Use of nickel-based alloys would, however, increase the weight.) The temperature of the gas at the rear of the compressor is determined by the inlet temperature to the engine (and therefore by the atmospheric temperature and flight speed) and by the overall pressure ratio and efficiency of the compressor, since

$$T_{03} = T_{02} (p_{03}/p_{02})^{(\gamma-1)/\gamma\eta},$$

where η is the polytropic efficiency. If T_{04}/T_{02} is held constant, and the nozzle remains choked with a constant throat area, the engine stays at a constant non-dimensional condition and the pressure ratios throughout the engine also stay constant. Under this condition the compressor delivery temperature is proportional to T_{02}, the engine inlet temperature.

There is a third non-dimensional parameter to be considered, which is related to the Mach number of flow in the blades, and is characterised by $N/\sqrt{c_p T_{02}}$, where N is the rotational speed of one of the shafts.[1] If T_{04}/T_{02} is constant, the engine is at a fixed non-dimensional condition and $N/\sqrt{c_p T_{02}}$ will also be constant. (Where there is more than one shaft there are as many rotational speeds as there are shafts; for constant engine condition there is a fixed ratio between the speeds of the shafts.) The turbomachinery aerodynamic performance, particularly of the compressor, is sensitive to $N/\sqrt{c_p T_{02}}$; if this parameter becomes too high the efficiency falls precipitously (see Sections 11.3 and 11.4) and there is a risk of the self induced aero-elastic vibration known as flutter.

---

[1] The parameter $N/\sqrt{c_p T_{02}}$ is derived from a non-dimensional term $ND/\sqrt{c_p T_{02}}$, which is proportional to the relative Mach number into the first set of rotor blades. It is normally understood that T_0 is the stagnation temperature *into* the component or engine.

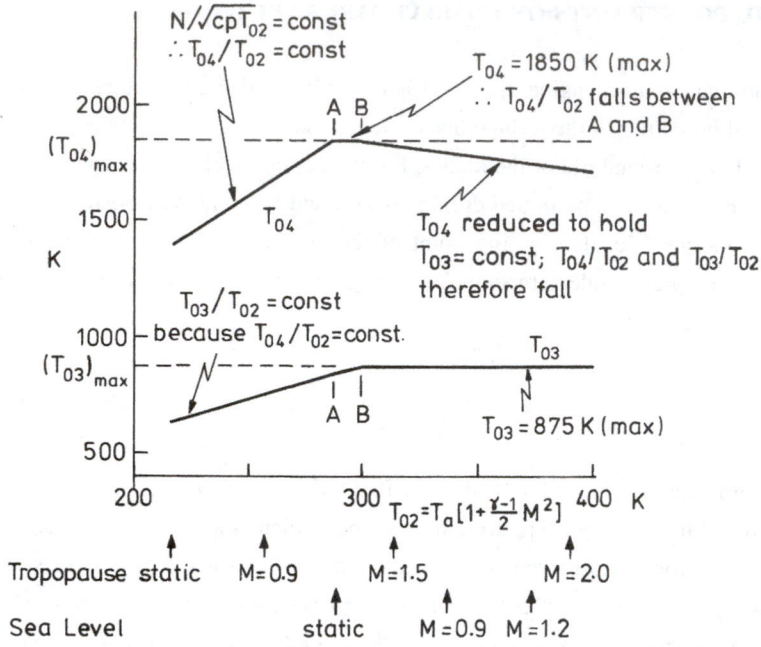

Figure 15.9. Turbine inlet temperature T_{04} and compressor delivery temperature T_{03} versus engine inlet temperature T_{02}. Engine overall pressure ratio $p_{03}/p_{02} = 30$ at the design point when $T_{02} = 288$ K. Compressor polytropic efficiency 0.90.

The LP compressor experiences much larger changes in $N_L/\sqrt{c_p T_{02}}$ as T_{04}/T_{02} is varied than the corresponding changes in $N_H/\sqrt{c_p T_{023}}$ for the HP compressor. (Note that for fixed T_{04}/T_{02} the ratios T_{023}/T_{02} and N_H/N_L are both constant.)

The maximum allowable value of N is primarily limited by the mechanical stresses in the discs holding the rotor blades. If N were held at its limit a higher than acceptable value of $N/\sqrt{c_p T_{02}}$ could occur if the inlet temperature were lower than the value used in design; this could occur, for example, with an engine designed for sea-level static conditions and flown at low Mach number at high altitude or operation at sea-level static conditions with temperature well below 288 K. By holding T_{04}/T_{02} at is its sea-level design value when T_{02} is below the design value, this over-speeding is prevented. When the inlet temperature T_{02} is higher than the value used in the design it is unlikely that the design value of $N/\sqrt{c_p T_{02}}$ will be exceeded because this would require T_{04} to increase too; what normally happens as T_{02} rises is that T_{04}/T_{02} and $N/\sqrt{c_p T_{02}}$ are both reduced because the limits on T_{04} and N are reached.

The effects of the constraints imposed on T_{04}, T_{03} and $N/\sqrt{c_p T_0}$ are illustrated in Figure 15.9 for an engine with an overall pressure ratio of 30 at design. The abscissa shows the inlet stagnation temperature and the ordinate shows the turbine inlet temperature and the compressor delivery temperature. The point marked **A** is designated the design point and has been selected here to be at $T_{02} = 288$ K, the sea-level temperature for the standard atmosphere; at this condition

$T_{04} = 1850$ K is selected, the maximum allowed. Also the design is such that at **A** non-dimensional speed $N/\sqrt{c_p T_0}$ and all the pressure ratios have their maximum values; this is true too along the line to the left of **A** because along this line the engine is at the same non-dimensional condition. Constant T_{04}/T_{02} is achieved along the line to the left of **A**, as inlet temperature T_{02} is reduced, by reducing the fuel supply to lower T_{04} and thereby maintain the temperature ratio constant. Operation along the line to the left of **A** could be during take-off on a cold day ($T_a < 288$ K) or could be a high altitude and low flight speed. To the right of **A** the same T_{04}/T_{02} cannot be maintained without exceeding the limit on T_{04}. Therefore moving to the right of **A** T_{04} is held constant whilst T_{04}/T_{02} is reduced and so too the other engine non-dimensional pressure ratios and variables such as $N/\sqrt{c_p T_0}$.

To the left of **A** the compressor delivery temperature varies in proportion to T_{02} since the pressure ratios are held constant. Between **A** and **B** however the compressor delivery temperature increases but more slowly than in proportion to T_{02} because the pressure ratio is decreasing in line with the decrease in T_{04}/T_{02}. At point **B** the compressor delivery temperature has reached its upper limit. If T_{02} is increased further the overall pressure ratio must be decreased to reduce T_{03}/T_{02}; reducing the pressure ratio is achieved by lowering the turbine delivery temperature T_{04} below the maximum. In other words from point **B** to the right the engine performance is limited by compressor delivery temperature.

In the case shown the range of engine inlet temperatures which cause the engine output to be limited by the turbine inlet temperature, the region from **A** to **B**, is quite small. Much larger are the regions where it is non-dimensional rotational speed (to the left of **A**) or compressor delivery temperature (to the right of **B**) which restricts thrust. **A** is at the value of T_{02} where maximum T_{04} and maximum $N/\sqrt{c_p T_0}$ both occur. The designer is able to select the value of T_{02} where **A** occurs, thereby fixing the design point of the engine. (Equivalently, point **A** is at the value of T_{02} where maximum T_{04} and maximum T_{04}/T_{02} coincide.) The position of **B** is less arbitrary because it follows directly from the choice of overall pressure ratio compressor efficiency and compressor disc capability.

It would seem sensible to put **A** where the highest value of T_{04}/T_{02} will be achieved, that is where T_{02} is low – but there is a snag. If the engine is going to have to operate efficiently at higher values of T_{02} it is going to be operating off-design. The further to the left is point **A** (i.e. the lower value of T_{02} selected for the design point) the further off-design some high-speed flight conditions will be. Selection of point **A** is therefore an important design choice which reflects the most important duties expected of the engine.

Exercise
15.7 Find the overall pressure ratio at which the points A and B in Figure 15.9 coincide. Take polytropic efficiency for the compression system to be 0.9.   (**Ans:** 33.1)

For an aircraft flying at the tropopause, $T_a = 216.65$, determine the Mach number at which T_{02} is equal to 288.15 K, the ambient temperature at sea level?   (**Ans:** 1.285)

15.5 ENGINE RATING

Military engines now use electronic control systems to ensure that operation does not exceed one of the many restrictions. (FADEC – full authority digital electronic control is the abbreviation for the control system used on recent engines.) Some of the selection remains, of course, in the hands of the pilot, but nevertheless there are restrictions on the time for which certain engine operating conditions may be maintained. The terms frequently used to refer to critical engine operating points are:

Combat rating

This is operation with the afterburner lit (if one is installed) and the engine itself operating at its maximum allowed condition (which may correspond to maximum turbine inlet temperature, maximum compressor delivery temperature or maximum $N/\sqrt{c_p T_0}$). Operation is normally limited to a short period, say 2.5 minutes.

Maximum dry rating

This gives the maximum performance without the afterburner. In the simplest mode of operating the afterburner the engine condition upstream of jet pipe would be the same as that in 'Combat Rating'.

Intermediate or military rating

In this condition operation is allowed for much longer, about 30 minutes. The afterburner is not normally lit for this condition.

Maximum continuous rating

This is the condition at which the engine may be operated without restriction on operating time.

SUMMARY OF CHAPTER 15

Just as there is no single aircraft type to fulfil all the various roles in an optimum way, so too with the engine the final choice must be a compromise. The process of selection for the engine, and for the aircraft, will involve numerous simulations to establish how well each of the many combinations accomplishes the various missions attempted. A good combination will be able to accomplish most, but it may be that a small number will need some compromise, such as additional fuel supplies from tanker aircraft. In general it can be said that the optimum engine will have high specific thrust without use of the afterburner (i.e. 'dry') if the combat phases of a mission use a large fraction of the total fuel; such a mission would be of the intercept type. Conversely an aircraft with a primary mission for air superiority (the classic fighter) is expected to use less than half its fuel in the combat phase and fuel consumption during cruise and loiter will be more important; for such an aircraft a relatively small engine giving adequate 'dry' thrust for many phases of the mission, but using the afterburner for the short time of combat, will be best.

The requirement to have high thrust–weight ratio directs the engine towards a low bypass ratio with a high jet velocity; as a way of categorising engines the specific thrust is rather better than the bypass ratio. With the exception of the very specialised engine for the Harrier, all combat engines mix the core and bypass streams before the final propulsive nozzle, obtaining thereby a small increase in thrust. In most cases an afterburner is installed between the mixing plane and the final nozzle. If there is an afterburner it is essential that the final nozzle has a variable throat area; for supersonic flight it is common, though not universal, to have a convergent–divergent nozzle, though the extent of the divergent portion is usually less than that to give isentropic expansion down to ambient pressure.

For aircraft with a major mission which involves flight at Mach numbers approaching or exceeding 2.0 a variable inlet is common so as to obtain small shock losses from a succession of (up to 3) oblique shocks. For aircraft having their primary role at a lower speed the optimum intake is likely to involve something simpler and lighter without variable ramps or vents.

A crucial decision in the design and specification of an engine is the choice of design point at which the non-dimensional variables such as T_{04}/T_{02}, overall pressure ratio, fan pressure ratio and non-dimensional rotational speeds occur simultaneously with the highest allowed turbine inlet temperature, T_{04}. If this occurs at a relatively high inlet temperature it implies that the maximum thrust capability of the engine is not being achieved at low flight speeds and high altitudes; if it occurs at too low an inlet temperature the engine may be a long way off-design, with much reduced pressure ratio, at high flight speeds. It turns out that for many designs there is only a small region in which limited by the maximum allowable turbine inlet temperature. The subsequent chapters will take the required thrusts and operating points from Chapter 14 and use these with engines of the type discussed in this chapter.

CHAPTER 16

DESIGN POINT FOR A COMBAT ENGINE

16.0 INTRODUCTION

In this chapter we will consider three separate engine designs corresponding to distinct operating conditions. For convenience here the three design points are at the tropopause (altitude 11 km; standard atmosphere temperature 216.65 K and pressure 22.7 kPa) for Mach numbers of 0.9, 1.5 and 2.0. The thrusts required for these conditions were determined in Exercise 14.4a. At each condition a separate engine is designed – this is quite different from designing the engine for a single condition and then considering the same engine operation at different conditions, referred to off-design, which is the topic of Chapter 17.

For this exercise all design points will correspond to the engine being required to produce maximum thrust, even though the ultimate suitability of an engine for its mission may depend on performance, particularly fuel consumption, at conditions for which the thrust is very much less than maximum. The designs will first be for engines without an afterburner (operation 'dry') and then with an afterburner; the afterburner will be assumed to raise the temperature of the exhaust without altering the operating condition of the remainder of the engine so the stagnation pressure entering the propulsive nozzle is unchanged.

The engines considered will all be of the mixed turbofan type – such an engine was shown in Figure 15.1 with a sketch showing the station numbering system adopted. Note that the numbering shows station 13 downstream of the fan in the bypass and station 23 downstream of the fan for the core flow; in the present simplified treatment it will be assumed that $p_{023} = p_{013}$ and that $T_{023} = T_{013}$. There are small losses associated with mixing of core exhaust and bypass stream, but these will be neglected here. As a result, if the fan pressure ratio is fixed then so too is the pressure at outlet from the turbine and the conditions through the core are also determined. Fixing the pressure ratio across the fan is a more direct way of specifying properties inside the engine than, for example, the bypass ratio; the pressure ratio is also the dominant term in the expression for jet velocity and therefore for gross and net thrust.

In comparing different designs some parameters need to be held constant. In the present case a uniform technology standard will be assumed, so maximum temperatures and component efficiencies are maintained equal for all the designs. This is the subject of the next section.

Table 16.1   Parameters and constraints assumed throughout Chapters 16–18

Turbine inlet temperature	$T_{04} \not> 1850\,\text{K}$
Compressor outlet temperature	$T_{03} \not> 875\,\text{K}$
LP Compressor *polytropic* efficiency[1]	$\eta_{pc} = 0.85$
HP Compressor *polytropic* efficiency	$\eta_{pc} = 0.90$
HP & LP Turbine *polytropic* efficiency	$\eta_{pt} = 0.875$
Cooling air supplied downstream of HPT NGV	12% of HP compressor inlet air

16.1 TECHNOLOGY STANDARD

The standard of technology described in this section will be used throughout the treatment of the fighter engine in Chapters 16–18. The standard adopted here is believed to be broadly in line with what designers inside major companies are using at the time of writing, but Section 16.7 explores the effect of changes in many of the parameters assumed. The most basic parameters defining the technology standard are set out in Table 16.1, but other aspects of the general treatment follow in the remainder of the section.

The turbine entry temperature is actually the temperature downstream of the HP turbine nozzles and upstream of the rotor, after the cooling air for the nozzles has mixed out. The maximum compressor delivery temperature is appropriate for blades and discs made of titanium alloy; using nickel based alloy a temperature perhaps 100 K higher might be acceptable, but these would be heavier and the maximum rotational speed would be lower for reasons of mechanical stress. The chosen values of both T_{04} and T_{03} are close to the current maxima for this type of engine.

For convenience the *polytropic* efficiencies will be used in this and the following chapter, but the conversion between polytropic efficiency and isentropic (sometimes called adiabatic) efficiency is explained in Section 11.4. Note that the efficiencies are generally lower than those for the civil transport aircraft. These efficiency values are representative for maximum thrust operation; at reduced thrust the Mach numbers inside the engine are lower and the component efficiencies will then generally be higher.

It is essential to include the cooling flows in cycle analysis if appropriate magnitudes are to be found for the specific thrust. It will be assumed that 20% of the air which enters the HP compressor is used for turbine cooling. Of this, 8% is used to cool the HP nozzles and its effect is included in the temperature quoted as turbine entry temperature. Another 8% is used to cool the HP rotor; in the simple cycle analysis here this air, at compressor delivery pressure and temperature p_{03} and T_{03}, is taken to mix with the gas leaving the HP rotor where its only effect is to lower the

---

[1] The polytropic efficiency used for the LP compressor may appear low by modern standards but at maximum speed this value is not unreasonable and at part speed about $\eta = 0.90$ would be more representative. This issue is returned to in Chapter 18.

temperature. A further 4% is used to cool the LP turbine and this is mixed out at LP turbine rotor exit.

The turbine inlet temperature can be increased by better cooling technology and by using more cooling air. For the compressor, however, the exit temperature can only be allowed to rise if there are advances in material science, since there is no air at adequate pressure and lower temperature with which to cool the compressor.

A more detailed analysis would include stagnation pressure loss in the combustor, in the bypass duct and in the jet pipe. In each the stagnation pressure falls by of the order of 4% of the local stagnation pressure. There is also a loss in stagnation pressure associated with the temperature rise in the afterburner. For the present purpose all these losses can be neglected in the interest of simplicity without seriously distorting the estimates. Loss of stagnation pressure in the intake is neglected at subsonic flight speeds, but included for supersonic flight by the relation from MIL-E-5007/8 introduced in Section 15.2.

The nozzle will be assumed to provide a fully reversible expansion to the ambient pressure, though this is at best an oversimplification even with a convergent–divergent nozzle. Many engines operate with only a convergent final nozzle and some significant thrust loss is inevitable at high pressure ratios; this is more serious for high-speed flight because of the large rise in pressure in the inlet.

When the afterburner is lit the maximum flow of fuel is determined by the ability to consume the oxygen and the temperature approaches the stoichiometric value; here a temperature of 2200 K will be assumed. Some engines in service allow a gradual or staged increase in fuel supplied to the afterburner, giving a variable degree of thrust boost, but it will be presumed here that whenever the afterburner is used this maximum temperature is produced. As noted above, any additional loss in stagnation pressure associated with combustion in the afterburner is neglected.

In treating combat engines the value of specific heat will be obtained by specifying γ and the gas constant R. It is a very satisfactory approximation to take $R = 287 \, \text{J} \, \text{kg}^{-1} \text{K}^{-1}$ throughout. For unburned air it will be assumed that $\gamma = 1.40$, giving $c_p = 1005 \, \text{J} \, \text{kg}^{-1} \text{K}^{-1}$, whilst for products of combustion $\gamma = 1.30$ giving $c_{pe} = 1244 \, \text{J} \, \text{kg}^{-1} \text{K}^{-1}$. (Both these values represent simplifications, with γ being a function of temperature, as shown in Figure 11.1.)

Exercise

Note that many of these calculations are repeated with a change in flight Mach number. The use of a programme or spreadsheet has much to recommend it.

16.1 An aircraft flies at the tropopause, 11,000 m, at which the ambient temperature is 216.65 K. Find the maximum overall pressure ratio for the engine at flight Mach numbers of 0.9, 1.5 and 2.0 given that the maximum allowable temperature at outlet from the compressor is 875 K. Assume that the *polytropic* efficiency of the combined compressors is 0.875. (**Ans:** 45.4; 23.0; 11.9)

Mixing

Whenever mixing takes place there is a loss in stagnation pressure, though there may be a rise in static pressure. The process depends on the mass flow, momentum and energy of each stream.

Mixing takes place between the turbine cooling air and the main flow, but this only involves a small fraction of the total air. A much more significant mixing process takes place between the core and the bypass streams downstream of the LP turbine. A detailed calculation could be performed for mixing of core and bypass, though the loss in pressure is only a few per cent and for the present it is assumed that the mixing takes place without loss in stagnation pressure. As discussed in Section 15.2 it is assumed that the stagnation pressure of the gas downstream of the LP turbine is equal to the stagnation pressure downstream of the LP compressor, that is

$$p_{05} = p_{013},\qquad\qquad\qquad (16.1)$$

and this in turn is equal to the stagnation pressure at the nozzle throat p_{08} and the nozzle exit p_{09}. The assumptions for stagnation pressure in the bypass and in the duct leading to the nozzle exit can be justified only by recognising that the losses rise steeply with Mach number and the Mach number of the flow leaving the turbine and in the bypass duct is typically less than 0.5.

The core stream at T_{05} will be assumed to mix with the cooler bypass stream at T_{013} to a uniform temperature T_{06} prior to entering the propulsive nozzle. In 'dry' operation (that is, without the afterburner in use) the jet temperature is given approximately by

$$c_{pe}T_{05} + bpr\, c_p T_{013} = (1 + bpr)\, c_{pm} T_{06},\qquad\qquad (16.2)$$

where c_p, c_{pe} and c_{pm} are the specific heats of the air prior to combustion (i.e. in the bypass stream), of the gas leaving the turbine and of the flow mixed out before it enters the propulsive nozzle, as in Equation 15.4. (This is covered in more detail in Equation 16.10.)

When operating with afterburner lit the temperature at the nozzle throat $T_{08} = T_{0ab}$ is effectively fixed by the fuel flow; it will be assumed that the combustion leads to a negligible loss in pressure in the jet pipe and that $\gamma = 1.30$, $c_p = 1244$ J kg^{-1}K^{-1} for the gas. When operating with the afterburner lit the gas temperatures are so high that some of the bypass stream is used to cool the walls of the jet pipe and the walls of the nozzle; this is neglected here and T_{0ab} is taken to be the temperature when the cooling air has mixed out and the flow is uniform.

16.2 GENERAL ENGINE SPECIFICATION

With a mixed turbofan of the type shown in Figure 15.1 with fixed geometry and a fixed level of technology, two parameters are sufficient to specify the whole engine type. One is the overall pressure ratio p_{03}/p_{02} and the other is the pressure ratio between inlet to the compressor and inlet to the final nozzle, p_{08}/p_{02}. This second pressure ratio is, neglecting pressure losses in the duct and mixing, equal to the LP compressor (fan) ratio p_{013}/p_{02}. Thus specifying p_{03}/p_{02} and p_{013}/p_{02}, together with T_{04}/T_{02}, is sufficient to determine the engine performance, more specifically its specific thrust and specific fuel consumption.

For a given overall pressure ratio p_{03}/p_{02} and temperature ratio T_{04}/T_{02} there is a fixed amount of power to be delivered to the fan. If the chosen fan pressure ratio is raised, the mass flow rate compressed in it must be reduced, in other words the bypass ratio must fall. A small change in pressure ratio can produce a large change in bypass ratio because an increase in pressure ratio not only increases the work per unit mass flow required by the fan but it also increases the back pressure to the LP turbine and therefore reduces its power output.

Overall calculations

LP and HP compressors

With overall pressure ratio and fan pressure ratio selected as input parameters, the calculations of the engine cycle become comparatively straightforward. The stagnation conditions at entry to the compressor, station 2, are determined by the altitude, flight Mach number and losses in the intake (the losses are normally only substantial at supersonic flight speeds). The fan pressure ratio fixes the stagnation pressure at entry to the HP compressor p_{023}. As noted above it is assumed that the fan delivers equal stagnation pressure and stagnation temperature to both the bypass duct and core,

$$p_{023} = p_{013} \quad \text{and} \quad T_{023} = T_{013}.$$

This is a simplification which is not necessary but which simplifies the analysis somewhat. For a known fan efficiency η, the corresponding temperature and pressure ratio for the core flow in the fan is

$$T_{023}/T_{02} = (p_{023}/p_{02})^{(\gamma-1)/\eta\gamma}$$

with a similar expression relating the stagnation temperature and pressure at combustor inlet T_{03} and p_{03}.

Combustor

The fuel flow to the core combustor $\dot{m}_f$ required to produce a turbine inlet temperature T_{04} is given by the expression found in Section 11.5. Not all the air passes through the combustor, since some is used to cool the turbines. The mass flow rate of fuel should be added to the air entering the HP turbine. The air cooling the HP turbine nozzle is assumed to be mixed with the combustion products at nozzle exit, station 4, and therefore is accounted for in the definition of turbine inlet temperature T_{04}. The mass flow of gas leaving the HP turbine nozzle, station 4, is therefore $\dot{m}_{a4} + \dot{m}_f$ with specific heat c_{pe}. Since the turbine entry temperature is defined as the mixed-out temperature at turbine stator exit, the air used to cool these nozzle guide vanes is included in the balance

$$\dot{m}_f LCV = (\dot{m}_{a4} + \dot{m}_f)c_{pe}(T_{04} - 298) - \dot{m}_{a4}c_p(T_{03} - 298), \tag{16.3}$$

where $\dot{m}_{a4}$ is the air mass flow at turbine nozzle exit. This gives an overestimate of the mass flow of fuel because it assumes that the specific heats have values independent of temperature. The fuel mass flow should really be obtained from enthalpy at inlet and outlet. The combustion is not quite complete by the time the gases leave the combustor and in a more detailed treatment a combustion efficiency would multiply the calorific value. The combustion efficiency is likely to be in excess of 98% over most of the operating regime.

HP turbine and cooling

The HP turbine power must equal the HP compressor power, with $\dot{m}_a$ being the total core mass flow of air through the HP compressor, so that

$$\dot{m}_a c_p (T_{03} - T_{023}) = (\dot{m}_{a4} + \dot{m}_f) c_{pe} (T_{04} - T_{045}), \tag{16.4}$$

where T_{03} is known from the pressure ratio, compressor efficiency and inlet temperature, HP turbine exit temperature T_{045} can be found from Equation 16.4. Then, knowing T_{045}/T_{04}, the HP turbine pressure ratio can be calculated with the turbine polytropic efficiency and γ for the combustion products using the polytropic relation

$$p_{045}/p_{04} = (T_{045}/T_{04})^{\gamma/(\gamma-1)\eta}. \tag{16.5}$$

Both T_{045} and p_{045} are therefore known in terms of conditions out of the combustor.

   The mass flow of cooling air to the HP turbine rotor is equal to the difference in mass flow between nozzle HP turbine exit and HP rotor exit and is equal to $\dot{m}_{a45} - \dot{m}_{a4}$. The cooling air at compressor delivery temperature T_{03} is mixed at a constant pressure with the gas leaving the HP turbine rotor to give a mixed temperature $T_{045'}$,

$$(\dot{m}_{a45} + \dot{m}_f) c_{pe} T_{045'} = (\dot{m}_{a45} - \dot{m}_{a4}) c_p T_{03} + (\dot{m}_{a4} + \dot{m}_f) c_{pe} T_{045}. \tag{16.6}$$

Downstream of the mixing the specific heat capacity and γ are taken here to be the values for the burned gas, which is clearly an approximation but one which should suffice since the cooling air only represents a small fraction of the total gas flow.

LP turbine

The pressure and mixed-out temperature entering the LP turbine are p_{045} and $T_{045'}$. Across the LP turbine

$$T_{05}/T_{045'} = (p_{05}/p_{045})^{\eta(\gamma-1)/\gamma} = (p_{013}/p_{045})^{\eta(\gamma-1)/\gamma}. \tag{16.7}$$

The second equality in Equation 16.7 follows because for the mixed turbofan $p_{05} = p_{013}$, the pressure in the bypass duct downstream of the fan. Since p_{05} is fixed by the fan pressure ratio, the pressure ratio and temperature ratio across the LP turbine are determined, and so T_{05} can be found. The power from the LP turbine, passing a mass flow $(\dot{m}_{a45} + \dot{m}_f)$, must equal the power into the fan. The fan passes a mass flow $(1 + bpr)\dot{m}_a$ so that the power balance for the LP shaft is, with $T_{013} = T_{023}$,

$$(1 + bpr)\dot{m}_a c_p (T_{013} - T_{02}) = (\dot{m}_{a45} + \dot{m}_f) c_{pe} (T_{045'} - T_{05}). \tag{16.8}$$

There is then a further mixing process downstream of the LP turbine where the cooling air to the LP turbine is introduced back into the main stream, so

$$(\dot{m}_{a5} + \dot{m}_f) c_{pe} T_{05'} = (\dot{m}_{a5} - \dot{m}_{a45}) c_p T_{03} + (\dot{m}_{a45} + \dot{m}_f) c_{pe} T_{05}. \tag{16.9}$$

When the fan pressure ratio and overall pressure ratio are given as input parameters, application of Equations 16.3 to 16.9 allow direct calculation of all temperatures and pressures in the engine and the bypass ratio. With all the cooling flow mixed back with the main flow it follows that $\dot{m}_{a5} = \dot{m}_a$ the mass flow entering the core compressor.

Mixing in the jet pipe and the afterburner

Downstream of the LP turbine the core and the bypass stream mix out and when the afterburner is not lit the uniform temperature T_{06} is given by

$$\left(\dot{m}_a + \dot{m}_f\right)c_{pe}T_{05'} + \dot{m}_a \, bpr \, c_p T_{013} = \left\{\dot{m}_a \left(1 + bpr\right) + \dot{m}_f\right\}c_{pm}T_{06}, \tag{16.10}$$

where c_{pm} is the specific heat of the mixed flow.

When the afterburner is lit the flow does not mix out to any significant extent before burning. Nevertheless the mixed-out temperature T_{06} without the afterburner can be used in the energy balance to find the fuel flow to the afterburner $\dot{m}_{fab}$ needed to raise the temperature to the required level by the nozzle throat, $T_{08} = T_{0ab}$:

$$\dot{m}_{fab}LCV = \left\{\left(\dot{m}_a \left(1 + bpr\right) + \dot{m}_f\right) + \dot{m}_{fab}\right\}c_{pe} \left(T_{0ab} - 298\right)$$
$$- \left\{\dot{m}_a \left(1 + bpr\right) + \dot{m}_f\right\}c_{pm} \left(T_{06} - 298\right). \tag{16.11}$$

The chosen temperature at the nozzle may be as high as the stoichiometric temperature. The combustion process is not normally complete by the time the gases enter the nozzle, 90% is a reasonable estimate and in a more detailed treatment an afterburner combustion efficiency would be introduced.

16.3 THE SELECTION OF OVERALL PRESSURE RATIO

A range of calculations has been run for the three possible design conditions ($M = 0.9$, 1.5 and 2.0 at the tropopause) using the Equations 16.3 to 16.11 and some of these results are shown here in Figures 16.1 and 16.2. It is more appropriate to specify engines according to the fan pressure ratio than the bypass ratio. Nevertheless, *bpr* is used here because it characterises the type of engine and because the optimum pressure ratio is so different for the three Mach numbers used for the design points. The range of bypass ratios selected covers most of the design choices likely: zero (giving the case of a turbojet) 0.5 and 1.0. In all cases the technology level is held at that in Section 16.1.

The curves in Figure 16.1 show specific thrust (units m/s) and specific fuel consumption (*sfc*, units $\text{kg}\,\text{h}^{-1}\text{kg}^{-1}$) versus overall pressure ratio p_{03}/p_{02} for an engine flying at $M = 0.9$. A curve is drawn for each of the three arbitrarily chosen bypass ratios. The turbojet has, as expected, a higher specific thrust and a higher fuel consumption, whereas the highest bypass ratio is the lowest for both specific thrust and *sfc*. The effect of overall pressure ratio is similar at all three bypass ratios. Specific thrust is a maximum for a pressure ratio of about 15, whilst the fuel consumption shows a pronounced decrease with pressure ratio continuing beyond the upper level shown. Using

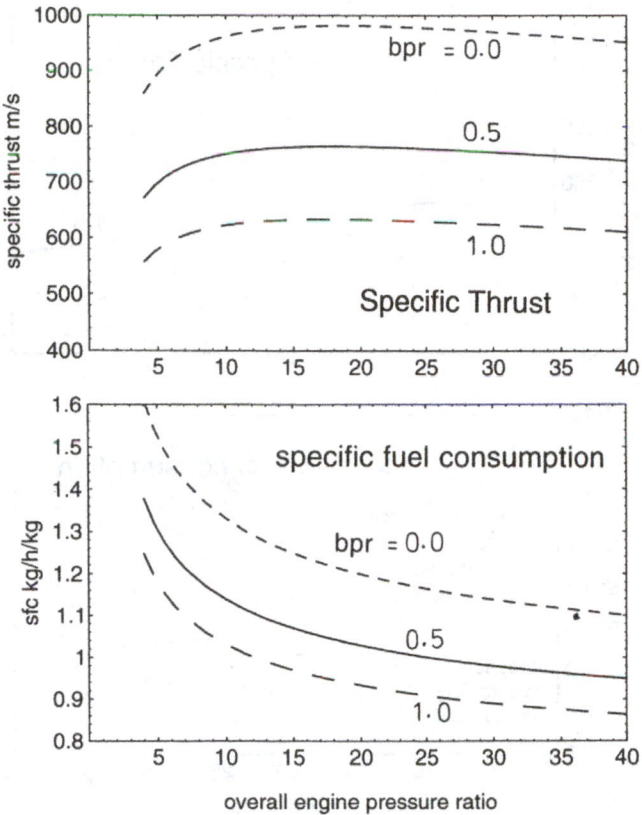

Figure 16.1. Specific thrust and specific fuel consumption (*sfc*) versus overall
pressure ratio for design point operation at $M = 0.9$ at tropopause. Dry.
(Constant technology standard – fan pressure ratio varies.)

an overall pressure ratio of, say, 30 gives a reduction in specific thrust of about 1.5% relative to the peak, but a corresponding reduction in *sfc* of about 10%.

Figure 16.2 shows specific thrust and fuel consumption at Mach numbers of 0.9, 1.5 and 2.0 for a fixed bypass ratio of 0.5. The overall pressure ratio for maximum specific thrust decreases as the Mach number increases, an effect which is understandable because of the increase in compression upstream of the engine as a result of the forward speed. If the Mach number were increased to 2.5 it would be found that the pressure ratio for maximum thrust is less than unity, indicating, that a ram-jet would give the highest specific thrust as anticipated in Chapter 15, though still with a high fuel consumption.

The selection of overall pressure ratio is, for the present purpose, somewhat arbitrary. An important balance exists between benefits of lower fuel consumption and higher thrust. The crosses in Figure 16.2 give upper bounds on *opr* and T_{03} because the design would give highest thrust possible. There is an additional factor, which is that higher pressure ratio means more turbomachinery with its attendant weight; the wish to achieve the desired pressure ratio with the least weight leads to lower component efficiencies than would be possible if the pressure ratios

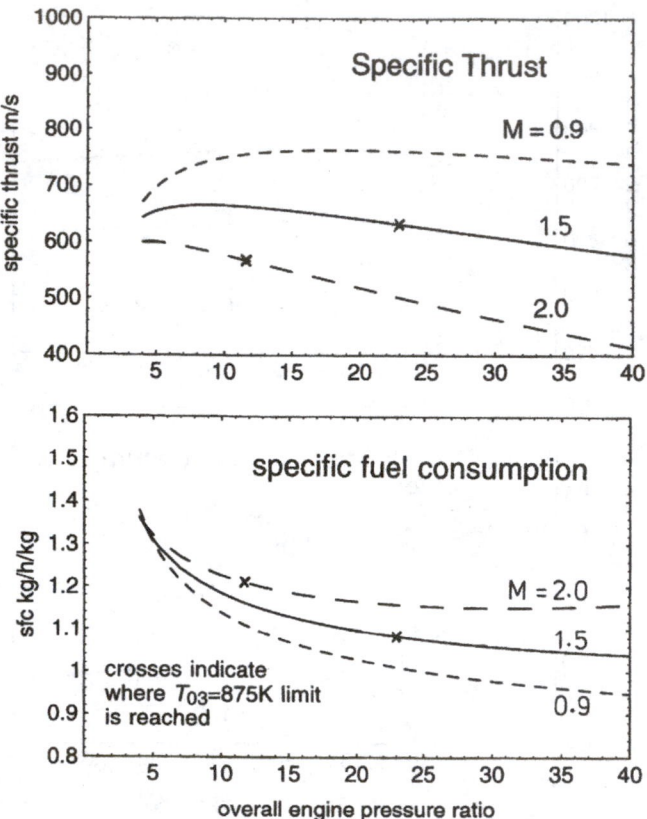

Figure 16.2. Specific thrust and specific fuel consumption (*sfc*) versus overall pressure ratio for design point operation, $bpr = 0.5$ throughout. Dry. (Constant technology standard – fan pressure ratio varies.)

could be reduced somewhat. As a design decision for the present work pressure ratios will be chosen as 30, 20 and 10 for design-point Mach numbers of 0.9, 1.5 and 2.0 respectively. These pressure ratios are also acceptable for the compressor delivery temperature, as Exercise 16.1 shows.

16.4 THE SELECTION OF FAN PRESSURE RATIO

For the mixed turbofan the selection of the fan pressure ratio fixes the pressure on the downstream side of the turbine; neglecting mixing losses and the pressure loss in the jet pipe, the stagnation pressure entering the propulsive nozzle is therefore the stagnation pressure leaving the fan. The jet velocity depends on the pressure ratio across the nozzle and the stagnation temperature entering the nozzle. For a given jet velocity, which implies a given specific thrust, the most efficient engine (i.e. the engine with the lowest specific fuel consumption) is the one with the lowest jet temperature. Raising the bypass ratio lowers the jet temperature so for a given jet velocity a benefit of high bypass ratio is realized as lower jet temperature. The interconnection of the fan pressure ratio,

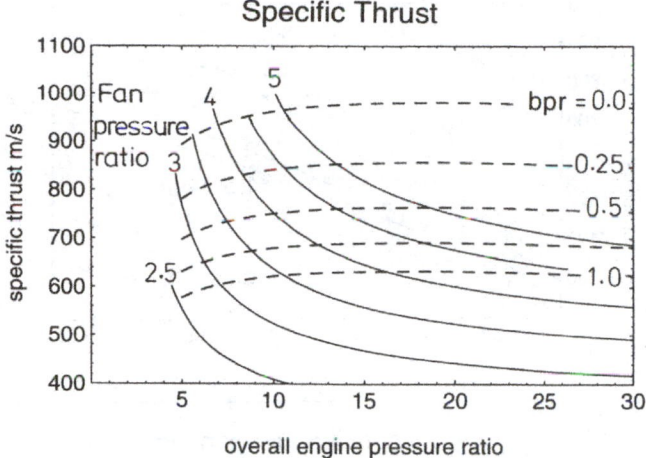

Figure 16.3. Specific thrust versus overall pressure ratio for design point operation at $M = 0.9$ at tropopause. Dry. (Constant technology standard.)

overall pressure ratio and bypass ratio on specific thrust are shown in Figure 16.3. With low overall pressure ratios the specific thrust changes rapidly with bypass ratio and thus changes rapidly with fan pressure ratio.

The selection of fan pressure ratio is complicated by many practical issues. A military fan should be simple and robust but it should also be light. Together these indicate few stages, typically no more than three, with the only variable stators (to cope with off-design operation) being the inlet guide vanes. Together these constraints make it hard to have fan pressure ratios more than about 5.

Specific thrust and specific fuel consumption versus fan pressure ratio are shown in Figure 16.4 for three different overall pressure ratios, all for a flight Mach number of 0.9. For the lowest overall pressure ratio the curve terminates at a fan pressure ratio of about 4.4, since at this condition the bypass ratio has fallen to zero and the engine has become a turbojet. At the overall pressure ratios of 20 and 30 bypass ratio tends to zero at higher fan pressure ratios than those shown in Figures 16.2 and 16.3. Although an increase in overall pressure ratio at a constant fan pressure ratio is beneficial, in that it reduces *sfc*, it also reduces the specific thrust. The reason for this is straightforward – the jet temperature falls as the overall pressure ratio is raised but with fan pressure ratio chosen the pressure into the propulsive nozzle is fixed. In other words the high specific thrust obtained with low overall pressure ratio is obtained at the cost of being less efficient in the conversion of fuel energy into kinetic energy.

Whilst holding overall pressure ratio constant, Figure 16.4 shows that specific thrust rises sharply with fan pressure ratio, a result to be anticipated from the equation for jet velocity. Because a high jet velocity gives a lower propulsive efficiency, the fuel consumption also rises with fan pressure ratio, as also shown in Figure 16.4. The choice of fan pressure ratio, as for the overall pressure ratio, is a compromise between achieving the highest specific thrust and lowest *sfc*. For the fan there is no easily specified optimum because specific thrust and *sfc* rise continuously with

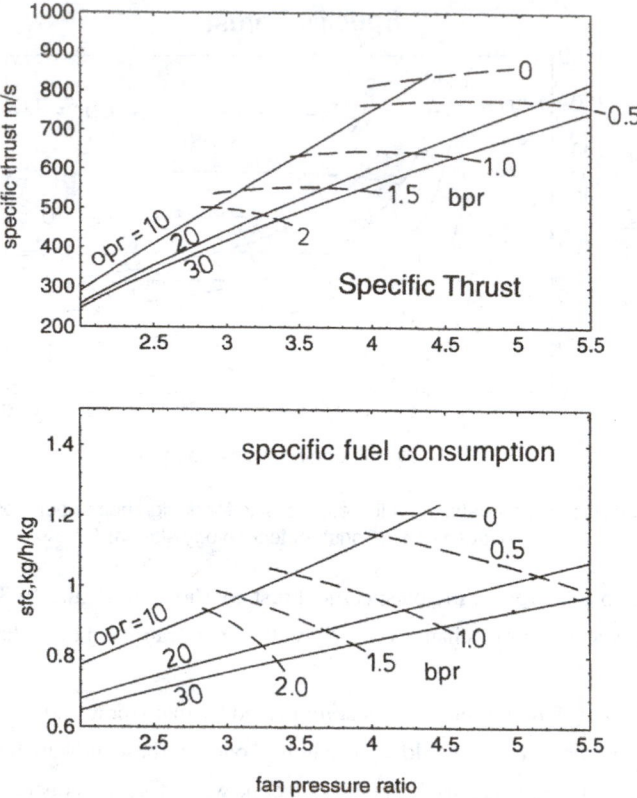

Figure 16.4. Specific thrust and specific fuel consumption (*sfc*) versus fan pressure ratio
for design point operation at $M = 0.9$ at tropopause. Dry. Curves show lines of constant overall
pressure ratio and bypass ratio. (Constant technology standard.)

fan pressure ratio until the engine becomes a turbojet. The final compromise can only be arrived
at when engine weight can be compared with the fuel consumed, the latter depending on mission
details.

Figure 16.5 shows specific thrust and specific fuel consumption versus fan pressure ratio for
the three design conditions $M = 0.9$, 1.5 and 2.0 at the tropopause with overall pressure ratios of
30, 20 and 10 respectively adopted earlier. For the two higher Mach numbers the curves terminate
when the bypass ratio goes to zero whereas for $M = 0.9$ the bypass ratio is still close to 0.5 at the
highest fan pressure ratio of 5.5. The specific thrust varies sharply with fan pressure ratio, increasing
more than threefold over the range shown for $M = 0.9$, whereas the fuel consumption increases
by about 50% in the same range. At the higher Mach numbers the change in specific thrust with
fan pressure ratio is steeper whilst the slope of the *sfc* with *fpr* is similar for all three values of
M. The precise balance of advantage will, as mentioned earlier, depend on the mission, since this
determines the weight of fuel used. For the purpose of this design exercise it seems appropriate to
select somewhat arbitrarily fan pressure ratios of 4.5, 4.0 and 3.0 for flight Mach numbers of 0.9,
1.5 and 2.0 respectively.

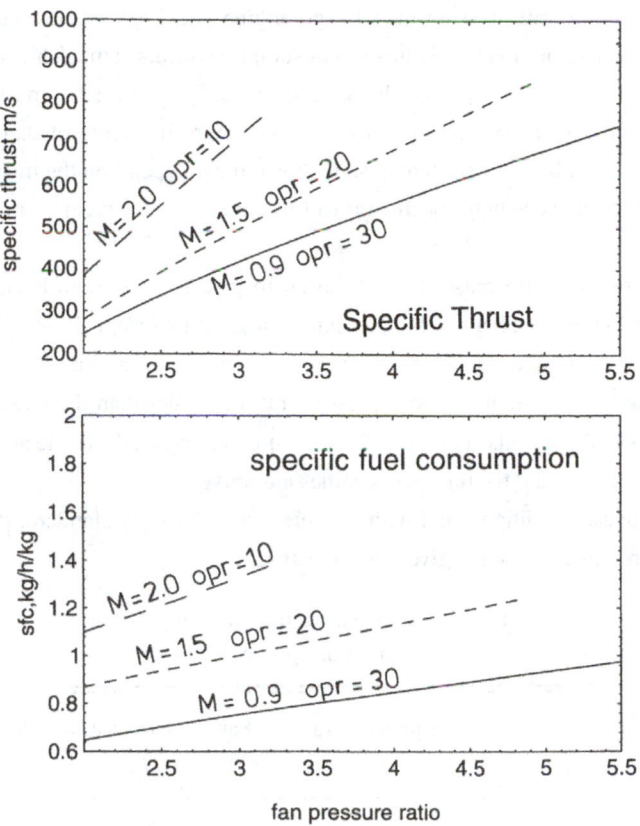

Figure 16.5. Specific thrust and specific fuel consumption (*sfc*) versus fan pressure ratio for design point operation at tropopause. Dry. Curves show combinations of overall pressure ratio (*opr*) and flight Mach number from Table 16.2. (Constant technology standard.)

SUMMARY OF SECTIONS 16.1 TO 16.4

It is appropriate to compare designs using a common technology level, giving maximum turbine entry and compressor delivery temperature and component efficiencies. Simple balances for mass flow and power allow conditions in the engine to be found. Although many sources of loss may be neglected in a simple treatment such as this, it is essential to include the effect of the cooling air flow in the power balance if realistic levels of specific thrust are to be obtained. (This is explored below in Section 16.7.) Because of the mixing in the jet pipe after the turbine and before the final nozzle, an iterative calculation of the engine cycle is required if *bpr* is specified whereas if fan pressure ratio is specified a direct calculation is possible.

The design decisions at this stage are the determination of overall pressure ratio and fan pressure ratio. With engine and turbine inlet temperatures and component efficiencies known, the bypass ratio is determined. As flight Mach number increases the inlet stagnation pressure rises steeply and the overall pressure ratio in the engine which gives maximum specific thrust therefore decreases. The minimum specific fuel consumption occurs at much higher overall pressure ratios than those for maximum specific thrust. The overall pressure ratio must be kept below a value which will produce a compressor delivery temperature in excess of that laid down in the technology level, 875 K here. The actual design choice must depend on the mission to decide the balance of advantage between high specific thrust (giving low engine weight) and low specific fuel consumption.

The fan pressure ratio must also be chosen to give the optimum balance between high specific thrust, calling for the highest fan pressure ratio, and low *sfc*, calling for low fan pressure ratio. As the flight Mach number increases the allowable range of fan pressure ratios narrows but the rate of increase in specific thrust with pressure ratio is larger than the rate of increase in *sfc*. This points to choosing the highest possible fan pressure ratio being desirable at high flight speeds; at Mach numbers close to 2.0 the turbojet becomes attractive.

For tropopause conditions, and with the selected technology standard, the design choices adopted for the present exercise are given in Table 16.2.

Table 16.2   Selected overall and fan pressure ratios at design point

M	Overall pressure ratio p_{03}/p_{02}	Fan pressure ratio p_{023}/p_{02}
0.9	30	4.5
1.5	20	4.0
2.0	10	3.0

Exercise
The use of a programme or spreadsheet has much to recommend it.

16.2* Calculate the fan delivery temperature T_{013} and the compressor delivery temperature T_{03} for the three flight conditions and pressure ratios set out in Table 16.2 (at the tropopause, $M = 0.9, 1.5, 2.0$). Use the efficiencies set out for the technology standard in Table 16.1.

(**Ans:** $T_{013} = 417.3, 500.5, 564.2$ K; $T_{03} = 762.1, 834.3, 826.8$ K)

16.5 THE SIZE OF THE ENGINE FOR 'MAXIMUM DRY' OPERATION

In Chapter 15 certain design points were selected for the engine: $M = 0.9, 1.5$ and 2.0 at the tropopause. After having adopted the particular technology standard laid out in Section 16.1, and after selecting the overall pressure ratio and the fan pressure ratio in Sections 16.3 and 16.4, the operating cycle of the engine is defined and with it the specific thrust and specific fuel consumption. These are for the engine producing its maximum thrust at the flight condition without the use of the afterburner, the 'maximum dry' condition. The mass flow of air and fuel, and the size of the engine at the design point may then be found.

As pointed out in the introduction to this chapter, the three design points are for distinct 'paper' engines and do *not* correspond to the operating conditions which the same engine would adopt at the different flying conditions – this has to be considered in Chapter 17. For the design point consideration the parameters such as the pressure ratios are taken to be fixed independently, whereas when one engine is operated at different flight speeds off -design the pressure ratios become dependent variables.

Exercises

The use of a programme or spreadsheet has much to recommend it.

16.3 Three design cases were given in Table 16.2 and addressed in Exercise 16.2. Adopt the standard of technology listed in Table 16.1 with the HP turbine entry temperature $T_{04} = 1850$ K being defined as the mean temperature at HP stator (nozzle) exit after the nozzle cooling air has been mixed. A further 12% of the air compressed will be added downstream to cool other parts. For each of the three design cases:

a Find the mass flow rate of fuel, $\dot{m}_f$, per unit mass of air entering the engine core, taking $LCV = 43$ MJ/kg. Assume $\gamma = 1.30$, $c_{pe} = 1244$ J kg^{-1}K^{-1} for the products of combustion.
(**Ans:** $\dot{m}_f = 0.0314, 0.0299, 0.0300$)

b Using an energy balance find the temperature at outlet from the HP turbine, when for each unit of air compressed, the mass flow out of the HP turbine nozzle vanes is $(0.88 + \dot{m}_f)$. Using the turbine efficiency given in the technology level, $\eta_p = 0.875$, find the pressure at HP turbine outlet.
(**Ans:** $T_{045} = 1544.4, 1553.7, 1617.0$ K; $p_{045}/p_{04} = 0.409, 0.421, 0.513$)

c The HP rotor cooling air is assumed to mix at constant pressure between the HP rotor and the LP nozzle. If the HP rotor cooling air is 8% of the total compressed, and its temperature is that of the compressor delivery, find the temperature entering the LP turbine $T_{045'}$.
(**Ans:** $T_{045'} = 1467.0, 1482.6, 1540.2$ K)

d The pressure ratio across the LP turbine can be found from the value of p_{045}/p_{04} found in Exercise 16.3b and the specified pressure ratio across the fan, p_{023}/p_{02}, since it is assumed $p_{05} = p_{013} = p_{023}$. Find the temperature T_{05} after the LP turbine. The cooling air to the LP turbine is then mixed with the flow after the turbine (i.e. after T_{05} has been found) to give the mixed-out temperature $T_{05'}$.
(**Ans:** $T_{05} = 1200.1, 1275.5, 1381.9$ K; $T_{05'} = 1177.4, 1252.1, 1354.2$ K)

e Use the LP turbine work to find the bypass ratio by matching it to the work into the fan.
(**Ans:** $bpr = 0.997, 0.361, 0.114$)

f For unit mass flow of air entering the HP compressor the mass flow in the core stream out of the LP turbine is $(1 + \dot{m}_f)$ and the corresponding bypass mass flow is equal to bpr. Using a simple mass weighting system find the mean specific heat c_{pm} for the exhaust stream, and thence the mean value of γ. Use c_{pm} to find the temperature entering the propulsive nozzle.
(**Ans:** $c_{pm} = 1126, 1182, 1222$ J kg^{-1}K^{-1}; $\gamma = 1.342, 1.320, 1.308$, $T_{08} = 844.1, 1086.2, 1289.3$ K)

g Using the results of Exercises 15.3, 16.1 and 16.3d, determine the pressure ratio across the propulsive nozzle, p_{08}/p_a. Take *fpr* from Table 16.2 and use the flight Mach number to get p_{01}/p_a. For $M = 2$ use Equation 15.6 to get the pressure loss in the intake, for lower Mach numbers assume the intake is loss free.
(**Ans:** 7.61, 14.24, 21.71)

h Find the jet velocity assuming isentropic expansion in the propulsive nozzle. Hence find the specific thrust (based on *net* thrust and including the mass flow rate of fuel) and the specific fuel consumption.

(**Ans:** V_j = 876,1104, 1273 m/s; Spec. thrust = 624.5, 686.5, 717.0 m/s;

sfc = 0.888, 1.127,1.325 kg h^{-1}kg^{-1})

16.4 a From Exercise 16.3, which gives the specific thrust at the Maximum Dry condition, determine the mass flow rate of air necessary through each engine to create the thrust for steady level flight at the tropopause for Mach numbers of 0.9, 1.5 and 2.0. (The necessary thrusts are given in Exercise 14.4a.)

(**Ans:** $\dot{m}_a$ = 12.8, 37.9, 61.9 kg/s)

b In the inlet duct just ahead of each engine the flow is purely axial and the Mach number is 0.7. (This is equivalent to saying that the LP compressor will be at the same non-dimensional mass flow at each design-point flight Mach number.) The non-dimensional mass flow $\bar{m}_2 = \dot{m}\sqrt{c_p T_{02}}/(A_2 p_{02})$ is given in terms of γ and Mach number in Section 6.2 – use this to find the area of the duct for each flight Mach number in part (a). Assuming that the diameter of the fan hub is 0.4 times the fan tip diameter, find the engine inlet diameter. (**Ans:** fan inlet diameter D = 0.47 m, 0.58 m, 0.55 m)

16.5 Repeat the calculations of Exercise 16.4 for the aircraft designed to make sustained 3g turns at the tropopause at Mach numbers of 0.9 and 1.5. (The necessary thrusts are given in Exercise 14.4c.)

(**Ans:** $\dot{m}$ = 57.3 kg/s, 57.2 kg/s; fan inlet diameter = 0.985 m, 0.71 m)

16.6* a If the engine were designed for take off at standard sea-level conditions (T_a = 288.15K, p_a = 101.3 kPa) and the same technology level, find the bypass ratio, the specific thrust and the *sfc* operating 'dry' whilst stationary. Take the fan pressure ratio and the overall pressure ratio to be 4.5 and 30 respectively.

(**Ans:** 0.471, 865 m/s, 0.805 kg h^{-1}kg^{-1})

b The static sea-level thrust–weight ratio of the New Fighter is selected to be 0.66 in the 'dry' condition. The mass of the aircraft (18 tonne) and the number of engines (2) is given in Section 13.4. Find the mass flow of air required and the inlet diameter of the engine on the same basis as in Exercises 16.4 and 16.5.

(**Ans:** 67.4 kg/s, 0.681 m)

16.6 THE EFFECT OF THE AFTERBURNER

The choice of specific thrust determines the jet velocity and therefore the kinetic energy of the jet. The kinetic energy determines the minimum work which must be supplied to the air for the thrust, but because the static temperature is invariably higher than that of the atmosphere, the fuel consumption will be greater than this hypothetical minimum. Whilst operating in the 'dry' condition the temperature of the exhaust gas can vary and efficient propulsion coincides with comparatively low exhaust temperature; the benefit of high bypass ratio arises because it lowers the temperature of the jet.

When the afterburner is used the exhaust temperature is determined by the amount of fuel burned in the engine and afterburner. In consequence the efficiency of the engine components does not affect the thrust or fuel consumption in the same way, nor does the bypass ratio. With the afterburner in use the specific thrust and specific fuel consumption are uniquely determined by the fan pressure ratio $p_{08}/p_a = p_{013}/p_a = p_{023}/p_a$ and exhaust temperature $T_{08} = T_{0ab}$ (neglecting losses in the nozzle itself and incomplete combustion of afterburner fuel).

If c_{pe} and γ_e are the specific heat and ratio of specific heats for the products (taken here to be 1244 J kg^{-1}K^{-1} and 1.30 respectively) then the jet velocity is given by

$$V_{jab} = \sqrt{2c_{pe}T_{0ab}\left\{1 - (p_a/p_{08})^{(\gamma_e-1)/\gamma_e}\right\}}.$$

With the temperature at exit from the afterburner, T_{0ab}, fixed, the only variable is, as anticipated above, the pressure ratio across the nozzle. It is essential that the propulsive nozzle throat area increases when the afterburner is lit because the mass flow through the engine should not reduce but the rise in temperature reduces the mass flow per unit area through the choked nozzle. Very often this is arranged so that the pressure in the jet pipe remains unaltered; under this condition the intake, compressors, combustor and turbines are unaware of the afterburner being used. The fan pressure ratio and mass flow therefore do not change between the 'maximum dry' condition (with the afterburner unlit) and the 'combat' condition, with the afterburner lit. Apart from the change in gas properties, the increase in jet velocity with the afterburner is equal to the square root of the temperature ratio T_{0ab}/T_{06}, the ratio of temperature with the afterburner lit to the mixed temperature ahead of it, which is also the temperature without the afterburner lit.

Although the mass flow of air remains unaltered when the afterburner is lit, the mass flow of fuel rises rapidly, and the additional fuel flow $\dot{m}_{fab}$ per unit mass of air through the core $\dot{m}_a$ is then given by Equation 16.9. The specific thrust, the net thrust per unit mass flow of air through the engine, is given by

$$\frac{F_N}{(1+bpr)\dot{m}_a} = \frac{V_{jab}\left\{(1+bpr)\dot{m}_a + \dot{m}_f + \dot{m}_{fab}\right\}}{(1+bpr)\dot{m}_a} - V, \tag{16.12}$$

where $\dot{m}_f$ is the mass flow of fuel in the main combustor and V is the flight speed. Because the net thrust is proportional to the difference between the jet velocity and the flight speed, the proportional increase in net thrust and specific thrust due to the afterburner is substantially greater than the proportional increase in the jet velocity.

Figure 16.6 shows the specific thrust and specific fuel consumption for the engine at the three design points with and without afterburning; the 'dry' results are the same as those shown in Figure 16.5. Of immediate note is the large increase in specific thrust and in specific fuel consumption when the afterburner is used. Whereas for the 'dry' engine the specific thrust increases quite rapidly with fan pressure ratio, the increase is comparatively slow for the afterburning engine; as a result the boost from the afterburner is noticeably bigger for engines with low fan pressure ratios. This is because the temperature of the exhaust gases is lower for the low pressure ratio fan in the 'dry' condition so the increase in temperature of the exhaust when the afterburner is used is that much greater. With afterburning the specific thrust is almost equal at design for the three flight speeds.

For the 'dry' engine the specific fuel consumption increases when the fan pressure ratio is raised, corresponding to higher jet velocity and specific thrust. With the afterburning engine the opposite is true; the specific fuel consumption falls with fan pressure ratio. The fall may be

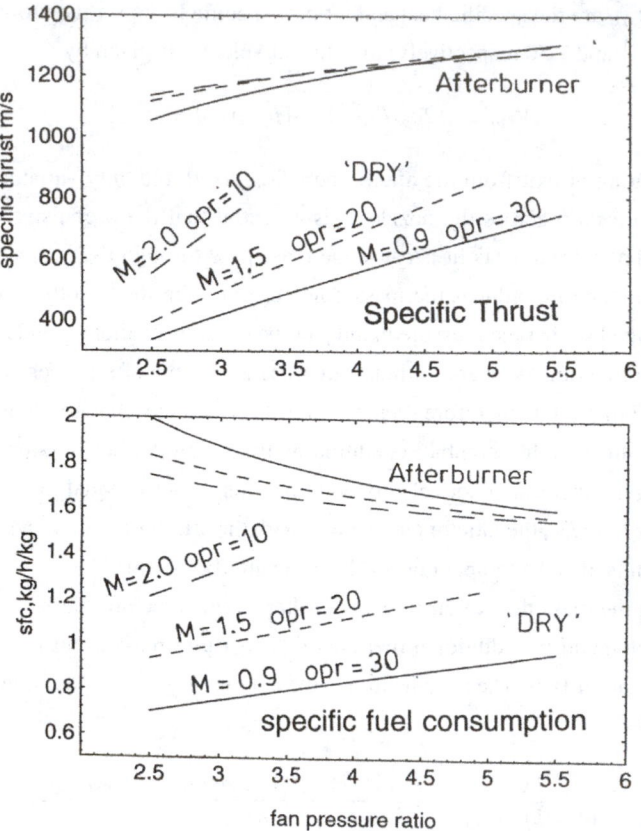

Figure 16.6. Specific thrust and specific fuel consumption (*sfc*) versus fan pressure ratio for design point operation at tropopause with and without afterburner. Curves show combinations of overall pressure ratio (opr) and flight Mach number. (Constant technology standard.)

understood by realising that the stagnation temperature of the jet is fixed with the afterburner lit, whereas the jet velocity, and therefore the thrust, increases as the pressure ratio increases. As a result the thrust for the same fuel input increases with pressure ratio, that is the *sfc* falls. It may also be noted that for $M = 2.0$ the difference between the *sfc* 'dry' and with the afterburner is relatively small.

Exercises

The use of a programme or spreadsheet has much to recommend it.

16.7 a For the same three design conditions used in Exercise 16.3 find the specific thrust and specific fuel consumption with the afterburner producing an exit temperature of 2200 K.

(**Ans:** Specific thrust = 1250, 1233, 1168 m/s; *sfc* = 1.68, 1.66, 1.69 $kg\,h^{-1}kg^{-1}$)

b* Find the specific thrust and *sfc* for the engine of Exercise 16.6 during take off using the afterburner.

(**Ans:** specific thrust = 1341 m/s, *sfc* = 2.27 $kg\,h^{-1}kg^{-1}$)

16.8 Determine the mass flow of air and the fan inlet diameter (on the same basis as in Exercises 16.4 and 16.6) when the afterburner is used to create the necessary thrust.

    a For 1g flight at $M = 0.9$, 1.5 and 2.0   (**Ans:** $\dot{m} = 6.4$, 21.0, 37.9 kg/s; $D = 0.33$, 0.43, 0.43 m)

    b For 3g flight at $M = 0.9$ and 1.5.   (**Ans:** $\dot{m} = 28.6$, 31.8 kg/s; $D = 0.70$, 0.53 m)

    c* For take off with a thrust/weight ratio equal to unity.   (**Ans:** $\dot{m} = 65.9$ kg/s; $D = 0.67$ m)

16.9 For the New Fighter Aircraft, Chapter 13, assume that the maximum fuel available during the high-thrust part of combat to be 2 tonne (i.e. 1 tonne per engine). Using the results of Exercises 16.4 and 16.7, with the specified thrust from Exercise 14.4, calculate the maximum time the thrust can be produced with and without afterburning at $M = 2.0$ for 1g flight and at $M = 0.9$ and 1.5 at 3g flight.

                        (**Ans:** 1g, $M = 2.0$; dry 10.0 min, afterburning 7.9 min: 3g, $M = 0.9$; dry 18.5 min, afterburning 9.8 min: 3g, $M = 1.5$; dry 13.2 min, afterburning 9.0 min)

16.7 THE NEED FOR A VARIABLE CYCLE

It is clear that any engine will be required to operate for some, at least, of the time at conditions for which it was not designed, that is off-design. It is most plausible, for example, that a short period of combat at maximum thrust will be after a long period of cruise or loiter and be followed by a further long period of cruise. A fixed-cycle engine necessarily is something of a compromise, providing less than the highest possible specific thrust when high thrust is needed in order to achieve adequately low specific fuel consumption for long periods when high thrust is not needed. Although the engine may be able to produce the thrust required off-design, it is not doing so in the most efficient way. As this chapter has shown, high specific thrust is best achieved with relatively high fan pressure ratio and low bypass ratio, whereas low *sfc* is achieved with low *fpr* and relatively high *bpr*. Furthermore, to achieve low *sfc* one needs a high overall pressure ratio, but at high speed there is a limit on this because the compressor exit temperature becomes too high. The maximum *opr* which is allowable for flight at $M = 2.0$ would lead to an inefficient engine at subsonic flight conditions.

The way around this is to have variable geometry, sometimes called a variable cycle. This is now an active area of study and development, and correspondingly rather secret. It appears that two approaches are being pursued. One is to vary the overall pressure ratio, to give high *opr* for subsonic operation and low *opr* for supersonic; this probably requires a variation in the area of the turbine nozzle guide vanes. The other is to vary the bypass ratio by flaps downstream of the fan. Beyond noting the desirability of the variable-cycle engine, this will not be pursued further here.

SUMMARY OF SECTIONS 16.5 AND 16.6

The afterburner produces a substantial rise in thrust, with the gross thrust increasing approximately as the square root of the ratio of the exhaust temperature with and without the afterburner. The fuel consumption is much higher with the afterburner, but the difference decreases as the Mach number increases and as the pressure ratio across the propulsive nozzle increases.

When operating with the afterburner lit, performance is determined entirely by the ratio of fan discharge stagnation pressure to ambient static pressure and the temperature into the nozzle; the efficiencies and losses in the engine become irrelevant. Provided that the fan pressure ratio can be achieved the overall pressure ratio and the turbine inlet temperature have no effect on performance.

The current focus of study and development in industry for combat engines is on the variable-cycle engine together with engines capable of operating with the aerodynamic problems associated with the convoluted intakes and exhausts needed to achieve low observables.

16.8 THE EFFECT OF ALTERATION IN ASSUMED PARAMETERS

It has inevitably been necessary to assume parameter values, including the technology level, to enable calculations to be performed. This immediately begs the question, how sensitive are the conclusions to these assumptions? This section will attempt to answer some of the questions which arise. The calculations will centre on operation 'dry', without the afterburner lit, since this is much more sensitive to the assumed values of engine variables.

The comparisons will be made for the design case at $M = 0.9$ for which the overall pressure ratio p_{03}/p_{02} is taken to be 30 and the fan pressure ratio p_{013}/p_{02} to be 4.5. With this combination, and with the technology level of Section 16.1, the bypass ratio is 0.997 (Exercise 16.3). In Table 16.3 one parameter is altered at a time, with the others at their datum values. It can be seen, for example, that reducing the turbine inlet temperature from 1850K to 1750K increases specific thrust $F_N/\dot{m}$ by 1.2%, increases *sfc* by 2.0% and reduces bypass ratio by 0.23. The predicted increase in specific thrust as temperature decreases, and elsewhere in the table when efficiency falls, is counter-intuitive. The reason for this is the stipulation that the pressure ratio of the fan and of the whole engine are constant. A reduction in work output from the core is therefore translated into a reduction in bypass ratio. Taking as an example the 100K reduction in turbine inlet temperature, the specific thrust has increased to 1.012 times its previous value whilst at the same time the bypass ratio has been reduced by 0.23 from 0.997 to 0.767. As a result of the drop in bypass ratio the total mass flow through the engine, for the same mass flow through the core, is reduced and because of this the total net thrust has been decreased in the ratio $1.012 \times (1 + 0.767) \div (1 + 0.997) = 0.896$, that is a 10.4% drop in thrust.

In all cases in Table 16.3 the changes to specific thrust and fuel consumption are comparatively small and the most significant effect is to the bypass ratio. A reduction in bypass ratio leads to a reduction in thrust whenever component efficiencies are lowered or losses increased. A change in the efficiency of the LP compressor has the largest effect.

Table 16.3   Effect on specific thrust and *sfc* of variation in parameters at design point, mass flow through the core held constant and pressure ratios held constant

	$\Delta(F_N/\dot{m})$ %	$\Delta(sfc)$ %	$\Delta(bpr)$
Reduce turbine inlet temp. from 1850 to 1750 K	1.2	2.0	−0.23
Increase LP compressor efficiency from 0.85 to 0.90	−4.0	−4.5	0.21
Reduce HP compressor efficiency from 0.90 to 0.85	3.4	3.8	−0.17
Reduce HP turbine efficiency from 0.875 to 0.825	2.4	2.7	−0.10
Reduce LP turbine efficiency from 0.875 to 0.825	2.6	2.8	−0.10
Stagnation pressure loss in combustor 5% of p_{03}	2.3	2.5	−0.09
Stagnation pressure loss in bypass 5% of p_{023}	−3.5	−0.9	0.09
Stagnation pressure loss in jet pipe 5% of p_{05}	−1.4	−1.4	0.00
Turbine cooling air increased by 50%	2.5	3.7	−0.25
Dispense with turbine cooling air	−3.5	−5.1	0.48

The magnitudes of the changes in specific thrust, specific fuel consumption and bypass ratio in the table are generally small, compared with the relatively large changes in efficiency or loss which are considered. The trends and general magnitudes found in the chapter may therefore be considered robust. It is noteworthy that the effect of including the pressure loss in the combustor, bypass duct or jet pipe are all relatively small, supporting the decision to simplify the treatment by not including arbitrary estimates for these in the body of calculations. The effect of altering turbine cooling air, however, is large; omitting the cooling air would lead to net thrust being over-estimated by 43%, a very substantial error. It is also possible to see from the table that a 'trade' exists between raising turbine inlet temperature and increasing cooling air: from the figures given in Table 16.3 a 100K reduction in T_{04} gives a 10.4% reduction in thrust, whilst increasing the cooling air by 50% gives a 10.3% reduction in thrust. This is an important balance which designers have to take into account.

Figure 16.7 considers the effect of inlet loss for flight at $M = 1.5$ and 2.0. Curves are shown for both speeds for three cases: when the inlet is assumed reversible, when there is a normal shock and where the loss is given by MIL-E-5007/8, Equation 15.6. The loss with MIL-E-5007/8 is appropriate for a sophisticated variable inlet decelerating the flow with several shock waves. At a Mach number of 2 the specific thrust with a normal shock is about 6% below that with an isentropic deceleration, and there is also a significant increase in *sfc*, particularly for low fan pressure ratios. An intake conforming to MIL-E-5007/8 introduces a thrust loss of only about 2% compared with the isentropic case for $M = 2.0$. At $M = 1.5$ the loss associated with the normal shock is small, justifying the use of simple fixed inlets for aircraft whose primary mission does not exceed this speed range, for example the F-16.

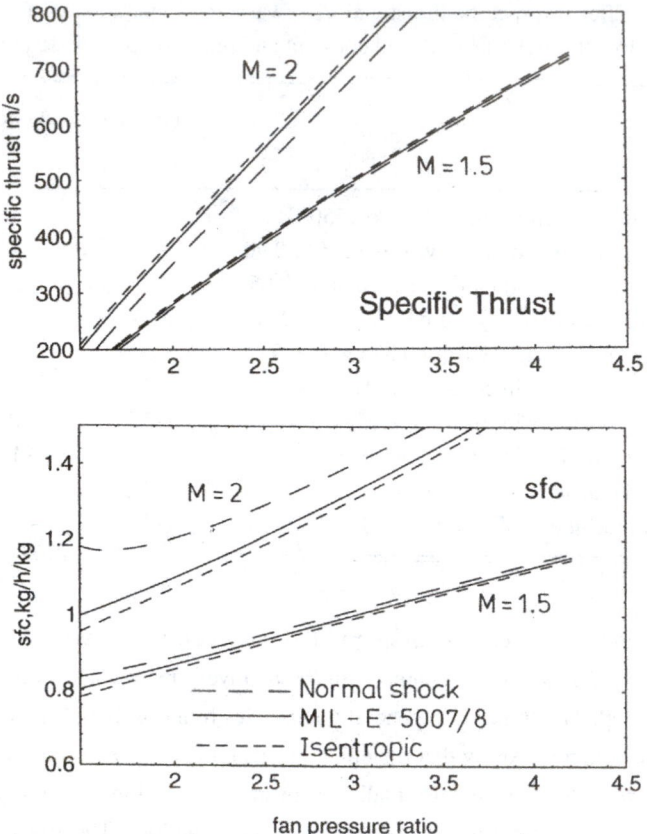

Figure 16.7. Specific thrust and specific fuel consumption (*sfc*) versus fan pressure ratio for design point operation at tropopause showing effect of intake loss. (Pressure ratios from Table 16.2. Constant technology standard apart from intake loss.)

SUMMARY OF CHAPTER 16

Calculations have been carried out for design at three flight Mach numbers at the tropopause and for take off at sea level. In a design calculation the overall pressure ratio and either the fan pressure ratio or bypass ratio are selected, so the calculations for any particular engine only apply when it is operating at the one condition for which it is designed. In terms of engine non-dimensional operating point the crucial parameter is the ratio of turbine inlet temperature to compressor inlet temperature, T_{04}/T_{02} and this ratio would alter if one of the resulting engine designs were used at a different condition.

The calculations were carried out at the same technology level (turbine inlet temperature, maximum compressor delivery temperature, compressor and turbine efficiencies) given in Table 16.1, which includes a constant proportion of the core air used for turbine cooling. The values used are representative of many aircraft recently entering service. Calculations were also performed to show that modest changes in any of the assumed values used would not seriously distort the results, either in the magnitudes of specific thrust and *sfc*, or in their trends. The most important additional input, relative to the more elementary calculations carried out for the civil engine in Chapters 4 and 12, is the mass flow used to cool the turbine; if this is not included the work supplied to the fan is far too large. Losses in stagnation pressure in the combustor, bypass duct and jet pipe, at levels which are plausible, do not significantly alter the magnitudes of specific thrust or specific fuel consumption.

The afterburner can significantly increase the specific thrust, whilst increasing the specific fuel consumption by a rather higher proportion. The increase in both specific thrust and specific fuel consumption with afterburning is proportionally much smaller for an engine designed for $M = 2.0$ than one designed for $M = 0.9$, mainly because the exhaust temperature of the 'dry' engine at $M = 2.0$ is already high. For the engine at $M = 0.9$ at the tropopause the specific thrust is almost exactly doubled with the afterburner; for the engine sized to allow a $3g$ turn at this Mach number the inlet diameter can be decreased from about 1 m to about 0.7 m with an afterburner. If the engine weight were proportional to linear dimension cubed this would represent an afterburning engine weighing only 35% of the comparable 'dry' engine; to offset this huge reduction is the weight of the afterburner and the long jet pipe and, of course, the additional weight of fuel burned when afterburning is used.

An engine specifically designed for high-speed propulsion, $M = 2.0$ for example, will have a very low bypass ratio and such an engine is sometimes colloquially described as a "leaky turbojet". (The bypass ratio for the design exercises here for $M = 2.0$ is only 0.11.) For $M = 0.9$ the optimum bypass ratio is close to unity at the current technology level. If the technology level were lower the bypass ratio and/or the fan pressure ratio would be lower too.

CHAPTER 17 COMBAT ENGINES OFF-DESIGN

17.0 INTRODUCTION

The engine for a high-speed aircraft is required to operate over a wide range of conditions and some of these have been discussed in Chapters 13, 14 and 15. Of particular importance is the variation in inlet stagnation temperature, which can vary from around 216 K up to nearly 400 K for a Mach-2 aircraft. The turbine inlet temperature is limited by metallurgical considerations, and as a result the ratio of turbine inlet temperature to engine inlet temperature T_{04}/T_{02} alters substantially, even when the engine is producing the maximum thrust it is capable of for its flight condition. In contrast, for the subsonic civil aircraft the value of T_{04}/T_{02} changes comparatively little between take-off, climb and cruise, the conditions critical in terms of thrust and fuel consumption. It is normally only when a civil aircraft is descending to land or is forced to circle an airport that T_{04}/T_{02} is reduced radically.

In Chapter 8 the dynamic scaling and dimensional analysis of engines was considered. There the engine non-dimensional operating point was held constant, for example T_{04}/T_{02} was kept constant, so the engine remained 'on-design'. To designate the engine operating condition the value of $N/\sqrt{c_p T_0}$ or any of the pressure ratios or non-dimensional mass flows could also be used, but T_{04}/T_{02} has the intuitive advantage since engine thrust is altered by varying fuel flow rate to change T_{04}. In Chapter 12 the more challenging issue of a civil engine operating away from its design condition was addressed, i.e. the case when $T_{04}/T_{02} \neq$ constant, and the subject of the present chapter is the similar case for military engines. The principal difference in approach from that of Chapter 12 arises from the practice in the combat engine of mixing the core and bypass flows.

The first part of this chapter develops the theory along the lines used in Chapter 12 for the civil transport engine off-design. This is first applied to the Combat engine when it is required to produce maximum thrust: in the terminology of Section 15.3 either at Combat rating with the afterburner in use, or at Maximum Dry without the afterburner. Later some consideration will be given to the engine operating at reduced thrust, as it must for most of the flight mission. To simplify discussion two alternative design cases will be considered. In case 1 the design condition is sea-level static, the condition at the start of the take-off or on a static test bed, with the inlet temperature at 288 K. In case 2 the design condition is taken to be $M = 0.9$ at the tropopause, one of the important operating conditions of the envelope, when the engine inlet temperature is only about 252 K. What follows is a simple treatment intended to demonstrate principles. Some sweeping simplifications are adopted, of which one of the most important is the assumption that the polytropic efficiencies of the compressors and turbines remain constant when operating away from design.

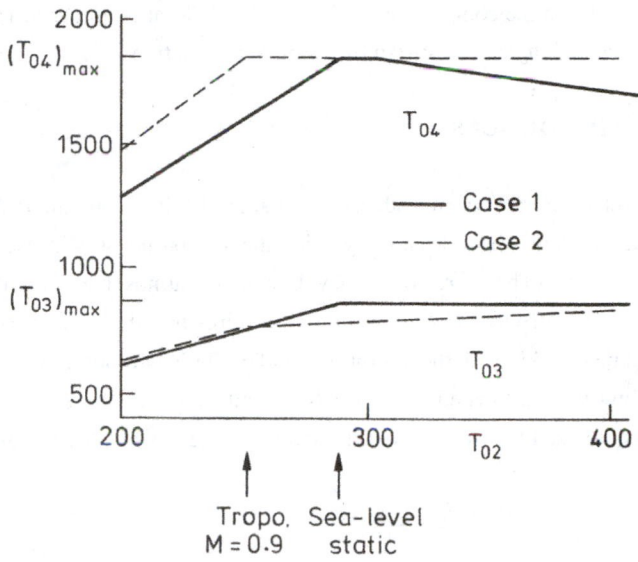

Figure 17.1. Turbine inlet temperature T_{04} and compressor delivery temperature T_{03} as a function of engine inlet temperature T_{02}. Design point for case 1 – sea-level static; design point for case 2 – $M = 0.9$ at tropopause.

17.1 THE SIGNIFICANCE OF OFF-DESIGN OPERATION

Table 15.1 gives some of the engine inlet stagnation temperature at important operating points. The way that the inlet temperature affects the engine operation and thrust is not always intuitively obvious, and depends on the design of the engine, as will be shown in this chapter. Operation off-design is fixed by the three major constraints on the engine, see Figure 15.9; and Figure 17.1 shows a corresponding plot of temperatures for the two design cases which are to be considered in this chapter.

At low inlet temperature the constraint is on $N/\sqrt{c_p T_0}$,[1] pressure ratios and T_{04}/T_{02} are constant at their design values. As engine inlet temperature T_{02} is raised the turbine inlet temperature T_{04} reaches its maximum allowable value, which we designate the design condition, and further increase in inlet temperature necessitates progressively lower values of T_{04}/T_{02} to keep T_{04} constant; the engine is therefore no longer operating at a constant non-dimensional condition so $N/\sqrt{c_p T_0}$ and the pressure ratios in the engine all fall. Further increase in inlet temperature can lead to the compressor delivery temperature T_{03} reaching its upper limit which is shown in Figure 17.1 for case 1 just below $T_{02} = 300\,\text{K}$. If T_{02} is increased still further for case 1, T_{04} must be lowered to

---

[1] All modern engines will have at least two shafts and there is a value of $N/\sqrt{c_p T_0}$ for each shaft, with the T_0 referring to the inlet temperature to the compressor on that shaft. At any non-dimensional operating point there will be a fixed ratio between $N/\sqrt{c_p T_0}$ for each shaft; for brevity in the text only a single $N/\sqrt{c_p T_0}$ is therefore referred to.

reduce the pressure ratio in the compressors and hence hold T_{03} at its upper limit. The trends and magnitudes in this rematching can be shown by a simple calculation.

17.2 ALTERNATIVE DESIGNS

When the New Fighter Aircraft was introduced in Chapter 13, it was required that the thrust-to-weight ratio at take-off should be at least unity when the afterburner was in use and at least 0.66 when the engine was dry. (These are for sea-level, standard atmosphere conditions, 288 K and 101.3 kPa). Given a take-off weight of 18 tonne this requires that the static sea-level thrust from each engine should be 88.3 kN with the afterburner and 58.3 kN without. In Chapter 14 the thrust requirements in flight were estimated (Exercise 14.4) and these are shown in Table 14.1. These thrust requirements are used to size the two alternative engine designs (case 1 and case 2) in this chapter.

Case 1 design

The design point is taken to be at sea-level static. Resting on the experience of Chapter 16 for the design point, it seems appropriate to aim for a fan pressure ratio p_{013}/p_{02} of 4.5 and an overall pressure ratio p_{03}/p_{02} of 30. Using a simple design calculation, Exercises 16.6 and 16.7, and the technology level set out in Section 16.1, the values in Table 17.1 result. These are for the engine at its highest non-dimensional operating point, in other words T_{04}/T_{02}, $N/\sqrt{c_p T_0}$ and all the pressure ratios are at their design values which are close to maximum values.

Table 17.1   Case 1: Sea-level static design point fan pr $= 4.50$,
overall pr $= 30$; $T_{02} = T_a = 288$ K, $T_{04} = 1850$ K, $T_{04}/T_{02} = 6.42$;
technology level as in Section 16.1

	T_{03} K	bpr	V_j m/s	Spec. thrust m/s	sfc kg h^{-1}kg^{-1}
Dry	872.5	0.471	848	865	0.805
Afterburning	872.5	0.471	1267	1341	1.543

The compressor delivery temperature in Table 17.1 for design point is only 2.5 K below the limit set for T_{03} of 875 K. For sea-level operation on a day only slightly warmer than the ISA value of 288 K it would therefore be necessary to reduce the turbine inlet temperature, and therefore T_{04}/T_{02}, below the design values. (If T_{04}/T_{02} is reduced the pressure ratios and non-dimensional rotational speeds are reduced as well.)

The temperatures for the compressor delivery T_{03} and turbine entry T_{04} are shown in Figure 17.1 (similar to Figure 15.8) for both the case 1 and case 2 engines. For inlet temperature below the design value, such as would occur at low Mach numbers and increased altitude, the engine can be kept at the design point by holding the ratio T_{04}/T_{02} constant, that is by reducing turbine inlet temperature. For example, at the tropopause $T_a = 216.7$ K and an inlet Mach number of 0.9 the inlet temperature T_{02} is only 251.7 K; although this flight condition is not the design condition, the engine can nevertheless be at the same non-dimensional condition as design, but with T_{04} reduced to 1616 K. Presure ratios and non-dimensional rotational speeds would be at the design values too, of course. For low inlet temperatures the value of T_{04}/T_{02} could be raised beyond its design value without exceeding a temperature limit, but the limits on pressure ratio or $N/\sqrt{c_p T_0}$ might be exceeded. The limit for operation at low inlet temperature is usually expressed in terms of $N/\sqrt{c_p T_0}$ because of its importance for aeroelasticity.

At the tropopause the Mach number must be raised to 1.28 to bring T_{02} back to the sea-level static temperature of 288 K, at which point the case 1 engine is at its design point and $T_{04} = T_{04max}$. For the case 1 engine T_{04}/T_{02} must fall if $T_{02} > 288$ K, because T_{04} has reached its maximum value, and the engine can then only operate below its non-dimensional design point. A further increase in Mach number to 1.35 leads to the compressor delivery temperature reaching its maximum of 875 K, so for Mach numbers above 1.35 T_{04} must be reduced (not just T_{04}/T_{02} reduced) to prevent the compressor deliver temperature exceeding its limit. In summary the case 1 the engine must be 'throttled back' (i.e. operate at lower non-dimensional conditions) at flight Mach numbers above 1.28 at the tropopause.

Case 2 design

The design is taken here as the engine to give a $3g$ turn at $M = 0.9$ at the tropopause. From Table 14.1 the thrust requirement per engine for this condition is 35.8 kN. From Chapter 16 for the design point, it is again appropriate to select a fan pressure ratio p_{013}/p_{02} of 4.5 and an overall pressure ratio p_{03}/p_{03} of 30; the case 2 engine is one specified in Exercises 16.3 and 16.5 and the design-point parameters are set out in Table 17.2. For the case 2 engine at the design point $T_{04}/T_{02} = 1850/251.7 = 7.35$, considerably higher than the value for case 1, which was 6.42.

Table 17.2   Case 2: Tropopause, $M = 0.9$ design point fan pr $= 4.50$, overall pr $= 30$; $T_a = 216.7$, $T_{02} = 251.7$ K, $T_{04} = 1850$ K, $T_{04}/T_{02} = 7.35$; technology level as in Section 16.1

	T_{03} K	bpr	V_j m/s	Spec. thrust m/s	sfc kg h^{-1}kg^{-1}
Dry	762.2	0.996	876	625	0.888
Afterburning	762.2	0.996	1430	1250	1.680

The specific thrust and thrust requirement allow the mass flow to be found, so that to produce this thrust dry the mass flow is equal to $(35.8 \times 10^3)/625 = 57.3$ kg/s and correspondingly, 28.6 kg/s with afterburning. An engine passing only 28.6 kg/s would be small by modern standards (only a little bigger than the Viper, see Section 12.2) reflecting the fact that a sustained $3g$ turn at $M = 0.9$ would not be regarded as a condition requiring a high thrust and afterburning would never normally be used to achieve this.

If this is the design point it requires that the non-dimensional rotational speeds and the pressure ratios are at their maximum and cannot normally be increased much. If T_{02} were reduced, which would occur if flight speed were reduced, the turbine inlet temperature would then be reduced too to stay at the design point $T_{04}/T_{02} = $ constant. If engine inlet temperature were increased the ratio T_{04}/T_{02} must fall, since T_{04} is already at its maximum. For case 2 at sea-level static conditions, for example, $T_{04}/T_{02} = 1850/288 = 6.42$ (as in case 1) but for case 2 the overall pressure ratio will have been reduced to only 21.4 because the design value is $T_{04}/T_{02} = 7.35$. This will be demonstrated quantitatively below. The temperatures are shown in Figure 17.1 alongside case 1. The primary limit on performance in case 2 is imposed by turbine entry temperature. Because the engine is 'throttled back' for $T_{02} > 251$ K (the temperature corresponding to $M = 0.9$ at the tropopause) the compressor pressure ratio and temperature rise never reach their upper limits. This contrasts with the case 1 engine for which the operating limit over most of the range is compressor delivery temperature.

17.3 A MODEL FOR THE TWO-SHAFT TURBOFAN ENGINE

The approach is so similar to that described in Chapter 12 that much of the explanation can be omitted here. The major difference is the mixing of the core and bypass stream ahead of the final propulsive nozzle. In line with the justification in Chapter 15 it will be assumed that the core and bypass streams have equal stagnation pressure where they mix and that this mixing takes place without loss in stagnation pressure.

The mixing of the core and bypass streams downstream of the LP turbine determines that, if the losses in the bypass duct are small enough to neglect, the outlet pressure from the fan and the LP turbine are equal, in other words

$$p_{05} = p_{013}.$$

Furthermore, with small losses in the jet pipe the fan discharge pressure is equal to the stagnation pressure at the nozzle throat, p_{08}. For the design-point calculations we could specify the fan pressure ratio and thus the outlet conditions from the turbine. It is then straightforward to find the LP turbine work. Off-design it is less straightforward because the fan pressure ratio is not normally known in advance. It will be realised that a reduction in p_{013} reduces the power input needed to the fan, if the bypass ratio is constant, and simultaneously increases the pressure ratio across the LP turbine and the power output from the LP turbine. The problem lends itself to a simple numerical iteration.

The efficiencies for the compressor and turbine used in Chapter 16 are retained here, though at lower values of T_{04}/T_{02} and $N/\sqrt{c_p T_0}$ than at design point the efficiencies might be expected to rise somewhat. The proportion of the core flow used to cool the turbine also remains as it was for the design point: it is assumed that 20% of the core flow does not go through the combustor but 8% is used to cool the HP nozzles and is mixed back in before the stipulation of T_{04} (so this does not need to be addressed specially); a further 8% is used to cool the HP rotor and is mixed in with the HP turbine outlet flow; and 4% is used in the LP turbine and mixed in downstream of the LP rotor.

Choked turbines

Just as in Chapter 12 the HP and LP turbine are assumed to be choked and their polytropic efficiencies are assumed to remain constant. The HP turbine is then constrained so that

$$(T_{045}/T_{04}) = (p_{045}/p_{04})^{\eta(\gamma-1)/\gamma} = \text{constant}. \tag{17.1}$$

The mass flow of air entering the core compressor is $\dot{m}_a$, but some of this is used for turbine cooling so the air mass flow into the HP turbine rotor is $\dot{m}_{a4}$ and the total mass flow into the turbine, including the mass flow rate of fuel, is $\dot{m}_{a4} + \dot{m}_f$. It then follows that the HP turbine power is given by

$$\dot{W}_{HP} = (\dot{m}_{a4} + \dot{m}_f)c_{pe}(T_{04} - T_{045})$$
$$= (\dot{m}_{a4} + \dot{m}_f)c_{pe}T_{04}(1 - T_{045}/T_{04}).$$

In line with the treatment in Chapter 12 it is convenient to define a coefficient k_{HP} for the HP turbine by

$$T_{04} - T_{045} = k_{HP}T_{04}, \tag{17.2}$$

where the coefficient can found from the turbine performance at the design point.

By equating HP turbine power to the HP compressor power it follows that the HP compressor temperature rise is given by

$$T_{03} - T_{023} = \{(\dot{m}_{a4} + \dot{m}_f)c_{pe}/\dot{m}_ac_p\}k_{HP}T_{04}. \tag{17.3}$$

If turbine inlet temperature T_{04} is given, the temperature rise in the HP compressor is fixed; with a known or assumed polytropic efficiency for the HP compressor the pressure ratio depends only on the temperature of the air leaving the LP compressor, T_{023}, into the core. The temperature out of the LP compressor in the bypass stream T_{013} is assumed equal to T_{023}. With the mixed turbofan the nozzle is also choked for most conditions of interest, but the balance of flow between the core and bypass is unknown and in this case it is not possible to find an expression for LP turbine work equivalent to Equation 17.2.

Power balance for the LP shaft

Since $T_{023} = T_{013}$, the power balance across the LP shaft may be written as

$$\dot{m}_a(1 + bpr)c_p(T_{023} - T_{02}) = (\dot{m}_{a45} + \dot{m}_f)c_{pe}(T_{045'} - T_{05}), \tag{17.4}$$

where $\dot{m}_{a45}$ is the mass flow of air entering the LP turbine with temperature $T_{045'}$ after mixing the cooling flow after the HP turbine.

For a fixed LP compressor efficiency,

$$p_{013}/p_{02} = (T_{013}/T_{02})^{\eta\gamma/(\gamma-1)},$$

with a similar expression for the core flow out of the fan. Likewise for a fixed LP turbine efficiency

$$p_{05}/p_{045} = (T_{05}/T_{045'})^{\gamma/\eta(\gamma-1)}.$$

Using these equations and $p_{05} = p_{013}$ the power balance Equation 17.4 may be solved to yield the combination of LP compressor pressure ratio and bypass ratio appropriate.

Mass flow

There are two mass flows of air to be found, that which goes through the core, $\dot{m}_a$, and that which goes through the whole engine, $(1 + bpr)\,\dot{m}_a$. The conditions which determine these are the choking of the HP turbine nozzle and the choking of the final propulsive nozzle, respectively. In both cases the mass flow rate of fuel should also be included; when the engine is dry this is approximately 2.5% of the mass flow of core air, whereas when the afterburner is in use it can amount to 10% of the air through the core.

At turbine inlet the non-dimensional mass flow rate is given by

$$\overline{m}_4 = \frac{\dot{m}\sqrt{c_{pe}T_{04}}}{A_4 p_{04}} = 1.389,$$

based on $\gamma = 1.30$, where the relevant mass flow $\dot{m} = (k\dot{m}_a + \dot{m}_f)$. Here k is the fraction of air compressed in the HP compressor which leaves the HP turbine nozzle guide vanes (88% in the present calculation), the remaining 12% being used to cool the HP rotor and LP turbine. Constancy of $\overline{m}_4$ is what determines the mass flow of air through the core.

At the throat of the propulsive nozzle, with area A_8, a corresponding relation applies

$$\overline{m}_8 = \frac{\dot{m}\sqrt{c_{pm}T_{08}}}{A_8 p_{08}} = \text{constant},$$

where the relevant mass flow $\dot{m} = \{\dot{m}_a\,(1 + bpr) + \dot{m}_f\}$. It is the requirement for constant $\overline{m}_8$ which determines the overall mass flow and thence the bypass ratio. The value of the constant depends upon c_{pm} and γ_m. The appropriate value for specific heat c_{pm} needs to be found; if the engine is dry the value of c_{pm} may be estimated by Equation 15.4, after which γ_m may be found. If the afterburner is in use $\gamma = 1.30$ and $c_{pe} = 1244\,\mathrm{J\,kg^{-1}K^{-1}}$ are assumed appropriate. To use Equation 15.4 requires the bypass ratio to be known and bypass ratio is something which has to be found. The process is therefore iterative.

Method of solution

For a given engine the design-point operation is used to fix k_{HP}, which relates the HP turbine inlet temperature to the drop in temperature across the turbine, as well as $\overline{m}_4$ the non-dimensional mass flow rate into the HP turbine and the non-dimensional mass flow rate into the final nozzle $\overline{m}_8$. The efficiencies and losses are assumed here to remain at the design values. For a particular value of engine inlet temperature T_{02} the turbine inlet temperature T_{04} is selected.

The calculation begins by assuming a value for the fan pressure rise p_{013}/p_{02} and from this the fan temperature rise in the bypass stream and in the core stream T_{023} can be found. Similarly the stagnation pressure downstream of the LP turbine is fixed. The temperature rise in the HP compressor is found from $k_{HP}T_{04}$ and, given T_{023}, the overall pressure ratio p_{03}/p_{02} is determined. From the turbine temperature drop $k_{HP}T_{04}$ the HP turbine pressure ratio can be found. Then, since the pressure downstream of the LP turbine is known from the fan pressure ratio, the pressure ratio and power per unit mass flow rate for the LP turbine can be found. The LP turbine power fixes the bypass ratio and from this the non-dimensional mass flow in the propulsive nozzle $\overline{m}_8$ can be found. The mass flow rate through the core has been determined by the choking of the HP turbine nozzle vanes. If $\overline{m}_8$ is not equal to the value at design the fan pressure rise must be altered and the calculation repeated until there is agreement.

17.4 VARIATION WITH TURBINE INLET TEMPERATURE

One of the things which complicates the general treatment of engines off-design is the dependence of some of the parameters such as thrust and *sfc* on both engine operating condition, characterised by the ratio T_{04}/T_{02} and flight Mach number. Because for the combat engines the propulsive nozzle

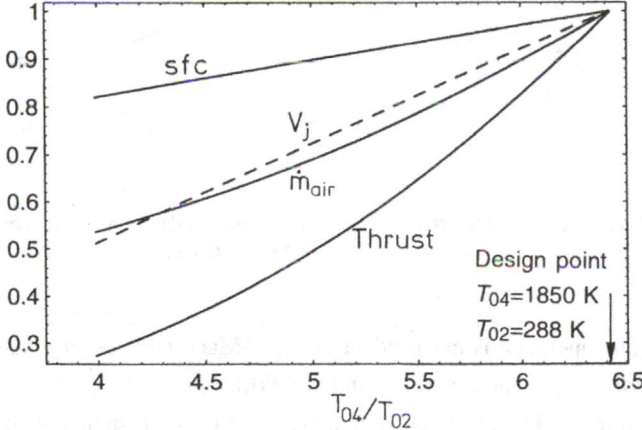

Figure 17.2. Case 1 engine performance as a function of T_{04}/T_{02}, normalised by values at design point (sea-level static, $T_{04} = 1850$ K). Constant component efficiencies.

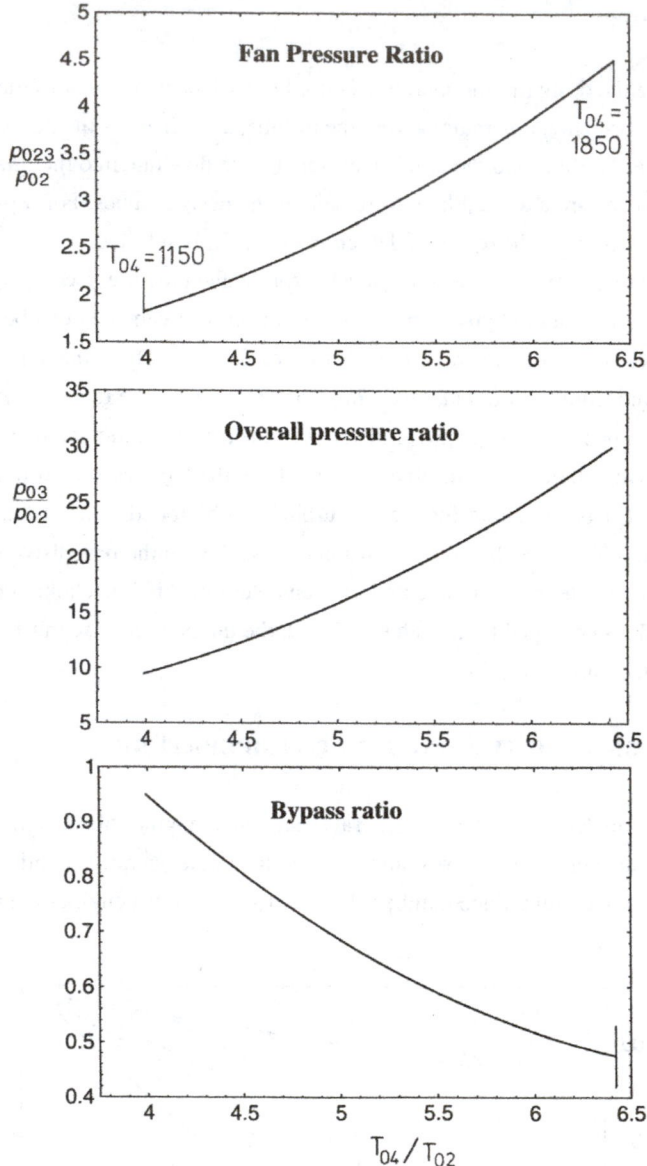

Figure 17.3. Case 1 engine pressure ratios and bypass ratio as a function of T_{04}/T_{02}. Constant component efficiencies.

is choked the engine operation is independent of flight Mach number, or ambient pressure, and depends only on inlet stagnation conditions and fuel flow. Figures 17.2, 17.3 and 17.4, as well as 17.6 and 17.9 all refer to the case 1 engine and are valid for both static conditions and in-flight conditions because the choked propulsive nozzle. The method for getting gross and net thrust at different conditions is discussed in Chapter 8. Figure 17.2 shows the variation in various gross

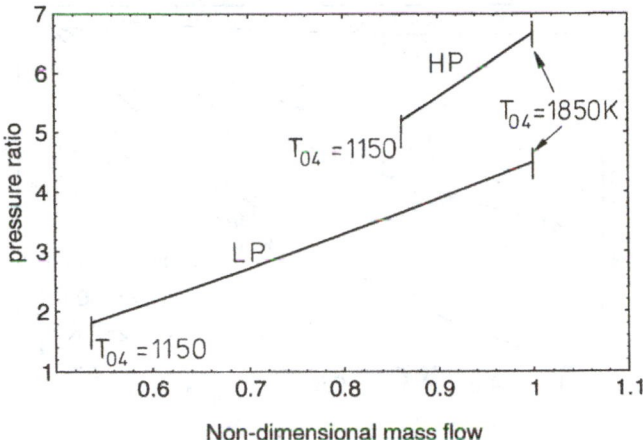

Figure 17.4. Case 1 engine compressor working lines (pressure ratio versus non-dimensional mass flow rate). Extreme turbine inlet temperatures T_{04} shown for engine inlet temperature $T_{02} = 288$ K. Constant component efficiencies.

parameters of the case 1 engine (design point for sea-level static design) operating on a sea-level static test bed as the turbine inlet temperature is varied between 1850 K (the design-point value) and 1150 K.

Figure 17.3 shows, for the same case, the variation in fan pressure ratio, overall pressure ratio and bypass ratio. At the lowest value of turbine inlet temperature the fan pressure is such that the final nozzle will be on the point of unchoking and below this pressure ratio and an approach similar to that described in Chapter 12 is required. Of particular note is the steep rise in bypass ratio as T_{04}/T_{02} is reduced, an effect seen also for the unmixed engine, Figure 12.10, though the effect is larger in the present case.

As was discussed in Chapter 12, the major difficulty encountered when engines are operated at reduced pressure ratio is the matching of the compressors. Figure 17.4 shows the working line for the LP and HP compressors in terms of the pressure ratio and non-dimensional mass flow at compressor inlet. The non-dimensional mass flow is normalised by the value at the design condition. The two working lines end to the right at the temperature ratio for design point which occurs at sea-level static conditions with maximum turbine inlet temperature. The HP compressor operating point varies relatively little over the range of T_{04}/T_{02} encountered, whereas the pressure ratio and the non-dimensional mass flow for the LP compressor are reduced to less than half the value at design when $T_{04} = 1150$ K.

For the case 2 engine (design point for $M = 0.9$ at the tropopause) the corresponding variation in major parameters to is shown in Figure 17.5. The most significant difference, compared with case 1, is the larger extent to which it is throttled back for potentially important operating conditions.

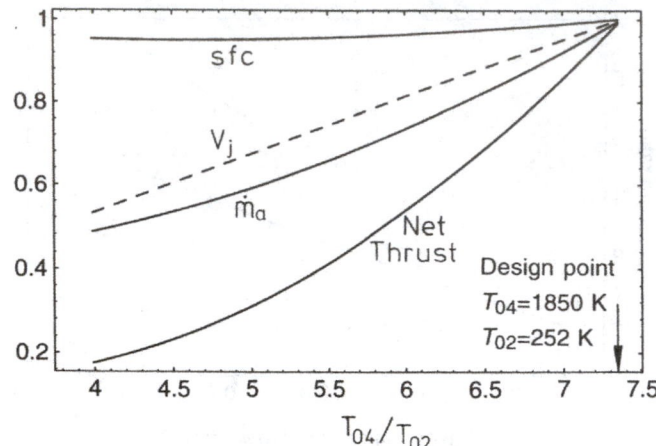

Figure 17.5. Case 2 engine performance as a function of T_{04}/T_{02}, normalised by values at the design point ($M = 0.9$ at the tropopause, $T_{04} = 1850$ K). Constant component efficiencies.

17.5 DIMENSIONAL REASONING AND DYNAMIC SCALING

So long as the propulsive nozzle is choked, which covers all important operating conditions, the engine is unaware of the external static pressure. The inputs to the engine are then the inlet stagnation pressure and temperature, p_{02} and T_{02} and the fuel flow $\dot{m}_f$. The temperature ratio T_{04}/T_{02}, which is varied by changing the fuel flow, may be used as a surrogate for the non-dimensional group based on fuel flow. It then follows that *all* the pressure ratios and temperature ratios inside the engine are determined by T_{04}/T_{02}. So too are the non-dimensional rotational speeds $N/\sqrt{c_p T_0}$ and the non-dimensional mass flow into the engine $\dot{m}\sqrt{c_p T_{02}}/(A_2 p_{02})$.

It has already become clear that one of the most important and commonly encountered constraints on the operation of the engine is the compressor delivery temperature, T_{03}. From what has been written above it should be clear that T_{03}/T_{02} is a unique function of T_{04}/T_{02} for a given engine; so too is, for example T_{08}/T_{02}, where T_{08} is the stagnation temperature of the gas entering the final propulsive nozzle. In the case of the mixed turbofan a numerical iteration is needed to obtain all the pressure ratios and temperature ratios off-design, as outlined in Section 17.2.

Figure 17.6 shows non-dimensional groups for the case 1 engine which were obtained after iterative calculations and which facilitate hand calculations of thrust in the exercises of this chapter. Shown as a function of T_{04}/T_{02} are the fan pressure ratio, the ratio of compressor delivery temperature to compressor inlet temperature (a simple function of overall pressure ratio if the efficiency is constant) and the exhaust stagnation temperature ratio T_{08}/T_{02}.

The procedure for using the curves in Figure 17.6 to find the engine operating point and thrust is as follows:

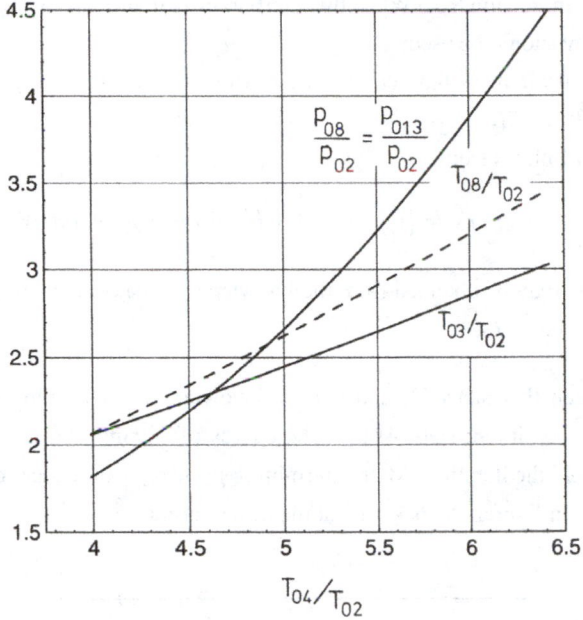

Figure 17.6. Case 1 engine pressure ratio and temperature ratios as a function of T_{04}/T_{02}.
Constant component efficiencies.

1. Determine the inlet stagnation temperature and pressure T_{02} and p_{02} from the ambient static conditions using the expression in terms of Mach number and allowing for inlet loss, if any;

2. Determine the maximum value of T_{03}/T_{02} which can be allowed, assumed here to be $T_{03} = 875\,\text{K}$. This effectively finds the maximum overall pressure ratio and using Figure 17.6 this provides one upper bound on the value of T_{04}/T_{02};

3. For the given maximum allowable T_{04} determine T_{04}/T_{02}. If this ratio is in excess of the value at the design point it means that the limit is not on T_{04} itself but on $N/\sqrt{c_p T_0}$. T_{04} must be reduced to give the design value of T_{04}/T_{02};

4. Using the lower of the values for T_{04}/T_{02} from ii) and iii) find fan pressure ratio, p_{013}/p_{02} on Figure 17.6. The pressure ratio across the propulsive nozzle is given by

$$p_{08}/p_a = (p_{08}/p_{02})\,(p_{02}/p_a) = (p_{013}/p_{02})\,(p_{02}/p_a)\,;$$

5. The jet velocity can be obtained, on the assumption that the expansion is reversible, from

$$V_j = \sqrt{2c_{pm}T_{08}\left\{1 - (p_a/p_{08})^{(\gamma-1)/\gamma}\right\}}.$$

The appropriate values of c_{pm} and γ_m for the exhaust gas must be used. (If a simple convergent nozzle were used there is some loss of thrust and the expression of gross thrust in terms of exit velocity and exit static pressure must be used ($\dot{m}V_9 + p_9 A_9$), as discussed in Section 8.3.) When the afterburner is in use $T_{08} = T_{0ab}$ and suitable values of c_p and γ should be used;

6. The overall air mass flow can be found from $\overline{m}_8$, whilst the core mass flow can be obtained from $\overline{m}_4$;

7. Net thrust follows from

$$F_N = \left\{ \dot{m}_f + \dot{m}_a(1 + bpr)V_j - \dot{m}_a(1 + bpr)V \right\},$$

and the *sfc* may be obtained from the knowledge of calorific value and the temperature difference $T_{04} - T_{03}$.

It will be seen that steps (2) and (3) are involved in determining what constrains the maximum operating condition; step (4) uses the curves of Figure 17.6 to find the fan pressure ratio and thereby avoid the iteration. After determining p_{013}/p_{02} a procedure can be used which is identical to that used in Exercises 16.4–16.7 at the design point.

Exercises

Note that many of these questions ask for repeat calculations with a change in Mach number. The use of a programme or spreadsheet has much to recommend it.

17.1 The case 1 engine introduced in Section 17.1 was calculated in Exercise 16.6 as a design based on sea-level static conditions. Use the design-point calculation of Exercise 16.6 to find k_{HP}, $\overline{m}_4$ and $\overline{m}_8$ for the engine.   (**Ans:** $k_{HP} = 0.190$, $\overline{m}_4 A_4 = 0.455 \ 10^{-3}$, $\overline{m}_8 A_8 = 3.55 \ 10^{-3}$)

17.2 The case 1 engine is to operate at the tropopause for $M = 0.9, 1.5$ and 2.0. Variations in engine parameters are shown in Figure 17.6: use this to find the maximum allowable turbine inlet temperature at the tropopause for each Mach number. (At $M = 0.9$ the engine is restricted by $N/\sqrt{c_p T_0}$ and so T_{04}/T_{02} must be at the design value; for $M = 1.5$ and 2.0 it is the compressor delivery temperature which restricts operation.)   (**Ans:** $T_{04\,max} = 1616, 1828.5, 1732.5$ K)
        Use Figure 17.6 to determine the pressure ratio p_{013}/p_{02} across the LP compressor and the corresponding compressor exit temperature T_{013}. Use the value of T_{03} to find the overall compression ratio. The efficiencies given in the technology level in Table 16.1 should be assumed to be still valid.
(**Ans:** $p_{013}/p_{02}=4.5, 3.64, 2.18$; $T_{023}=417, 485, 507$ K; $p_{03}/p_{013}=6.66, 6.41, 5.58$; $T_{03}=762, 875, 875$ K)

17.3 Using results of Ex 17.2, at each of the three flight Mach numbers for the Maximum Dry condition:

a Find the fuel mass flow rate per unit mass flow of air entering the HP compressor.
                    (**Ans:** $\dot{m}_f = 0.0250, 0.0283, 0.0257$)

b Use the maximum turbine inlet temperatures and corresponding pressures found in Exercise 17.2 to determine the ratio μ of the core mass flow to core mass flow at design point (that is, sea-level static and $T_{04} = 1850$ K). Use $\overline{m}_4$ from Exercise 17.1.   (**Ans:** $\mu = 0.407, 0.626, 0.684$)

c Find the temperature drop across the HP turbine and thence the exit temperature T_{045}. Use the ratio T_{045}/T_{04} to find the pressure at LP turbine entry.
                    (**Ans:** $T_{045} = 1310, 1482, 1404$ K; $p_{045}/p_a = 17.7, 29.3, 31.1$)

d 8% of the flow into the compressor is supplied to cool the LP turbine. It is delivered at compressor delivery temperature and pressure. Find the temperature into the LP turbine, $T_{045'}$
                    (**Ans:** $T_{045'} = 1253, 1419, 1347$ K)

e Knowing the pressure ratio across the LP turbine, find the temperature at exit T_{05}.
(**Ans:** $T_{05} = 1057, 1203, 1175$ K)

f 4% of the flow into the compressor is supplied to cool the HP turbine. It is delivered at compressor delivery temperature and pressure. Find the mixed-out temperature leaving the LP turbine, $T_{05'}$.
(**Ans:** $T_{05'} = 1039, 1184, 1157$ K)

g From the temperature drop in the LP turbine and, knowing the temperature rise in the LP compressor, find the bypass ratio.
(**Ans:** $bpr = 0.449, 0.546, 0.803$)

h Knowing the bypass ratio find the value of c_{pm} and γ_m after the core and bypass flows mix.
(**Ans:** $c_{pm} = 1171, 1161, 1139$ J kg^{-1}K^{-1}; $\gamma = 1.324, 1.328, 1.337$)

i Find the stagnation temperature into the propulsive nozzle T_{08} after the core and bypass streams mix.
(**Ans:** $T_{08} = 877, 974, 905$ K)

j Knowing jet pipe stagnation pressure and temperature find the jet velocity and the specific thrust.
(**Ans:** $V_j = 897, 1030, 1016$ m/s; spec. thrust $= 636, 606, 441$ m/s)

k Using fuel mass flow and the specific thrust find the specific fuel consumption *sfc*.
(**Ans:** $sfc = 0.958, 1.067, 1.142$ kg h^{-1}kg^{-1})

l The ratio of total mass flow entering the engine compared to that at the design point is given by $mr = \mu(1 + bpr)/(1 + bpr_{des})$. Find the value of *mr*.   (**Ans:** mass flow ratio $= 0.401\ 0.658, 0.839$)

m Determine the ratio of net thrust to that at design point, that is

$$F_N/F_{Ndesign} = mr \times (spec\ thrust)/(spec\ thrust\ at\ design) \qquad \text{(\textbf{Ans:} 0.295, 0.462, 0.428)}$$

Notes: (1) It might seem that the choking condition for the final nozzle is not needed, but this only appears to be the case because use is made of results in Figure 17.6.

(2) The bypass ratio for the case 1 engine at $M = 0.9$ at the tropopause found in section (g) is not equal to the value at design 0.471. This is surprising since the engine is expected to be at the same non-dimensional operating point at design and at this flight condition. The explanation for this is that the fuel flow per unit mass flow of air needed at the $M = 0.9$ tropopause case is about 20% smaller than at sea-level static since the rise in temperature in the combustor, which is proportional to T_{02}, is smaller. This gives a smaller mass flow through the turbine which is sufficient to alter the bypass ratio by this amount.

17.4 For the case 1 engine calculate performance at the Combat condition, with the afterburner raising the final stagnation temperature to 2200 K for flight at the tropopause for $M = 0.9, 1.5$ and 2.0. Assume that the engine operates at the same conditions as in the Maximum Dry condition. (The solution can therefore begin from Exercise 17.3h.) At the three Mach numbers find the jet velocity, specific thrust, *sfc* and net thrust (as a fraction of the thrust with the afterburner at sea-level static conditions, Exercise 16.7b).
(**Ans:** $V_j = 1431, 1563, 1600$ m/s; spec.thrust $= 1250, 1210, 1098$ m/s; $sfc = 1.66, 1.67, 1.78$ kg h^{-1}kg^{-1};
ratio of net thrust to that at design with the afterburner in use 0.374, 0.594, 0.687)

17.6 CASE 1 AND 2 ENGINES AT MAXIMUM DRY AND COMBAT RATING

It may be recalled that the case 1 engine took sea-level static conditions, with $T_a = 288$ K to define the design point, whilst for case 2 the design point was taken to be $M = 0.9$ at the tropopause. Figures 17.7 and 17.8 show comparative results for the operation of the case 1 and case 2 engines at the tropopause for a range of Mach numbers. The discontinuities in the curves correspond to the different constraints coming into play. Also shown by bold crosses are the results for the three

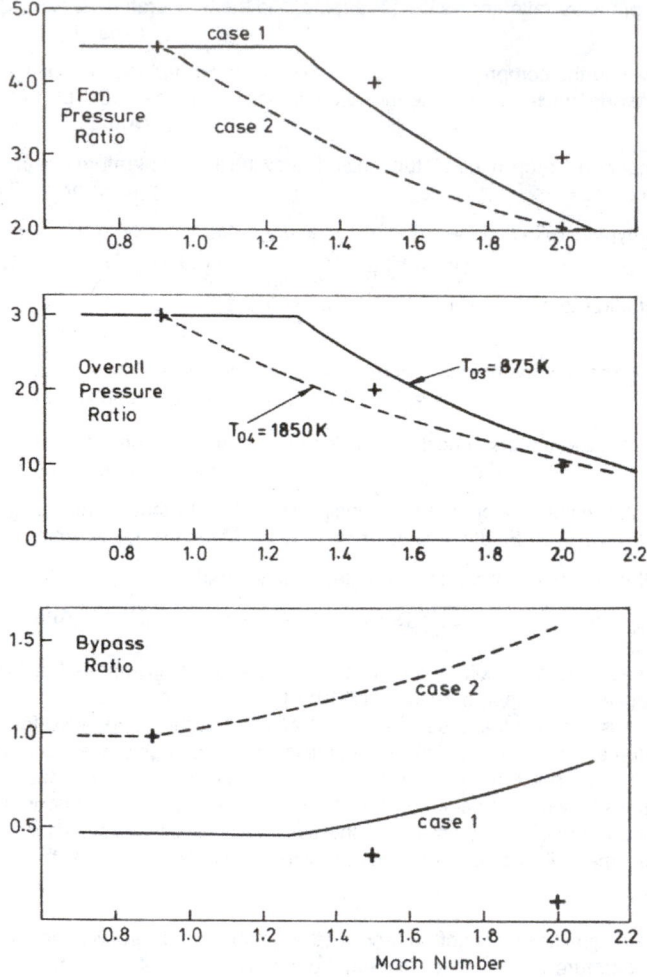

Figure 17.7. Case 1 and Case 2 engines at Maximum Dry condition as a function of Mach number for flight at tropopause. Comparison of p_{023}/p_{02}, p_{03}/p_{02} and *bpr*. Crosses show *design-point* engines from Chapter 16, Exercises 16.3 and 16.7.

design-point studies of Chapter 16 at the tropopause for $M = 0.9$, 1.5 and 2.0. (In Figures 17.7 and 17.8 $M = 0.9$, of course, coincides with the design point for the case 2 engine.)

In Figure 17.7 the pressure ratio across the fan and the overall pressure ratio are shown. For case 1 that the engine can be at its non-dimensional design point at $M = 0.9$ with the turbine inlet temperature only 1616 K because the inlet temperature $T_{02} = 251.7$ K is below the design value for the sea-level test bed. Both case 1 and case 2 engines are at their design conditions at the tropopause and $M = 0.9$, so the pressure ratios are the same for both engines. For higher Mach numbers the pressure ratios for the case 2 are below those at design point whereas the pressure ratio for case 1 is maintained for a flight Mach number of 1.28 when T_{04} reaches 1850 K. For case 1, the pressure

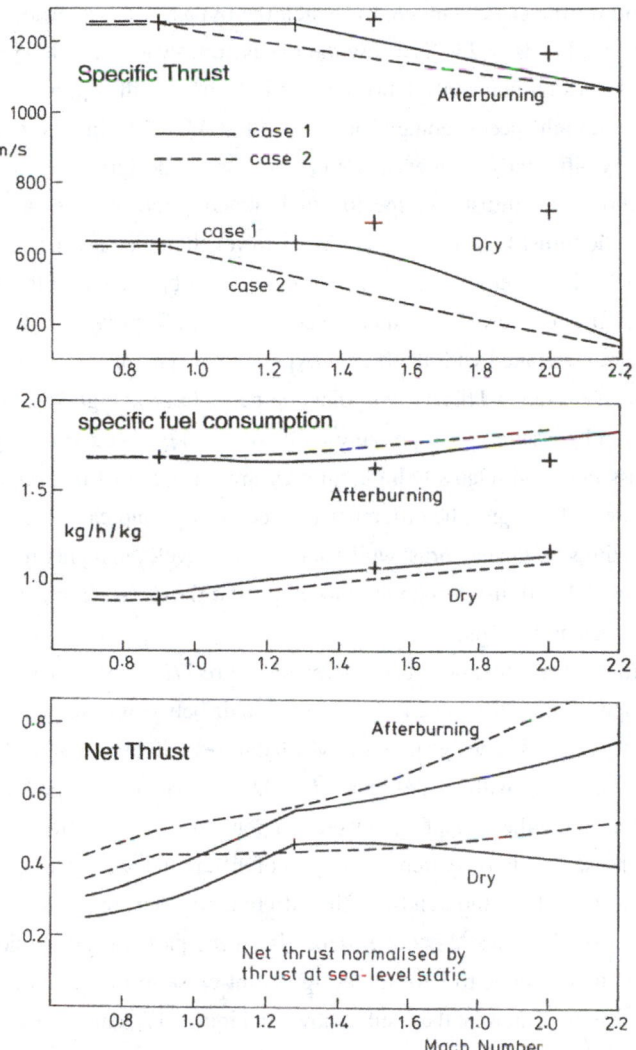

Figure 17.8. Case 1 and Case 2 engines at Maximum Dry and Combat conditions as a function of Mach number for flight at tropopause. Comparison of specific thrust, net thrust and *sfc*. Crosses show *design-point* engines from Chapter 16, Exercises 16.3 and 16.7.

ratios do approximately follow the trends selected for the design-point calculations of Chapter 16, shown as crosses.

At its design point at $M = 0.9$ at the tropopause the core of the case 2 engine is producing more work per unit mass of air compressed because the turbine inlet temperature is much higher, 1850 K rather than 1616 K for case 1. he fan pressure ratio is equal for both case 1 and case 2 at this condition, this extra work appears as a higher bypass ratio for case 2; in other words for the same mass flow of air through the core more air is passing through the engine. For both engines the bypass ratio increases as the engine is 'throttled back', that is, as T_{04}/T_{02} is reduced at higher

Mach number, and for the case 2 engine, for which the bypass ratio is always higher, a a bypass ratio of 1.57 is reached at $M = 2.0$. The way the bypass ratio varies off-design is the opposite of the way it varies the design-point calculations which indicate that the optimum engine would have *lower* bypass ratio at high speeds, going as low as 0.11 at $M = 2.0$. In this respect the off-design engine behaves very differently from what would be chosen at design.

Specific thrust, net thrust and specific fuel consumption are shown in Figure 17.8. As expected, the specific thrust for the case 2 engine is lower than the case 1 engine except at Mach numbers below 0.9. The lower specific thrust for case 2 can be related to the lower fan pressure ratio for $M > 0.9$ (it is entirely due to this effect when the afterburner is in use) and the lower jet pipe temperature associated with the higher bypass ratio. The difference between the specific thrust for engines off-design and the design-point engines is large at high Mach number because of the big difference in bypass ratio and pressure ratio. As already noted, on-design engines at high M have low bypass ratio and relatively high fan pressure ratio, whilst the trend is in the opposite direction for engines off-design. The difference between case 1 and case 2 is less straightforward for the ratio of net thrust to the net thrust whilst static at sea level; the higher mass flow for the case 2 engine leads to a higher thrust despite the lower specific thrust. With the afterburner the case 2 engine generally gives more thrust.

The specific fuel consumption curves given in Figure 17.8 show remarkably little difference between cases 1 and 2; the difference is even quite small between cases 1 and 2 off-design and values calculated for engines at design points at Mach numbers of 0.9, 1.5 and 2.0 at the tropopause. (Case 2, of course, coincides with the design at $M = 0.9$.) This is because pressure ratio plays such a large part in determining the *sfc* and these were not generally very different.

Case 1 and case 2 set fairly extreme examples of different design assumptions. As the results in Figures 17.7 and 17.8 show, the trends for the different engines are not in any sense obvious or intuitive. Whether case 1 or case 2 is better depends on the parts of the mission which are most critical; it is generally desirable to have the design point close to the condition which is critical. Perhaps the most striking feature of the results shown in Figures 17.7 and 17.8 is the relatively small difference between the performance of two engines when the difference in their design point is so pronounced. This brings an important general conclusion; if the differences between the engines is relatively small the almost endless search for an optimum engine may be fruitless. The differences between different missions, and the lack of knowledge about what form any mission would actually take when the aircraft and engine are being designed over many years with a service life of 20 or more years, could far outweigh the small differences between engines.

Exercises

17.5 In Exercise 16.8 the mass flow of air needed for the case 1 engine with the afterburner to provide a thrust–weight ratio of unity at take-off at sea level on a standard day was calculated to be 65.9 kg/s. Assuming that the aircraft weight for subsequent manoeuvres is 15 tonne, values of drag were calculated in Exercise 14.4 and are given in Table 14.1. Use these with the off-design thrust ratios in Exercises 17.3k and 17.4 to calculate the specific excess power (*SEP*) for the aircraft. If $SEP < 0$ find the amount by which the mass flow of air through the engine must be increased to achieve the sustained flight condition. Do this for the engine at Maximum Dry and at Combat (i.e. with the afterburner) for the following:

a 11 km altitude and $M = 0.9$, 1.5 and 2.0 in straight and level (i.e. 1g) steady flight;

b 11 km at $M = 0.9$ and 1.5 for *sustained* 3g turns;

(**Ans:** a) dry: $SEP = 33.2$ m/s, 5.4 m/s, increase $\dot{m}_a$ by 78%;
a/b: $SEP = 90.2$ m/s, 158.8 m/s, 131.5 m/s;
b) dry: increase $\dot{m}_a$ by 108%, 46%; a/b: increase $\dot{m}_a$ by 16%, $SEP = 78.8$ m/s)

Note: The results show that 'supercruise' (supersonic flight without the afterburner) would be possible for the case 1 engine at $M = 1.5$ at the tropopause, but 'supercruise' at $M = 2.0$ would require a much bigger engine. For 3g sustained turns at the tropopause – this is impossible with the case 1 engine at $M = 0.9$ even with the afterburner, and possible only with the afterburner at $M = 1.5$.

17.6 For $M = 0.9$ at sea level a calculation for the case 1 engine shows that the ratio of net thrust to that at sea-level static conditions is 0.798 when dry and 0.979 with the afterburner. Exercise 14.4b gave the drag at this speed and altitude for 5g and 9g turns. Determine the specific excess power for both turn rates dry and with afterburner – if the SEP is negative determine the increase in engine size which would be necessary. The mass for take-off and manoeuvre are 18 tonne and 15 tonne respectively.

(**Ans:** dry; 5g turn, $SEP = 27.8$ m/s, 9g turn, increase $\dot{m}_a$ by 80%:
a/b; 5g turn, $SEP = 194$ m/s, 9g turn $SEP = 12.5$ m/s)

Note: Close to sea level the case 1 engine allows 5g sustained turns with the engine dry, but 9g sustained turns are possible only with the afterburner.

17.7 OPERATION AT REDUCED THRUST

In Section 14.3 some attention was directed to the operation of the aircraft when the drag is much less than the maximum and the amount of engine thrust needed is very small. In Exercise 14.6 the altitude for loiter (to maximise the time spent flying in a vicinity) at a Mach number of 0.6 was found to be about 5.6 km, at which condition the thrust from each engine was only required to be 7.1 kN. In Exercise 14.7 the altitude for maximum range at $M = 0.8$ was found to be about 5.7 km at which condition the thrust should be 8.1 kN. Since minimising fuel consumption is the primary goal for both these flight conditions the engine must obviously operate dry.

To determine the condition at which the engine should operate to produce the required thrust, more specifically the temperature at entry to the turbine, it is necessary carry out an iterative analysis of the type described earlier in Section 17.4. It is also possible to use the ideas of Section 17.5 to generalise a small number of results and to use a graphical solution to remove the need to carry out an iteration.

For operation with a choked final nozzle which expands the exhaust reversibly the nozzle divergent section must alter with pressure ratio across the nozzle, p_{08}/p_a. The pressure ratio is determined by the engine operating condition and by flight Mach number, the latter determining the pressure in the intake. The change in the nozzle area ratio represents a change in the geometry of the entire engine and to obtain simple non-dimensional relations one must stop at the nozzle throat. The primary relations for determining thrust will then be the pressure ratio $p_{08}/p_{02} = p_{013}/p_{02}$ and the temperature ratio T_{08}/T_{02} in terms of the familiar ratio of turbine inlet temperature to compressor inlet temperature T_{04}/T_{02}; these are shown in Figure 17.6. Knowing p_{02}/p_a from the Mach number it is possible to calculate the jet velocity; but to facilitate this it is convenient to have the mean specific heat and the mean specific heat ratio plotted versus T_{04}/T_{02}. To obtain the specific thrust the additional mass flow attributable to fuel is needed and the ratio

$$\frac{\dot{m}_f + \dot{m}_a \left(1 + bpr\right)}{\dot{m}_a \left(1 + bpr\right)}$$

can also be plotted versus T_{04}/T_{02}. Specific thrust alone is not sufficient to determine net thrust since the total mass flow rate is needed, but non-dimensional mass flow $\overline{m}_2$ can also be given as a curve versus T_{04}/T_{02}. All of these curves are shown in Figure 17.9.

Exercise

17.7 It was calculated in Exercise 14.6 that the New Fighter Aircraft would need 7.1 kN thrust from each engine when flying at $M = 0.6$ at an altitude of 5.6 km. (At this altitude take the ambient pressure and temperature to be 50.5 kPa and 251.75 K.) Show that the case 1 engine will produce this thrust with a turbine inlet temperature of 1075 K. (This may be achieved using the curves in Figures 17.6 and 17.9.) If the bypass ratio is 0.958 find the specific fuel consumption.   (**Ans:** 0.87 kg h^{-1}kg^{-1})

SUMMARY OF CHAPTER 17

In a manner similar to that described in more detail in Chapter 12 it is possible to calculate fairly easily how a given engine will behave as either the inlet conditions or the fuel flow are altered. So far as the inlet conditions are concerned, the most important variable in determining engine operating condition is the stagnation temperature, which is a function of both altitude and Mach number. (It is also dependent on the climatic conditions, but throughout the book we have assumed only standard atmospheric conditions.) The inlet temperature, T_{02}, strongly affects engine operation and is a strong function of Mach number. At low engine inlet temperature the fuel flow is adjusted to keep T_{04}/T_{02} at its design value and so to prevent pressure ratios and non-dimensional rotational speed from getting too large. At higher values of T_{02} the turbine inlet temperature T_{04} reaches its limit and further increase in T_{02} leads to the ratio T_{04}/T_{02} decreasing (whilst T_{04} is held constant) and with it all the other non-dimensional variables. Still further increase in T_{02} leads to the compressor delivery temperature reaching its upper limit and then the overall pressure ratio has to be reduced, which means reducing T_{04}. At high flight speeds the engine is therefore inevitably operating at a condition equivalent to 'throttled back' at sea-level static, with all the pressure ratios, non-dimensional mass flows $\overline{m}$ and non-dimensional rotational speeds lowered.

For exactly equivalent reasons an engine operating on a stationary sea-level test bed (or taking off) when the inlet T_{02} is above the design value will be in a 'throttled-back' condition, that is, non-dimensional variables at values below design point. Throttling back has a much larger

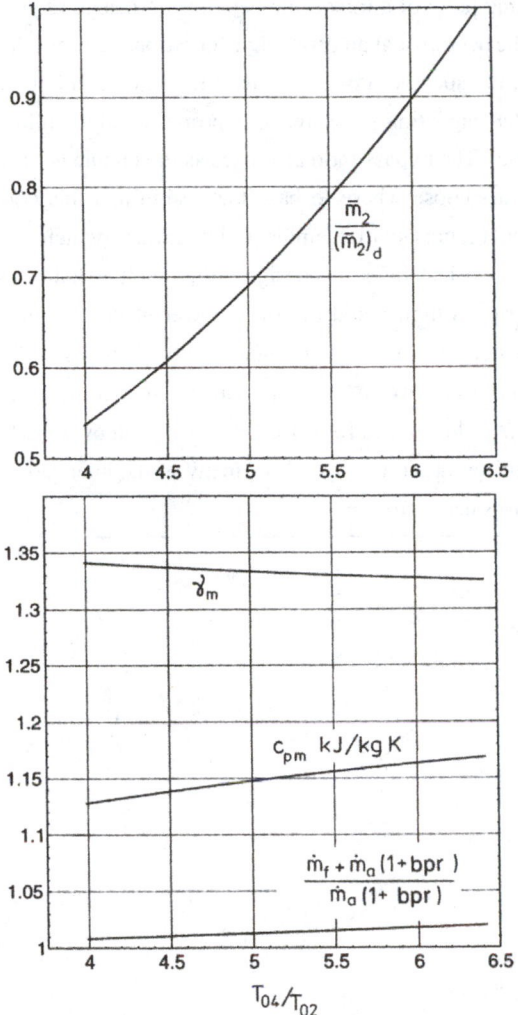

Figure 17.9. Case 1 engine properties versus temperature ratio. Inlet non-dimensional mass flow (normalised by design value). Also exhaust mass flow for dry operation normalised by inlet mass flow c_p and γ into the propulsive nozzle are shown as functions of T_{04}/T_{02}. Constant component efficiencies.

effect on the LP shaft than the HP, so the LP compressor has to accommodate large excursions in non-dimensional mass flow as its operating point changes, but the range for the HP may only be on the order of a quarter as large. The speed changes for the LP shaft are similarly much larger than for the HP shaft.

The calculations of the engine performance off-design are based here on two major simplifications. It is assumed that both the HP and LP turbines are choked, as is the final propulsive nozzle. It is also assumed that the compressor and turbine polytropic efficiencies remain constant.

To calculate the operating point of a mixed-flow engine off-design requires a simple iterative calculation. To calculate the net thrust at an off-design condition requires not only the the fan pressure ratio (or its equivalent, the specific thrust) but mass flow rate and bypass ratio. The mass varies in proportion to the inlet stagnation pressure, so is proportional to ambient pressure and a strong function of Mach number. The bypass ratio also rises as the engine is 'throttled back'.

The designer can choose where to base the design and this could be at sea-level static conditions with a selected representative ambient temperature, which was case 1 here. If a much lower inlet temperature is selected for the design point, such as that for case 2, ($M = 0.9$ at the tropopause), the net thrust is higher and the *sfc* is lower at the design point than for an engine designed at sea-level static. Away from the design point the benefits are less clear, with the engine operating more 'throttled back' over most of the range. The optimum design will emerge from a full analysis of many possible engines for off-design operation over the full mission (in fact, over a large number of candidate missions) to see how many of the intended goals are possible and the extent of the compromises necessary.

CHAPTER 18 TURBOMACHINERY FOR COMBAT ENGINES

18.0 INTRODUCTION

This chapter will look only briefly at the design of the compressors and turbines for combat engines, following on from Chapter 9 for the civil transport engine. The flow Mach numbers inside the compressors tend to be higher than for the subsonic transport and this makes the treatment of each blade row rather special; the design rules must take account of the presence of strong shock waves. A very important design consideration is how the compressor will behave at off-design conditions since combat engines are off-design for so much of their operation. The off-design problem arises because of the large density ratio between inlet and outlet of the compressor, and the reduction in this ratio when $N/\sqrt{c_p T_0}$ is decreased. The turbine stages do not suffer from this off-design problem. The turbines are required to produce large work output in relation to the blade speed, that is $\Delta h_0 / U^2$ must be high, but at off-design conditions for the engine the turbine condition is essentially unaltered from the condition at design point. This, it may be recalled from Chapters 12 and 17, is because the turbines and the propulsive nozzle are effectively choked, so the turbines are forced to operate at the same non-dimensional condition.

In this chapter the consideration will be based on the case 1 engine of Chapter 17, with fan pressure ratio 4.5 and HP compressor ratio 6.66 at design point, sea-level static. This was calculated as Exercise 16.6 with the technology standard given in Table 16.1. The polytropic efficiency chosen for the fan was 0.85, for the HP compressor it was 0.90 and for the turbine 0.875. The fan polytropic efficiency given here at the design point is lower than that given for other recent components in the book. In fact the fan efficiency would probably rise markedly when the engine was working below design point and $N/\sqrt{c_p T_0}$ is reduced, bringing down the Mach numbers inside the engine. At design the fan is "pushed" to pass the largest mass flow and give the largest pressure ratio leading to maximum thrust and some compromise in efficiency is possibly tolerable.

18.1 THE COMPRESSORS

It can be seen from Figure 15.1 that the LP compressor or fan has three stages and the HP compressor 5 stages and these numbers will be adopted here.

In Exercise 16.6 the mass flow of air into the engine to give the necessary thrust at take off was found to be 67.4 kg/s and with a Mach number in the duct upstream of the fan equal to 0.70

the fan tip diameter was fixed at 0.681m. (It was also assumed that the hub diameter was equal to 0.40 times the casing diameter.)

Exercises

18.1* If the fan tip diameter at inlet is equal to 0.681 m and the fan tip speed is 500 m/s find the rotational speed of the LP shaft. If the inlet flow is axial with Mach number 0.70 and the inlet stagnation temperature on a sea-level test bed is 288 K, find the relative velocity and relative Mach number at the tip of the first stage fan.   (**Ans:** 14,022 rpm, 549 m/s, $M = 1.69$)

18.2* The pressure ratio across the fan is 4.5. Assume that the stagnation temperature rise in each of the three stages are equal and that the stage polytropic efficiencies are the same and equal to 0.85. Find the stagnation temperature rise in each stage and the stagnation pressure ratios in each of the stages.
(**Ans:** 63.2 K; pr $= 1.804, 1.636, 1.525$)

18.3 At design the Mach number of the flow at inlet to the fan is 0.7. Through the LP compressor the Mach number is progressively reduced so that at outlet it is 0.3. (The flow is axial at both inlet and outlet.) Use the expression given in Section 6.2 to find $\overline{m}_2$ and then $\overline{m}_{23}$ and hence find the cross-sectional area of the flow at outlet assuming that the flow is uniform. If the hub diameter is increased by 20% from the front of the first stage to the rear of the last fan stage, find the hub and casing diameter at fan outlet.
(**Ans:** $\overline{m}_2 = 1.171$; $\overline{m}_{23} = 0.629$, $A_{out} = 0.163$ m^2, $D_{hub} = 0.327$ m, $D_{casing} = 0.561$ m)

18.4 In Exercise 17.2j the mass flow into the engine at $M = 2.0$ at the tropopause is 0.839 times that at sea-level static conditions. Find the value of $\overline{m}_2$ at this condition and verify that the flow Mach number into the fan is about 0.340. From Exercise 17.2 the pressure ratio across the fan at this condition is 2.18; use this to find value of $\overline{m}_{23}$ assuming the flow to be uniform and the fan efficiency to be unchanged. Verify that the Mach number at exit from the fan is about 0.346.   (**Ans:** $\overline{m}_2 = 0.703$; $\overline{m}_{23} = 0.717$)

For simplicity in the above exercises the flow is treated as uniform at fan exit. This is an oversimplification and the error increases at off-design conditions. It will be noted from Figure 17.7 that the bypass ratio increases from 0.47 at design to about 0.80 at $M = 2.0$ which leads to a radial shift of the streamlines passing through the fan so as to pass a smaller fraction of the total flow through the core.

The HP compressor varies its operating point relatively little over the entire operating range of the aircraft, as shown in Figure 17.4 and it is only for starting that any special measures to ensure satisfactory performance are needed. Typically bleed valves are used for this condition.

Exercises

18.5* From Figure 15.1 it is possible to estimate that the tip diameter of the first stage of the HP compressor is equal to 0.65 times the corresponding diameter of the fan. Assume that the flow into the HP compressor is uniform and axial with a Mach number at design of 0.5. If the relative Mach number onto the first rotor tip is 1.2 find the rotational speed of the HP shaft.   (**Ans:** $D_{tip} = 0.443$ m; $N_{HP} = 20,109$ rpm)

18.6 The pressure ratio across the HP compressor at design is 6.66. Assume that the stagnation temperature rises in each of the five stages are equal and that the stage efficiencies are the same and equal to 0.90. Find the stagnation temperature rise in each stage and the stagnation pressure rises in the first and last stages.   (**Ans:** 78.9 K; stage pr $= 1.619, 1.348$)

18.7 The flow is axial at HP compressor inlet and outlet with Mach numbers of 0.5 and 0.3 respectively; the corresponding values of $\overline{m}$ are 0.956 and 0.629. Find the inlet and outlet area necessary when the overall mass flow into the engine is 67.4 kg/s and the bypass ratio is 0.471. With the inlet tip diameter given from Exercise 18.5 find the inlet hub diameter. Assuming a constant casing diameter find the hub diameter at the HP compressor inlet and outlet and then the blade height h.

(**Ans:** area $= 0.0728, 0.0224$ m^2; $D_{hub,in} = 0.322$, $D_{hub,out} = 0.4095$ m; $h = 60, 17$ mm)

18.8 From Exercise 18.7 the mean radius at inlet and outlet is determined. Find the mean blade speed U_m for the first and last stages and thence the value of the loading parameter $\Delta h_0 / U_m^2$ for these stages.

(**Ans:** $U_m = 403, 448$ m/s; $\Delta h_0 / U_m^2 = 0.489, 0.394$)

18.2 THE TURBINES

The turbines of the combat engine bear a close resemblance to those of a civil engine. The non-dimensional loading $\Delta h_0 / U_m^2$ and flow deflection will normally be somewhat higher for the combat engine, with correspondingly lower efficiencies. The cooling flow may be higher, which generally leads to somewhat higher loss and lower efficiency. For the engine of the New Fighter Aircraft it will be assumed that there is a single stage HP turbine and a single stage LP turbine.

The high speeds and temperatures produce very high stresses in the turbine. Whereas the compressor can be made of lightweight titanium alloys, the turbine can only be made of more dense nickel alloys. The stress in the turbine is crucial in fixing the diameter of the turbine rotors. From Figure 15.1 it can be deduced that the *mean* diameters of the HP and LP turbines are approximately 0.675 and 0.725 times the fan inlet tip diameter respectively. With these estimates it is possible to deduce many of the aspects of the turbines.

Exercises

18.9* **a** Find the blade speed at mean radius U_m for the HP and LP turbines. Using the temperature drops given in Exercise 16.6 find the values for $\Delta h_0 / U_m^2$ for the HP and LP turbines. If the axial velocity is equal to 0.5 times the mean blade speed for each turbine use Figure 9.3 to estimate the stage efficiencies of each turbine. (**Ans:** $U_m = 484.3, 362.5$; $\Delta h_0 / U_m^2 = 1.86, 2.16$; $\eta_{HP} = 0.895$, $\eta_{LP} = 0.893$)

b On the assumption that the *absolute* velocity at outlet from each turbine is purely axial, use the Euler work equation (see Section 9.2) to calculate the *absolute* swirl velocity into each rotor at the mean radius. (**Ans:** $V_\theta = 901, 783$ m/s)

c Assuming that the axial velocity in the turbines is equal to 0.5 times the mean blade speed, find the resultant velocity leaving the HP and LP nozzle guide vanes and thence the *static* temperature. Use this to calculate the local speed of sound and hence find the corresponding Mach numbers assuming $\gamma = 1.30$. What are the flow deflections in the nozzle guide vanes? (**Ans:** $V = 933, 803$ m/s; $T = 1500, 1176$ K; $M = 1.247, 1.212$; $75.0°, 77.0°$)

18.10 For the mean diameter draw the velocity triangles at entry to and exit from the HP and LP rotor. Assume that the axial velocity is equal on both sides of each rotor. Hence find the deflection in each rotor at the mean diameter (**Ans:** $123°, 130°$)

18.11* **a** Find the area of the throat for the HP and LP turbine stators. The mass flow through the HP turbine throat is equal to 0.88 times the mass flow of air into the HP compressor plus the mass flow of fuel;

for the LP turbine the mass flow is 0.96 times the compressor flow plus the fuel flow. The mass of fuel is given in Exercise 16.6 to be 2.9% of the air into the HP compressor. The mass flow of air in the engine is given in Exercise 16.6b as 67.4 kg/s and the bypass ratio in Exercise 16.6a as 0.471. Assume $\gamma = 1.30$.

(**Ans:** 0.0150, 0.0406 m^2)

b As an approximation assume that the flow direction out of the nozzle guide vanes is the same as that at the throat. Hence using the mean diameters for the HP and LP turbines find the blade height at the throat.   (**Ans:** 40.1, 116 mm)

SUMMARY OF CHAPTER 18

This has been a brief chapter, resting firmly on Chapter 9. Use has been made of the drawing of an engine in Figure 15.1 to allow the relative diameters of the compressors and turbines tobe specified. Clearly the accuracy of this is low but it suffices for the purpose of this chapter. (Despite the inaccuracy of these estimates of diameter results are given to three significant figures for the sake of consistency and to aid checking of answers.) Once these diameters are fixed the ease or difficulty of designing satisfactory stages is more or less determined.

The diameters of the turbomachinery components largely determine the overall size and shape of the engine and if suitable information were accessible its weight could be estimated. Naturally the treatment here in Chapters 15–18 has skirted over many of the aerodynamic difficulties and all the mechanical ones. It is to be expected that every aspect of aerodynamic and mechanical design will be as aggressively pursued to optimise performance and minimise weight. It is one consequence of this aggressive push to optimise that the time taken to develop combat engines is so long and cost is so high.

Part 4

A Return to the

Civil Transport Engine

CHAPTER 19   A RETURN TO THE CIVIL TRANSPORT ENGINE

19.0 INTRODUCTION

When the engine for a new civil transport, the New Large Aircraft, was considered in Chapters 1 to 10 many assumptions were introduced to make the treatment as simple as possible. In the treatment of the engine for a New Fighter Aircraft in Chapters 13–18 the level of complexity was increased. The properties of the gas were allowed to be different before and after burning of the fuel and the effect of the mass flow of fuel added to the gas passing through the turbine was included. The effect of the cooling air supplied to the turbines was allowed for and the effect of the pressure loss in the combustor was accounted. It is appropriate to recalculate the performance of an engine for the civil aircraft with some of these effects included and that level of fidelity will apply to most of this chapter.

Another difference between the treatment for the civil engine in Chapters 1–10 and the treatment for the combat aircraft was the mixing of the core and bypass streams upstream of the final propulsive nozzle in the combat engine. Some large civil engines are mixed and this chapter therefore opens with a brief consideration of this option. Following this the consequences of different levels of fidelity in modelling will be addressed. A significant part of the chapter uses the most accurate model to look at the impact of cooling air, pressure drop in the combustor and component efficiency on the thrust and *sfc* of engines; this is done first for the engine on-design and then off-design. The chapter concludes with a brief consideration of propulsion for high-speed civil aircraft.

19.1 THE BENEFIT OF MIXING IN LARGE CIVIL ENGINES

It has been quite common for large civil engines to have a mixed exhaust, where the core and bypass streams are mixed before the final propulsive nozzle. Indeed, most of the Rolls-Royce RB211 engines were mixed and mixing is still common for smaller engines. Figure 19.1 shows two wide-body aircraft which entered service at almost the same time with engines made by Rolls-Royce: the A330 has Trent 700 engines with a mixed nozzle whilst the Boeing 777 has Trent 800 engines with separate nozzles. Both engines have proved to be successful. By a simple treatment (Exercise 19.1) it is possible to demonstrate the ideal advantage that the mixed configuration brings. The mixing can be idealized as a transfer of heat from the relatively hot core stream to the cool bypass stream before the gas expands in the nozzle. The benefits from doing this diminish as the bypass ratio increases, because the relatively smaller core stream has less energy to transfer.

Figure 19.1. Photographs of two contemporary aircraft. On the left is an Airbus-330 with a Rolls-Royce Trent 700 engine having a mixed-exhaust. On the right is a Boeing 777 with a Rolls-Royce Trent 800 engine having an unmixed exhaust.

Moreover, as the fan pressure ratio decreases, with consequent rise in bypass ratio, the specific thrust also decreases and the engine becomes relatively large for the same net thrust. This means that the additional weight and drag associated with the duct in which the mixing is supposed to take place becomes a disproportionately large penalty.

Exercise

19.1 a Compare the gross thrust from two engines, each having a bypass ratio of 6, one with mixed exhaust and the other unmixed. In each case the stagnation pressure at exit from the bypass and the core are equal to 1.65 times the inlet pressure p_{02} and, because of flight forward speed, $p_{02} = 1.50 p_a$, where p_a is the ambient pressure. The stagnation temperature of the core stream at exit from the LP turbine is 750 K, whilst the stagnation temperature of the bypass stream is 305 K. Calculate the gross thrust per unit mass flow through the core (ignoring the mass flow of fuel) for an unmixed engine when the two streams expand separately to ambient pressure. Then calculate gross thrust for a mixed engine when the two streams first mix to uniform temperature without loss in stagnation pressure before expanding to ambient pressure. Assume isentropic expansion in the nozzle in each case. Treat c_p as constant so that enthalpy $h = c_p T$, taking $\gamma = 1.400$ for the bypass, 1.342 for the core stream and $\gamma = 1.391$ for the mixed stream. **(Ans:** Unmixed $F_G = 2834 \, \mathrm{N \, kg^{-1} s^{-1}}$; Mixed $F_G = 2938 \, \mathrm{N \, kg^{-1} s^{-1}}$)
If the flight speed is 231 m/s, find the proportional difference in net thrust F_N. **(Ans:** 8.5%)

b Repeat the calculations of part a) for a fan pressure ratio of 1.5, and a bypass ratio of 10. Take stagnation the temperatures out of the bypass and core to be 280 K and 625 K respectively, with $\gamma = 1.400$ for the bypass, 1.356 for the core and 1.396 for the mixed out flow.
(Ans: Unmixed $F_G = 3922 \, \mathrm{N \, kg^{-1} s^{-1}}$; Mixed $F_G = 3991 \, \mathrm{N \, kg^{-1} s^{-1}}$;
Proportional change in net thrust 5.0%)

The increase in net thrust calculated in Exercise 19.1 looks significant, particularly for the bypass ratio of 6. In reality the benefits are much smaller than this idealised calculation and the increase in thrust calculated here is an overestimate for several reasons. The increase in thrust must be reduced to take account of the stagnation pressure reduction associated with the heat exchange between hot and cold gas (a feature of gas dynamics we will not follow up here) and the stagnation pressure loss associated with the viscous fluid mechanic effects of mixing. Furthermore, the mixing of core and bypass streams is incomplete, so not all of the potential benefit will be achieved. Set against the small increase in net thrust is the increased drag caused by the longer cowl around the outside of the engine nacelle and the increased powerplant weight. The residual increase in F_N could be less than 1% for the bypass ratio of 6 and with the engine of Exercise 19.1b the effect would be negative. All the newer large civil engines are unmixed.

19.2 THE EFFECT OF IMPROVED MODELLING AT THE DESIGN STAGE

The approaches to calculation

The intention throughout this book has been to use the simplest approach consistent with demonstrating the behaviour or estimating the effect being studied. In Part 1, essentially in Chapters 4, 5 and 7, the engine was designed with sweeping simplifications. When something had been learned of the performance of compressors and turbines in Chapter 11 it was possible in Chapter 12 to look at engines operating off-design and for this a slightly more advanced modelling approach was adopted. For looking at combat engines it was necessary to include some representation of turbine cooling and this can now be used to re-examine the civil engine. The various assumptions, which have been incorporated into the calculations described in earlier chapters, are summarized below.

Chapters 4, 5 and 7: The calculations did *not* include turbine cooling or the pressure drop in the combustor and no account was taken of the mass addition of fuel. *Isentropic* efficiencies of 0.90 were assumed for fan, compressor and turbine throughout. $\gamma = 1.40$ was used for air and combustion products, with $c_p = \gamma R/(\gamma - 1)$ calculated from γ.

Chapter 12: The calculations did not include turbine cooling or the combustor pressure drop and the mass addition of fuel was not included. *Polytropic* efficiencies of 0.90 were used throughout. $\gamma = 1.40$ was used for unburned air and $\gamma = 1.30$ for combustion products.

Chapters 15, 16 and 17: For the combat engine the cooling air and the mass flow of fuel were included, but not the pressure drop in the combustor. *Polytropic* efficiencies of 0.90 were used throughout. $\gamma = 1.40$ was used for unburned air and $\gamma = 1.30$ for combustion products.

Chapter 19: In the present chapter earlier approaches are first used to compare the approach of Chapters 4, 5 and 7 with Chapters 15, 16 and 17 to reveal those aspects which are most important. However, the underlying method set out in earlier equations is the same throughout. For Chapter 19 *polytropic* efficiencies have been used throughout. For some calculations $\gamma = 1.40$ for air and $\gamma = 1.30$ for combustion products were used, as in Chapter 12, 15, 16 and 17, but later calculations use gas properties dependent on local temperature which are sufficiently accurate that they cannot be a relevant source of error; the approach is set out in the appendix to this chapter. Such calculations are denoted by the term "Accurate γ", abbreviated to "Acγ" in the tables.

Table 19.1   On-design cycle calculations of varying sophistication; single stage HP turbine

Chap.	γ	Cooling	bpr	T_{03}	HPT p_{04}/p_{045}	LPT p_{045}/p_{05}	p_{05}/p_a	$T_{09} = T_{05}$	sfc	$\dot{m}_f/\dot{m}_2$
7	1.4	no	11.4	794	4.40	10.6	1.44	577	12.7	0.0171
12	1.4 &1.3	no	13.5	822	4.43	11.1	1.37	668	15.0	0.0235
19	1.4 &1.3	yes	11.7	822	5.15	9.09	1.38	650	15.3	0.0212
19	Acc.γ	yes	11.6	793	4.96	9.29	1.40	625	13.9	0.0191
GasTurb		yes	11.6	798	5.08	9.09	1.40	628	13.6	0.0186

Note: $M = 0.78$, 35,000 ft, $fpr = 1.5$, $opr = 45$, $T_{04}/T_{02} = 6.11$. When cooling presents 4% loss in pressure in the combustor is also included.

To compare the effects of the different approaches and approximations, Table 19.1 has been drawn up and shows relevant numbers calculated for the civil engine which began in Part 1 of the book. Throughout, the design point is specified as follows:

flight at $M = 0.78$ at 35,000 ft,

fan pressure ratio in the bypass stream $fpr = p_{013}/p_{02} = 1.5$.

fan inner region and booster, referred to as the LP compressor, $p_{023}/p_{02} = 2.5$

HP compressor $p_{03}/p_{023} = 18$

turbine inlet temperature $T_{04} = 1500$ K, giving $T_{04}/T_{02} = 6.11$.

For all of the results in Table 19.1 the bypass ratio has been adjusted in the calculations so that the jet velocity of the bypass stream is equal to that of the core stream. This is not a necessary requirement, and may not even lead to the best engine (lowest fuel burn or lowest powerplant weight), but it was shown in Chapter 7 that the effect of plausible values of this velocity ratio on *sfc* is small. (Equal velocities for core and bypass leads to a higher bypass ratio than for values typical of real engines.)

Some rows in Table 19.1 are for calculations where cooling air has been included; the cooling air amounts to 10% of the air entering the core compressor $\dot{m}_2$ and is withdrawn at exit from the HP compressor. It is used to cool the HP turbine rotor and accounted for in the cycle calculations by adding air at compressor delivery temperature back into the main flow downstream of the HP turbine rotor being cooled. The cooling air used to cool the HP turbine nozzle guide vanes is automatically included in the definition of turbine entry temperature T_{04}. In Table 19.1 the HP turbine is assumed to have one stage so that all the cooling air is accounted as reinserted downstream of the rotor; in some later calculations a two-stage HP turbine will be assumed with the cooling air reinserted in two places. The pressure drop in the combustor is assumed to be 4% of the pressure at compressor delivery, hence $p_{04} = 0.96\, p_{03}$.

Because the results in Table 19.1 are all on-design, the overall pressure ratio is constant and the fan pressure ratio are constant throughout. Because $fpr = 1.5$, and the velocities of the bypass and core jet streams are set equal, the jet velocity $V_j = 339$ m/s. The variable which shows the effect of the approximations most clearly is therefore the bypass ratio. The gross thrust is equal to $(1 + bpr) V_j$ per unit mass flow rate entering the core, $\dot{m}_2$.

In Table 19.1 the stagnation pressure ratios across the HP and LP turbines are shown, together with the pressure ratio across the core propulsive nozzle which, assuming no loss downstream of the LP turbine, is the stagnation pressure leaving the turbine divided by the exit static pressure, $p_{05}/p_a = p_{09}/p_a$. (The exit static pressure is assumed to be the atmospheric pressure for the unchoked core nozzle) at exit. It will be observed that the pressure ratio across the propulsive nozzle does not change much and this is because the jet velocity of the core stream is set to be equal to the velocity of the bypass stream and the bypass jet velocity is constant because the fan pressure ratio is constant. The core jet velocity has a relatively weak dependence on the temperature leaving the LP turbine, $T_{05} = T_{09}$. Even for the results for which the value of $\gamma = 1.4$ was used for the gas through the turbine, as in Chapter 7, T_{09} is only about 50 K less than the most accurate value. Consequently the core nozzle pressure ratio varies by no more than about 3%. Because the pressure ratio across the core nozzle does not change significantly and the engine overall pressure ratio is fixed, most of the reduction in the pressure ratio of the HP turbine is compensated by an increase in the pressure ratio of the LP turbine.

As already noted, in on-design calculations the pressure ratios are fixed so that the errors tend to accumulate, that is to become most apparent, in the pressure ratio across the LP turbine, p_{05}/p_{045}. If this pressure ratio is larger than the correct value, the power to the fan is too large and, for constant fan pressure ratio, this error appears as an enlarged bypass ratio. With fan pressure ratio specified the errors therefore appear in the bypass ratio. Bypass ratio is significantly overestimated by the method used in Chapter 12 because with temperature downstream of the LP turbine, T_{05}, nearly constant and with $\gamma = 1.3$ (and hence $c_p = 1244\ \mathrm{J\,kg^{-1}K^{-1}}$) the power per unit mass flow from the LP turbine is increased by almost 25% relative to Chapter 7 calculations. When turbine cooling air is included, as for Chapter 19, temperature and pressure ratios across the HP turbine are increased leaving less pressure ratio and hence less power for the LP turbine. The bypass ratio obtained with the Chapter 19 method therefore drops back to near the level of Chapter 7. The reasonable agreement in the calculated bypass ratio in Chapter 7 (no cooling and $\gamma = 1.4$) is therefore probably fortuitous.

In calculating the ratio of fuel mass flow rate to the mass flow rate of air entering the core, $\dot{m}_f/\dot{m}_2$, different approaches have been used. For Chapters 7 and 12, and for the simpler method of Chapter 19 (where we take $\gamma = 1.4$ and 1.3), the fuel was found from the temperature difference $T_{04} - T_{03}$ and specific heats, as in Equation 11.6. This is not satisfactory when c_p is a strong function of temperature and $\dot{m}_f$ is then overestimated; in the present case by about 11%. Therefore for the "Accurate γ" case in Table 19.1 fuel flow has been calculated using tabulated values of enthalpy for gas into and out of the combustor.

The bottom row of Table 19.1 shows the results of the most accurate method available, the commercial programme GasTurb. It should not be surprising that the "Accurate γ" results in Table 19.1 agree reasonably well with, GasTurb, since the physics is the same and the empirical inputs are as near as possible the same. What is surprising is that agreement is fair from even the simplest approach of Chapter 7, making it possible to deduce trends for design choices using it. As the results of Chapter 12 show, simply altering one set of inputs does not in general lead to better estimates and there is clear evidence of errors tending to cancel between different methods. So, for example, the results for Chapter 7 are generally more in line with GasTurb than those of Chapter 12 where more representative gas properties were used.

Turbine cooling flow

Calculations in the present chapter are shown for which $\gamma = 1.4$ was used for the air and $\gamma = 1.3$ for the combustion products and cooling air was included. The use of cooling air is the principal difference here from the calculations in Chapter 12 with the same values forγ. The comparison between the rows 2 and 3 of Table 19.1 shows the HP turbine pressure ratio is higher for the cases with cooling and the LP turbine pressure ratio is lower. This is because with turbine cooling the mass flow rate of highly energetic combustion products into HP turbine is reduced, yet the power required by the core compressor is unchanged. The temperature and pressure out of the HP turbine must therefore be lower, assuming that cooling has not changed the polytropic efficiency. The core thermal efficiency is reduced by the cooling air because work is done compressing the air used for cooling but less work is extracted in the turbine because the cooling air is at lower temperature and is mixed in downstream of the rotor. With the engine overall pressure ratio fixed, the pressure ratio across the LP turbine must therefore fall, giving less power from the LP turbine and, for the same fan pressure ratio, a lower bypass ratio.

19.3 THE EFFECT OF CHANGES AT THE DESIGN STAGE

In this section we will examine the effect of changes, such as changes to cooling air, reduction in combustion pressure drop and alterations to the efficiency of turbines and compressors. These calculations are equivalent to what might happen in a design office whilst the engine is still at the "paper" stage, when alterations do not alter the pressure ratios which are specified. So, for example, if the HP compressor efficiency were to be increased the power required to the compressor to achieve the specified pressure ratio would be less, so the power from the HP turbine would be lower. As a result the pressure and temperature drops in the turbine would be smaller and the pressure ratio and temperature drop across the LP turbine would be higher; for the reasons previously discussed, the bypass ratio would be increased. The calculations allow a sensitivity or a so called 'what-if' analysis to be conducted.

Variation in turbine cooling flow at the design stage

Here we will look at two variations in the cooling, one being the way it is taken and used, the other the quantity which is taken. For the results in Table 19.1 attributed to Chapter 19, turbine cooling

Table 19.2   Calculated effects of variation in cooling flow on engine parameters at the design stage

No HP stages	$\dot{m}_c/\dot{m}_2$	bpr	p_{04}/p_{045}	p_{045}/p_{05}	p_{05}/p_a	$T_{09} = T_{05}$	$F_N/\dot{m}_2$	sfc
1	0.1	11.69	4.95	9.33	1.40	625	1381	13.7
1	0.09	11.92	4.85	9.53	1.40	626	1405	13.6
2	0.1	11.86	4.69	9.79	1.41	619	1399	13.5
2	0.09	12.06	4.62	9.94	1.41	619	1420	13.4
2	0.05	12.83	4.38	10.57	1.40	630	1504	13.2

Note: $M = 0.78$, 35000 ft, $fpr = 1.5$, $opr = 45$, $T_{04}/T_{02} = 6.11$, $p_{04}/p_{03} = 0.96$. Datum has 10% cooling flow. Accurate γ.

air equal to 10% of the air entering the core compressor is taken at compressor discharge and used to cool the HP turbine rotor. As noted, taking compressor delivery air and essentially bypassing the HP turbine represents a loss in useful work for each unit mass of air compressed. It is more likely that in a real engine the HP turbine would have two stages and this would alter the cooling arrangement. For the present work with a two-stage turbine we assume half the air is used in the second (i.e. lower pressure) stage of the HP turbine. The pressure at this station is lower so the cooling air to the second stage is withdrawn half-way through the HP compressor. It is self-evident that drawing cooling air at higher pressure than needed is a source of lost power. The other half of the cooling air, for the first stage of the HP turbine, is taken from HP compressor delivery. Different cooling arrangements are explored in Table 19.2, all using the "Accurate γ" method for gas properties with $\dot{m}_c/\dot{m}_2$ being the ratio of cooling mass flow rate to the mass flow rate entering the HP compressor. Two cooling rates are considered for the datum engine which is the engine with one HP turbine stage used for Table 19.1. In the modified engine with two HP turbine stages cooling mass flow rate equally divided between the two stages and three different total levels of cooling are considered.

Because in Table 19.2 the overall pressure ratio is held constant, altering the cooling air affects the split in pressure ratio between the HP and LP turbines. The most pronounced change in the parameters shown is in the bypass ratio. Table 19.2 shows that with a realistic level of cooling for a civil engine, there are marked benefits in splitting the removal and reintroduction of the cooling air as in a two-stage HP turbine: a rise in thrust per unit mass flow through the core of 1.3% and a reduction of 1.5% in sfc is calculated. Note that in computing this improvement the turbine efficiency has been kept constant – in fact having two turbine stages is likely to lead to an increase in turbine efficiency because of lower loading coefficient.

Reducing the magnitude of cooling air by one tenth might be plausible with development, but as Table 19.2 shows, this gives only a modest reduction in sfc, worth about 0.6% for the single stage or the two-stage. Halving the magnitude of cooling air has a significant effect on sfc, reducing it about 1.8%, but such a large change in cooling air is unlikely.

Exercise

19.2 Use the results presented in Table 19.2 for a two-stage HP turbine, with 10% of the core compressor air used for cooling, to determine key flow areas in the engine at the design point. The engine is to produce a net thrust of 38.9 kN at $M = 0.78$ at 35,000 ft. It may be assumed that the fuel mass flow is equal to 2% of the mass of air entering the core giving an equivalence ratio of 0.34. It may be assumed that adequate accuracy is obtained by evaluating gas properties at the stagnation temperature, though static temperature is correct.

a Find the turbine nozzle throat area when $T_{04} = 1500$ K. Assume that $\gamma = 1.295$ and $c_p = 1259$ J kg^{-1}K^{-1}. Given the nozzle is choked, find $\overline{m}_4$ and then the nozzle area per unit mass flow into the core.

(**Ans:** $\overline{m}_4 = 1.393$, $A_4/\dot{m}_2 = 0.588.10^3$ m^2 kg^{-1}s)

b Find the core propulsive nozzle throat area when $T_{09} = 619$ K, $\gamma = 1.357$ and $c_p = 1091$ J kg^{-1}K^{-1}. Since the nozzle is unchoked find $\overline{m}_9$ and then the nozzle area per unit mass flow into the core.

(**Ans:,** $\overline{m}_9 = 1.225$, $A_9/\dot{m}_2 = 20.4.10^3$ m^2 kg^{-1}s)

c Find the bypass nozzle throat $\gamma = 1.4$. Since the nozzle is choked $\overline{m}_{19} = 1.281$. Find the bypass nozzle area per unit mass flow into the core.

(**Ans:** $A_{19}/\dot{m}_2 = 91.7.10^3$ m^2 kg^{-1}s)

Note: These areas are key requirements for off-design calculations. Recall that the pressure drop in the turbine must be compatible with passing the core flow through the HP turbine nozzles and the final propulsive nozzle. In addition the HP turbine pressure ratio is fixed (if gas properties and efficiency do not vary significantly) with HP turbine power proportional to T_{04}.

Changes of component efficiency at the design stage

In this section the effect of changes in component performance will be assessed when the pressure ratios are fixed. The ratio of turbine inlet temperature to engine inlet temperature T_{04}/T_{02} is also kept constant. In consequence a change in the performance of one component has to be compensated by an alteration in some other component.

The comparisons shown in Table 19.3 were carried out on the same engine as that used in Tables 19.1 and 19.2 with a two-stage HP turbine. 10% of the air entering the core compressor is used to cool the HP turbine, half coming from compressor discharge and half from midway along the compressor. The stagnation pressure drops by 4% in the combustor. Unless perturbed the polytropic efficiencies of the fan, compressors and turbines are 0.90. The "Accurate γ" for gas properties has been used.

For the changes in fan and the turbine efficiencies the proportional changes in net thrust F_N are almost equal to the inverse of proportional changes in *sfc*. This is because as the efficiency of these components is increased the fuel flow is unaffected. For the HP compressor, however, an increase in efficiency means a lower temperature into the combustor T_{03} so that, for the same turbine entry temperature T_{04}, the fuel supply increases and the proportional increase in net thrust per unit core mass flow is about double the reduction in *sfc*. The HP compressor is therefore particularly important in determining the thrust and the fuel consumption. For the HP turbine the influence of efficiency on F_N and *sfc* is smaller because higher efficiency means lower temperature into the LP turbine which produces less power.

Table 19.3   Design-stage calculations of variations in component performance.

Pressure ratios held constant	bpr	$\Delta F_N/F_N$	$\Delta(sfc)/sfc$ %
Datum engine (2-stage HP turbine)	11.86	0	0
Fan bypass η_p to 0.91	12.00	0.88	−0.88
HP compressor η_p to 0.91	12.09	1.75	−0.87
Core fan & booster η_p to 0.91	11.94	0.58	−0.29
HP turbine η_p to 0.91	11.94	0.58	−0.58
LP turbine η_p to 0.91	11.97	0.87	−0.86
Halve combustor pressure loss, $p_{04} = 0.98p_{03}$	11.95	0.71	−0.70
Increase turbine inlet temp. 10 K to 1510 K	12.06	1.57	−0.23

Note: One parameter varied whilst others held at datum value as in Table 19.2; $M = 0.78$, 35,000 ft, $fpr = 1.5$, $opr = 45$, $T_{04}/T_{02} = 6.11$, $p_{04}/p_{03} = 0.96$. Two-stage HP turbine. Datum with 10% cooling and 4% pressure loss in the combustor Accurate γ.

As Table 19.3 shows, halving the combustor loss would be beneficial, but the scope for reducing this is small, since a pressure drop in the combustor is needed to produce turbulent mixing in the combustor and also because the cooling air to the HP turbine nozzle guide vanes must have a higher pressure than the stagnation pressure of the main flow so that it can flow through the holes in the nozzle wall. The impact of turbine entry temperature is strong on net thrust per unit core mass flow, but relatively weak on the *sfc*.

The changes shown in Table 19.3 are small and the perturbations may be added and subtracted. So, for example, an increase in turbine entry temperature can be used to compensate for deficits in component efficiency. As shown, the thrust per unit core mass flow which would be lost from a deficit of 1% in HP compressor efficiency is 1.75%, whereas a rise in turbine entry temperature of 10 K gives a increase in thrust of 1.57%. Therefore the loss in thrust attributable to 1% loss in HP compressor efficiency can be restored by an in increase in T_{04} of $10 \times 1.75/1.57 = 11.1$ K. The resulting *sfc* would be higher than original design intent by $0.87-(0.23 \times 11.1/10) = 0.61\%$, however.

As noted above in connection with the cooling air, the propulsive efficiency is constant here and the variation in *sfc* comes from the alteration in core thermal efficiency and fan efficiency.

Changes in aircraft design speed

The performance of an engine is affected by the speed at which it flies. In this section we consider the effect of change in cruise Mach number on the design point. Later, in Section 19.4, we consider the off-design case when the flight Mach number is changed for an existing engine. Raising the Mach number increases the inlet stagnation temperature T_{02} and pressure,

Table 19.4   Effects of Design Mach number for Cruise for On-design Operation

	M	V (m/s)	T_{04} (K)	T_{04}/T_{02}	bpr	V_j (m/s)	$F_N/\dot{m}_{tot}$ (m/s)	$\Delta(sfc)/sfc$ %
Datum	0.78	231	1500	6.11	11.86	340	109	0
$T_{04}/T_{02} = $ const	0.70	207	1468	6.11	11.62	322	114	−5.8
	0.85	252	1531	6.11	12.09	356	104	5.3
$T_{04} = $ const	0.70	201	1500	6.24	12.30	322	114	−6.3
	0.85	252	1500	5.99	11.48	356	104	5.4

Note: 35,000 ft, *fpr* = 1.5, *opr* = 4, p_{04}/p_{03} = 0.96. Two-stage turbine. Cooling flow rate 10%. Accurate γ.

$p_{02}/p_a = \{1 + (\gamma - 1)M^2/2\}^{\gamma/(\gamma-1)}$. Raising flight Mach number increases the pressure throughout the engine, but it becomes significant by changing the pressure ratio across the core propulsive nozzle, which is not choked. The comparison can sensibly be carried out in two ways: at constant T_{04}/T_{02}, which means the core would approximate to constant non-dimensional condition, or constant T_{04}. Constant T_{04} is more like a real comparison since this temperature is limited by the level of material and cooling technology available to the designer. The effects are shown both ways in Table 19.4. It should be noted that the jet velocities are identical for the two approaches to T_{04} because the jet velocity is determined by the fan pressure ratio which is held constant for these on-design calculations. As forward speed V is increased the bypass pressure increases relative to ambient static pressure and the jet velocity V_j goes up.

Consider first the case where T_{04}/T_{02} is held constant as cruise Mach number is changed. In this case the core will be at nearly constant condition, so that the thermal efficiency of the core and the power output per unit mass flow will change comparatively little. The effect is best understood by considering specific thrust $F_N/\dot{m}_{tot}$, where $\dot{m}_{tot}$ is the total mass flow leaving the engine. The changes in the flight speed and jet velocity are of the same sign and similar in magnitude so the variation in specific thrust is relatively small with forward speed. The proportional variations in specific thrust are similar to the variations in specific fuel consumption. When T_{04} is held constant the variation in specific thrust is the same as for $T_{04}/T_{02} = $ constant but the variations in *sfc* are slightly larger. This is because the alteration in T_{04}/T_{02} changes the core efficiency and power output per unit core mass flow.

The changes in *sfc* with flight Mach number look attractive, but there is a snag. In the estimation of fuel burn from an aircraft, see Chapter 2, the relevant quantity is the range factor, $H = V(L/D)/sfc$, and the proportional reduction in *sfc* with reduction in speed is slightly smaller than the reduction in flight speed. Reducing cruise speed without modifying the aircraft (to take out empty weight or increase lift-drag ratio) will not lead to a fuel burn reduction.

19.4 OFF-DESIGN, THE EFFECT OF CHANGES TO AN IN-SERVICE ENGINE

Here we consider what happens when an existing engine experiences changes after it has been designed. In this case, which we call off-design, pressure ratios do *not* stay constant when parameters vary; for example, a reduction in HP compressor efficiency will lead to a reduction in the overall pressure ratio, lower power from the LP turbine and a reduced pressure ratio for the fan.

Two quite different types of alteration need to be considered. The first type of change is associated with the components of the engine, such as a change in compressor or turbine efficiency. The second type is change in aircraft or engine operation relative to the design case, such as change in flight speed or turbine entry temperature.

Changes associated with component efficiency arise in several ways. Here we specifically consider component polytropic efficiency for which the change could be the result of in-service operation (e.g. sand erosion, build-up of dirt, increase in tip clearance) or it could be failure of the delivered hardware to match the specification in the design. A change in efficiency in one component will lead to some consequent alteration in operation of other components in the engine. Failure to meet specification is obviously a problem most often encountered during the development of an engine, when the consequences need to be understood and accounted for.

The approach to determining off-design conditions is that set out in Chapter 12. The solution must begin with an on-design calculation. Taking choked flow at entry to the HP and LP turbines fixes the temperature and pressure ratio across the HP turbine and the dependence of the HP turbine power per unit mass flow on turbine inlet temperature. The on-design calculation also fixes the cross-sectional flow areas of the HP turbine nozzle, LP turbine nozzle, the core propulsive nozzle and the bypass stream nozzle. For the present simple calculations the polytropic efficiency of each component is assumed to remain constant at its set value unless explicitly altered, which means that large excursions in operating point will lead to inaccuracies. (In GasTurb this restriction to constant component efficiency is removed because maps of turbine and compressor performance are used, either representative ones stored in the programme or those supplied by the user.)

As noted earlier, the large overall pressure ratios makes the calculated results sensitive to the assumed values of gas properties; this is particularly true of the pressure ratio across the LP turbine and this becomes apparent in the computed bypass ratio. The effect is particularly noticeable for off-design calculation because the fan pressure ratio now depends on the power from the LP turbine. (For on-design calculations *fpr* was specified and the errors associated with inaccurate gas properties showed up in the calculated bypass ratio which did not affect the specific thrust or the propulsive efficiency.) For considering off-design operation of an engine with a low fan pressure ratio accurate gas properties, such as those derived using the empirical expression in the appendix to this chapter, are required to get the right trends.

Component efficiencies vary with altitude and forward speed because of alterations in Reynolds number, but this effect is small enough to neglect in the current treatment. Significant

Table 19.5   Off-design calculations of engine parameters holding turbine inlet temperature constant at design altitude and Mach number

	fan pr	bpr	Overall pr	$\Delta F_n/F_n$ %	$\Delta (sfc)/sfc$ %
Datum	1.500	11.86	45.0	0	0
Fan bypass $\eta_p = 0.91$	1.505	11.91	45.0	0.67	−0.75
HP compressor $\eta_p = 0.91$	1.519	11.48	46.9	4.75	−0.66
Core fan & booster comp. $\eta_p = 0.91$	1.507	11.73	45.6	1.51	−0.20
HP turbine $\eta_p = 0.91$	1.504	11.86	45.1	0.59	−0.49
LP turbine $\eta_p = 0.91$	1.505	11.87	45.1	0.95	−0.75
Increase turbine temp. $T_{04} = 1510$ K	1.510	11.75	45.8	2.41	−0.09

Note: Design is datum engine, $fpr = 1.5$, $opr = 45$, $T_{04} = 1500$ K, Two-stage turbine. as in Table 19.2. Where not perturbed: $T_{04} = 1500$ K and component polytropic efficiencies 90%. Pressure drop in combustor 4% and 10% cooling flow rate thoughout. Accurate γ.

variations in compressor efficiencies do occur when they are operating well away from design point, particularly at pressure ratios above those for the design point. This occurs at two important cases which will be considered: sea-level static, when the aircraft is still stationary, and top-of-climb, when the maximum climb rating is to be produced at the cruise altitude.

Off-design alteration in component efficiency

Calculations have been carried out to the engine first used for Table 19.2 with the two-stage HP turbine, for which the cooling air is drawn from two locations. Small perturbations are again applied in turn to component efficiencies: each component has in turn had its efficiency increased by 1% from 0.90 to 0.91. These changes are sufficiently small that the effects are essentially linear so they may be added and subtracted. Similarly, for these small changes the signs can be reversed to represent decreases in efficiency. In addition the turbine inlet temperature has been raised by 10 K with all the efficiencies at their unperturbed value of 0.9.

The largest change is in net thrust attributable to the increase in efficiency of the HP compressor. Recall that the turbine inlet temperature is constant. The higher efficiency leads to an increased overall pressure ratio, which increases core mass flow and the work out from all the turbines. More work from the LP turbine increases the booster pressure ratio and the fan pressure ratio and thus fan mass flow. For this reason too, the fan and booster efficiencies have a significant effect on thrust, though a relatively small one on *sfc*. The LP system, both the fan in the bypass stream and the LP turbine, have the largest impact on *sfc*.

Because the small changes in Table 19.5 are additive it can be seen that a shortfall in thrust because of a loss in component efficiency can be compensated by a rise in turbine entry temperature. For example, Table 19.5 shows that a 1% increase in fan efficiency gives a rise in thrust of 0.67%

Table 19.6   Off-design calculation of impact of change in flight Mach number

Design M	Operating M	fpr	opr	$\Delta F_n/F_n$ %	sfc $\mathrm{g\ s^{-1}kN^{-1}}$	$\Delta(sfc)/sfc$ %
0.78	0.78	1.50	45	0	13.5	0
0.78	0.70	1.52	47.4	4.7	12.7	−6.1
0.85	0.85	1.50	45	0	14.2	0
0.85	0.78	1.52	47.4	3.6	13.5	−5.0

Note: Design is datum engine of Table 19.2, Two-stage turbine$fpr = 1.5$, $opr = 45$, $T_{04} = 1500\,\mathrm{K}$, $p_{04}/p_{03} = 0.96$ Cooling flow rate 10%. Accurate γ.

and a decrease in *sfc* of 0.75%. A 10 K rise in T_{04} gives a thrust increase of 2.41% and a *sfc* decrease of 0.09%. Hence the net thrust loss from a deficiency in fan efficiency of 1% can be recovered by increasing the turbine inlet temperature by $10 \times 0.67/2.41 = 2.78$ K with a consequent increase in *sfc* relative to design intent of about $0.75 - 0.09 \times 0.67/2.41 = 0.72\%$

Altered aircraft operation

Cruise speed alteration

First we consider relatively small alterations around the design point, which is cruise at $M = 0.78$ at 35,000 ft. Aircraft have to obey requirements of air traffic control, which may require them to fly more slowly than their design speed. There is little scope for flying at a higher Mach number than that chosen for the design point because the aircraft drag then increases steeply. In Table 19.6 results are shown for two engine designs, one designed for cruise at $M = 0.78$, the choice for the New Efficient Aircraft, and the other designed for cruise at $M = 0.85$, the value of many of the recent aircraft designs, such as the Boeing 787 and Airbus A350. The cruise Mach number for each is reduced in the off-design calculation, with the 0.78 being reduced to 0.70 and 0.85 being reduced to 0.78. In all cases turbine inlet temperature is held constant

As the numbers in Table 19.6 show, for flight at $M = 0.78$ there is a small *sfc* advantage in designing the engine for $M = 0.78$ compared with operating an engine designed for $M = 0.85$ at this reduced Mach number. The changes in *sfc* with Mach number are very similar to those achieved for on-design Mach number reduction shown in Table 19.4. The net thrust goes up for both on-design and off-design when the Mach number is reduced because the ram drag falls more than the reduction in gross thrust. For the off-design decrease in Mach number the net thrust also goes up because the pressure ratios have increased and the increase in *opr* gives a small increase in the core thermal efficiency.

Climb and take-off

Next we consider two critical off-design conditions: the top-of-climb condition, when the aircraft is climbing just as it reaches the cruise altitude, and take-off, when the engine is at sea-level and

Table 19.7   Engine at design (cruise) and Off-design at top-of-climb and take-off

	T_{04} (K)	T_{04}/T_{02}	fpr	opr	HPT pr	LPT pr	T_{09} (K)	bpr	V_{jc} (m/s)	V_{jbp}	$F_n/F_{n.des}$
M = 0.78, 35 kft											
Design	1500	6.11	1.50	45	4.69	9.79	619	11.86	340	340	1
Climb	1550	6.31	1.56	49.9	4.70	10.2	635	11.23	375	349	1.143
Sea-level Static	1761	6.11	1.54	49.3	4.82	7.80	776	10.03	315	278	6.83

Note: Design is datum engine of Table 19.2 with two-stage HPT. Design point *fpr* = 1.5, *opr* = 45, T_{04} = 1500 K, Pressure loss in combustor 4%, cooling flow rate 10%, Accurate γ.

stationary (i.e. sea-level static). Top-of-climb is assumed to occur at the Mach number for cruise, the design value of 0.78 and initial cruise altitude of 35,000 ft. The results are shown in Table 19.7 where the design point parameters are repeated for comparison. For the climb condition the turbine inlet temperature has been raised by 50 K, increasing the temperature ratio T_{04}/T_{02} from 6.11 to 6.31. For take-off the temperature ratio has been kept at the design value of 6.11 so that T_{04} increases to 1761 K. As explained in Chapter 8, the constant non-dimensional scaling does not hold exactly for this engine when Mach number is varied, even though T_{04}/T_{02} is held constant, because the core propulsive nozzle is not choked and the turbine pressure ratios are therefore not held constant. Nevertheless the HP turbine hardly changes pressure ratio. As a result the pressure ratio of the HP compressor changes relatively little and most of the change is associated with the LP spool.

The climb results in Table 19.6 show that when the value of T_{04}/T_{02} is increased above the design value the bypass ratio is reduced and the core jet velocity has increased more than the bypass velocity (i.e. V_{jc}/V_{jb} is increased). This is a general result for all engines. For the sea-level static case the drop in bypass ratio is because the proportional drop in pressure ratio in the bypass duct is large when the "ram effect" is absent and, although the fan pressure ratio has increased, the pressure ratio across the nozzle p_{019}/p_a has gone down. The bypass nozzle therefore passes less flow relative to that through the core.

The 50 K rise in temperature to achieve climb results in a 14% increase in net thrust. As Exercise 19.3c shows, this would produce a rate of climb of 308 ft/min, which meets the minimum requirement but is less than customers often now require. Temperature rises greater than 50 K are possible but the overall pressure ratio then gets large and to allow this the design point efficiency of the compressors may be compromised.

The computed sea-level static thrust shown in Table 19.6 is equal to 6.8 times cruise thrust and may be compared with the requirement for take-off thrust to be 0.3 times maximum take-off weight given in Chapter 2. The reduction in weight from fuel burned in take-off and climb is no more than about 2% of take-off weight, so initial cruise thrust will be approximately the maximum take-off weight divided by lift-drag ratio, here $L/D = 21.6$. The ratio of required take-off thrust to initial cruise thrust is therefore 6.6. The computed take-off thrust in Table 19.6 at the design value of T_{04}/T_{02} should therefore be ample, confirming that take-off and cruise are at approximately the same non-dimensional condition for bypass engines with low fan pressure ratio.

The purpose of the present treatment is to remain simple so that the dependencies and interactions can be understood with the least possible amount of empirical information. It is important that the trends predicted are correct. In the calculations leading to Table 19.6 the compressor and turbine efficiencies have been held constant at 90%. This is a good approximation for the turbines but is a likely source of inaccuracy for the fan and compressors when the pressure ratios in the off-design operation are significantly different from the design values. When pressure ratios are above design the drop in efficiency of compressors and fans tends to be particularly rapid. A program like GasTurb does include maps of compressor performance and the drop in efficiency can be estimated: for the increase in pressure ratio with $T_{04} = 1550$ K, the climb condition, GasTurb predicts efficiencies of 87.5% for the fan and 89.6% for the HP compressor, compared with the values of 90% at design for each component.

Exercises

19.3 Use the results of Exercise 19.2 and the results presented in Table 19.6 to estimate engine performance at $M = 0.78$ at 35,000 ft when turbine entry temperature is increased to 1550 K, assuming component efficiencies are unchanged. As before assume that the fuel mass flow is equal to 2% of the mass of air entering the core and 10% of the core air is used for cooling the two-stage HP turbine.

a At HP turbine entry assume $T_{04} = 1550$ K and $c_p = 1265$ J kg^{-1}K^{-1}, $\gamma = 1.293$. Determine the core jet velocity and calculate ratio of core mass flow to core mass flow at design (i.e. when $T_{04} = 1500$ K).
(**Ans:** 375 m/s, 1.090)

b Calculate the bypass jet velocity and the bypass mass flow as a ratio of mass flow at design.
(**Ans:** 349 m/s, 1.029)

c Calculate the ratio of gross thrust and net thrust to those at design. Find the proportional increase in net thrust and the rate of climb would this produce if the lift-drag ratio is 21.
(**Ans:** 1.071, 1.143, 14.3%, 308 ft/min)

Note: The 50 K rise in turbine entry temperature would appear capable of giving the necessary minimum climb performance, normally taken as 300 ft/min.

19.4 Use the results of Exercise 19.2 and the results presented in Table 19.6 to estimate engine performance sea-level static when turbine entry temperature is increased to 1761 K so T_{04}/T_{02} is at the design value. It may be assumed that the component efficiencies remain constant. Assume that the fuel mass flow is equal to 2% of the mass of air entering the core.

a Take $c_p = 1287$ J kg^{-1}K^{-1}, $\gamma = 1.287$ for the HP turbine nozzle throat for the sea-level inlet temperature. Determine the core jet velocity ratio and the ratio of core mass flow to that at design (i.e. when $T_{04} = 1500$ K).
(**Ans:** 313 m/s, 1.113, 3.05)

b Calculate the bypass jet velocity, $\overline{m}_{19}$ for the bypass nozzle and thence the bypass mass flow as a ratio of mass flow at design.
(**Ans:** 278 m/s, 1.032, 2.59)

c Calculate the gross thrust per unit mass flow through the core. Calculate the ratio of gross thrust for sea-level static to net thrust at cruise.
(**Ans:** 9514 kN/kg, 6.8)

Note: The biggest change in the engine at sea-level static relative to cruise is to the LP turbine, mainly because the static pressure at core nozzle exit is then high relative to inlet stagnation pressure. The condition of the HP compressor and HP turbine at sea-level static is similar to that at design and even the fan pressure ratio is only slightly greater. The fan non-dimensional mass flow is lower than design by about 3% whilst the fan pressure ratio is higher than design by about 2% and together these can be a problem for fan stability.

19.5 HIGH-SPEED CIVIL TRANSPORT

From the Wright brothers' first flight to the time of the Boeing 707 entering service there was a continual increase in flight speeds. For civil transport the speeds have not increased significantly from the time of the Boeing 707 (with the notable, but uneconomic, exception of Concorde) and in some cases have fallen. The early big twins (e.g. Airbus A300 and Boeing 767) were designed for cruise near $M = 0.78$ rather than $M = 0.85$ for the four-engine jets, but more recent large twins have gone up to 0.85. The smaller twins (e.g. Boeing 737 and Airbus A320) are designed for cruise Mach numbers of below 0.80.

There has been long-standing interest in a supersonic transport which would be bigger than Concorde (Concorde could only carry about 100 passengers), with longer range (Concorde had a maximum range of 3500 nm, just adequate for Paris to Washington DC) and possibly with a higher cruise speed (Concorde cruised at $M = 2.0$). For $M = 2.0$ sonic boom prevents flights over populated regions, at least for large aircraft, and above $M = 2.2$ the skin has to be cooled or else titanium has to be used in place of aluminium. There are many difficulties, including airport noise, sonic boom, fuel consumption and the cost of the highly complicated aircraft. More recently concern for the environment puts another hurdle in the way of a new supersonic transport which would be less fuel efficient than a subsonic aircraft.

In the spring of 2001 Boeing announced that they were preparing designs for a medium sized aircraft (about 200 seats) with a range of 9000 nm and a cruise Mach number of 0.98; this was dubbed the 'sonic cruiser'. Not everyone in the aviation industry, however, believed that these targets were realisable. The favoured configuration appeared like a slender delta with forward (canard) wings. The aircraft was aimed at the business traveller who is prepared to pay more for speed. The increase in Mach number from 0.85 to 0.98 could take about one hour off a journey from London to Los Angeles. The engine companies (General Electric, Pratt & Whitney and Rolls-Royce) prepared engine proposals for propelling this aircraft.

The first matter to consider is whether the choice of Mach number was sensible – would it have altered significantly as the design refined? At first sight a Mach number of 0.98 was an odd choice, since aircraft drag rises very steeply from about $M = 0.85$. Beyond $M = 1.0$ the drag levels off and may even fall. At $M = 1.2$ perhaps three hours could be taken off the journey from London to Los Angeles yet at $M = 1.2$ sonic boom would not be an issue and the stagnation temperature would not be high enough to affect the strength of aluminium alloys.

The engineering problems that the sonic cruiser presented were large and consideration was given to employing a range of new technologies to improve engine performance (in particular, reduced fuel consumption and engine weight), improved aircraft aerodynamics and reduced airframe weight (largely by extensive use of composites).

With aircraft of conventional shape and Mach number it is possible to estimate the drag at cruise with considerable precision. This means that the engine thrust (and engine size) can be optimised with some confidence in advance. For a new shape of aircraft the drag will not be known with such confidence. Worse still, it is difficult to measure drag accurately in a wind tunnel at speeds close to sonic; the best that can be achieved is to use a small model in a large wind tunnel.

The initial responses from some airlines to the sonic cruiser proposal were enthusiastic, but over time there was a gradual cooling. This was partly because of the financial pressure that most airlines were under in 2002, but it was also because the economic benefits to the airlines turned out smaller than might at first be imagined. It turns out that it is hard to take advantage of the reduced journey time in many long flights because of constraints with airports at each end. As a result, Boeing cancelled the sonic cruiser project at the end of 2002, whilst announcing a new aircraft which has become the Boeing 787. The 787 exploits some of the advanced technology that had been directed towards the sonic cruiser and offers the benefits in lower operating costs and lower environmental impact. It is assumed that Airbus will have included similar technology to be competitive with the A350.

There is currently no serious plan for higher speed civil transport, though discussions continue for a supersonic business jet. This is aimed at a market sector which is insensitive to cost. Because the impact of sonic boom is strongly related to aircraft weight it is assumed that the relatively small business jets will be allowed to fly supersonically over populated areas. The issues of propulsion now have many features considered in Part 3 relating to combat engines, most notably that the inlet temperature T_{02} becomes high at cruise so the non-dimensional ratios in the engine are much lower at cruise than for subsonic operation.

It is the underlying assumption of this book that for civil transport to play its part in reducing greenhouse gas emissions the design specifications of civil aircraft need to change. There is a limit to what can be achieved through improvements in technology (principally lower engine sfc, higher aircraft L/D, lower aircraft empty weight) and great efforts have already gone into achieving the best possible values of these quantities. The major opening which remains is to alter the specification of the aircraft and two routes have been chosen for the New Efficient Aircraft used here: reduced full-payload range and reduced cruise Mach number. Given that there is a lower fuel-burn option it therefore seems irresponsible to be considering aircraft with still higher cruise Mach number because there would be an ineluctable increase in fuel burned per passenger-kilometre flown.

SUMMARY OF CHAPTER 19

When the civil engine was considered in the early chapters there were many simplifications introduced which can be removed with the increased level of complication introduced for treating the combat engine. It is also possible to consider the mixed high-bypass engine, such as that fitted to many current civil aircraft. There is an improvement in *sfc* if the core and bypass are mixed before the final nozzle. The benefits diminish as the fan pressure ratio decreases and for the values of *fpr* now considered for a new engine the benefits are probably less than the penalties arising from increased weight and drag for the mixed nacelle.

On-design calculations were performed for an unmixed engine at a constant fan pressure ratio and constant overall pressure ratio to assess the significance of the various ways of treating the engine, beginning with the simplest approach used in Chapter 7 and then including progressively more accurate gas properties, turbine cooling air and combustor pressure loss. It was shown that significant errors can occur when inaccurate properties are used for the gas, with the principle effect being due to the effect of gas temperature. Accurate prediction of performance also requires allowance to be made for turbine cooling air and combustor pressure loss. (Correct trends can be found with simple calculations even if the absolute values are in error.) Subsequent on-design calculations were performed when component efficiencies, cooling flow amount and combustor pressure loss were altered. These calculations showed that at constant pressure ratios and constant T_{04}/T_{02}, variations in component performance have a marked effect on bypass ratio. The alteration in bypass ratio changes the thrust per unit mass flow through the core and the specific fuel consumption. The HP compressor has the largest effect on both thrust per unit mass flow through the core and *sfc*; for a plausible design of engine the net thrust for the same core mass flow rate increases by about 1.9% and the *sfc* falls by about 0.9% for one per cent increase in compressor efficiency. Halving the cooling air flow rate, if it were possible, would lead to a reduction in *sfc* of about 0.7% and 0.7% greater net thrust from the same size of core.

For off-design consideration of a fixed configuration of engine with constant T_{04}/T_{02}, the pressure ratios are altered as the component efficiencies are decreased; such changes in efficiency might be as a result of in-service deterioration or of the failure of the compressor or turbine to achieve in operation the efficiency it was designed to give. Off-design calculations for the same datum engine showed that the largest alteration was produced by the change in LP system; 1% fan efficiency is worth 0.78% net thrust and 0.75% *sfc*. The HP compressor efficiency is still important and 1% loss in efficiency leads to a reduction in thrust of about 0.4% and an increase in *sfc* of about 0.7%. A rise in turbine inlet temperature raises net thrust by about 0.9% and decreases *sfc* slightly.

Reducing cruise Mach number was considered both on-design (with the pressure ratios held at the design values) and off-design, where the pressure ratios can change. The proportional changes in net thrust and sfc are similar on-design and off-design because the largest contributor to the change in each case is the alteration in ram drag.

For climb at cruise altitude an increase in turbine entry temperature of 50 K leads to an 11% increase in thrust which is capable of giving a climb rate of 308 ft/min. At this condition the fan pressure ratio is about 1.56 and the overall pressure ratio about 50. Take-off was considered with T_{04}/T_{02} at the design value so that $T_{04} = 1710$ K. At this condition the fan pressure ratio was found to be 1.54 and the overall pressure ratio 49. The static thrust was 6.8 times the net thrust at design, which is cruise at $M = 0.78$ at 35,000 ft and this would allow the engine to meet the requirements for take-off.

APPENDIX TO CHAPTER 19

The curves in Figure 11.1 show that any simple choice of gas properties is going to lead to inaccuracy if applied to compressions and expansions in which there are substantial temperature changes. The change in c_p and γ associated with combustion product composition is much less significant. For the air in the compressor γ falls from just below 1.40 to about 1.30 over the range of temperature involved, whereas for the turbine, with $\phi = 0.3$, γ on entry is around 1.28 increasing to about 1.33 at exit. An empirical relationship derived by Cumpsty and Marquis (2014) is very easy to use and gives much greater accuracy. The empirical relationships are for pure air and for stoichiometric combustion (equivalence ratio one) of fuel of composition C_nH_{2n}. The empirical result is divided for temperatures below 400 K and above 400 K

For temperatures below 400 K it is recommended that

$$c_p = 1005 \text{ J kg}^{-1}\text{K}^{-1} \quad \text{and} \quad \gamma = 1.40 \text{ is used for air}$$

and $c_p = 1063 \text{ J kg}^{-1}\text{K}^{-1}$ and $\gamma = 1.37$ is used for stoichiometric products

For temperatures between 400 and 2000 K it is recommended that

$$c_p = 967 + 469 \exp\{-1000/T\} \text{ J kg}^{-1}\text{K}^{-1} \text{ is used for air}$$

and $c_p = 1024 + 655 \exp\{-1000/K\} \text{ J kg}^{-1}\text{K}^{-1}$ for stoichiometric products.

By adding c_p in suitable proportions a value may be obtained for any equivalence ratio.

Since $c_p = c_v + R$ the ratio of specific heats is related to the specific heat at constant pressure by the relations

$$\gamma = c_p/(c_p - R) \quad \text{and} \quad c_p = \gamma R/(\gamma - 1)$$

where R is the gas constant per unit mass, with units $\text{J kg}^{-1}\text{K}^{-1}$. The gas constant R is related to the universal gas constant per mole by $R_0 = MR$, where M is the molar mass. For pure air $M = 28.97$ whilst for the stoichiometric products of burning a fuel of the form C_2H_{2n} the molar mass is 28.91. As a result R is well approximated by $R = 287 \text{ J kg}^{-1}\text{K}^{-1}$ for air and combustion products.

CHAPTER 20   TO CONCLUDE

20.0 INTRODUCTION

This chapter returns to some general issues related to both civil and military engines. These are topics which can be more satisfactorily addressed with the background of earlier chapters.

20.1 CIVIL AVIATION AND THE ENVIRONMENT

The choice of aircraft here, the New Efficient Aircraft, was predicated on the need to reduce fuel burn. The cruise Mach number was lower than current new large aircraft and the full payload maximum range, R1, was distinctly smaller than recent designs of aircraft offered for sale. The small reduction in speed would have little effect on the passenger and the vast majority of flights currently offered would be accommodated with the reduced range. The industry as a whole does not seem willing to embrace this approach to reducing fuel burn, even though fuel is now a major cost, perhaps 40% of the direct costs. No aircraft flying in 2015 has been designed with fuel cost approaching the values currently prevailing and expected in the future. Despite the relatively high cruise Mach numbers and long R1 ranges being offered, most people nevertheless agree that global warming mitigation demands a reduction in CO_2 emissions.

ICAO have come forward with a metric for fuel burn but, as described in Chapter 1, this is not well conceived if the object is to display the relative efficiency of different aircraft and aircraft types in moving people or freight. Now ICAO has to decide how to turn this metric into regulations for fuel burn. The earliest these regulations could begin to affect new aircraft types is probably 2020 with existing types several years later. It seems unlikely, however, that this will matter very much, for there is already major economic pressure to reduce mission fuel burn. The situation with fuel and CO_2 is quite unlike other areas in which ICAO is involved as setter of regulations, such as noise (discussed in the Appendix) and pollutants (like NO_x, discussed in Chapter 11). Without regulations of some sort there would be no incentive for an airline to reduce these noise and NO_x because there is no direct cost. This is not the case for CO_2 where there is a direct cost to the airline for fuel burned.

Civil aviation appears to be growing steadily and the opportunity to move large numbers of people around in Asia is likely to continue this for some time. This clearly conflicts with the need to contain CO_2 emissions. The industry has come out with goals of containing the growth in CO_2 emission, some from new technology, some from better operations and then some from bio-fuels.

New technology has achieved great improvements in the past, typically averaging about 0.8% per annum for the aircraft and engine over the last 40 years. Improvements in fuel burn can be expected in future, but it is generally agreed that it is getting harder to find them and the steep rate assumed by some to offset the growth in aviation exceeds the rate achieved in the past.

Real potential benefits from better operations exist – for example flying more direct routes and avoiding delays on the ground, with the engines idling, or, much worse, flying around awaiting an opportunity to land. Although the potential benefits are real, they are not easy to achieve and are limited in extent.

The most specious claim for saving aviation produced CO_2 is for the use of bio-fuels. Clearly bio-fuels can be used in aircraft, but this is not a proper saving in the sense that improved technology is. If a field produces oil seed which is processed and used to fuel aircraft, this same fuel is not being used to fuel trucks. But the truck operators could just as easily claim the reduction in net CO_2 from the use of bio-fuel and it is not logical, therefore, to credit the gain to aircraft alone. It may also be remarked that large land areas would be needed to make much impact on the total requirements of aviation fuel, large enough to have major impact on food supplies. If the goal is to reduce net CO_2 the best use of renewable fuels is substitution for heat, perhaps substituting for coal in thermal power plant.

Environmental issues will very likely come to play an increasing part in the choice of cycle for both the civil and military engine. The publication of the IPCC report on the effect of aviation in 1999 brought attention to emissions at high altitude; CO_2 is not the only greenhouse gas, but NO_x can be important and the effect of the contrails induced by the water vapour in the engine exhaust is another potentially important source. The release of combustion products around the tropopause, it appears, is several times more serious than the burning of a similar amount of fuel at sea level. (It does not make any difference where CO_2 is released because this persists in the atmosphere for hundreds of years and becomes well dispersed.) The creation of NO_x (oxides of nitrogen) is enhanced when the temperature of combustion increases, as it does when the compressor delivery temperature is raised. It may be that reducing the emissions, and not material limitations, is going to be the ultimate restriction on compressor delivery temperature and therefore on the overall pressure ratio.

Reducing noise will always favour going to lower fan pressure ratios, and to meet noise regulations a higher fuel burn may result than would be the case if noise could be ignored. Further in the future one could imagine aircraft geometries being adopted which are not optimal for aerodynamic performance but which minimise noise, for example, there are proposals for putting engines over the wings and using the wings to shield the ground. It is possible that airport noise rules will constrain the design of subsonic aircraft in ways not yet fully understood. No scheme for a supersonic transport with acceptable noise has yet been found which is capable of bearing serious examination.

Undoubtedly one of the great achievements has been the success of ETOPS, so that twin-engine aircraft can now operate safely almost everywhere in the world. Levels of reliability have been achieved which make failure of both engines so remote that the risk is small compared with other hazards of everyday life. In fact some time ago the primary driver for reducing engine in-flight shut-down frequency was economic rather than safety; an in-flight shut-down generally involves large disruption and cost, but negligible danger provided the pilot is adequately trained and behaves correctly.

20.2 SOME REAL ISSUES OF ENGINE DESIGN AND DEVELOPMENT

The cost of engine development is a strong disincentive to doing it very often. Whenever possible the engine from one aircraft application is adapted to another aircraft and even adapted for electrical power generation or other industrial use. Looking back, the General Electric CF6, designed for the DC10, was installed on the Boeing 747 and 767 and many Airbus designs, most recently the A-330. The Rolls-Royce RB211 was designed for the Lockheed L1011 and has found application on Boeing 747, 757 and 767 aircraft; as a more major derivative it has evolved into the Trent to power the Boeing 777 and the Airbus A330, A340, A380 and A350. In the cases of both the CF6 and the RB211 the engines have been modified considerably, but much of the original design architecture has remained. Going back further the CF6 core is itself a derivative of a core for a military transport aircraft. Perhaps more remarkably, the smaller CFM56 powering later versions of the Boeing 737 and many of the Airbus A320 aircraft uses a core developed from an engine which was originally designed to propel the supersonic B-1 bomber.

It would therefore be disingenuous to leave the impression that most engines have followed a logical design path from a specification in terms of the optimum new engine. The optimum provides a reference for comparison, but there is great pressure to adapt what already exists. Although the aircraft operator may be forced to use more fuel because of this, there are advantages to the operator because the use of existing developed technology reduces risk of delay and increases reliability whilst also containing the cost.

For the civil engine there is another aspect which confuses the logical development. The time taken to develop a new engine has, until recently, been longer than the time taken to design, test and produce a new civil aircraft, so the engine development needed to begin sooner. The engine manufacturer is a supplier to the airframe manufacturer and is dependent on the airframe maker to give the engine specifications; the specifications of the aircraft, and therefore the size of engine needed, may be altering right up to the time when the aircraft is complete. There is a tendency for aircraft to become heavier as the design progresses and problems are encountered; it is not unusual for engines to become heavier too, leading to the requirement for still more thrust. For the maker of engines this requires judgement and experience to reduce the risks that are entailed: the engines may be designed for a higher thrust than that originally asked for by the airframe maker on the expectation that thrust requirements will rise. If this conservatism is carried too far, however, the engines offered by the engine designers are too big and too heavy and as a result are likely to be uncompetitive.

For military aircraft there are similarities and differences. One similarity is the huge cost, so whenever possible a developed engine will be used in another application, perhaps in a wholly different aircraft. On other occasions only part of the engine, such as the core, may be re-used and incorporated with a new LP spool. When a new combat engine is to be designed and built the situation differs from the civil market. First a government is likely to be involved in stipulating the mission requirements, in making the decisions on engine parameters and type, and in deciding who shall build the engine. There are now marked differences between US practice and that in most of Europe. In the USA the trend has been to run competitions between engines produced by General Electric and Pratt & Whitney. In Europe the trend is to have a single engine designed and built by an international consortium; the exception here is France which has recently developed its own fighter aircraft and engines.

The huge costs of the combat aircraft and their engines lead to extensive studies to optimise the combination for a range of duties. As explained in Chapters 13 and 14, the different roles and different phases of any role for the combat aircraft impose conflicting requirements for air frame and engine. The optimisation and design takes a long time, not untypically more than a decade. The situation is complicated by the changing geo-political situation, so that the threat foreseen as dominant at the start of the process may have receded by the end. As Chapters 16 and 17 showed, a wide range of engine types could meet many of the goals and the search for the best possible may lead to a substantial increase in cost. Certainly the requirement for an aircraft to perform widely different roles necessarily leads to compromises which will make it less effective in each of the roles than an aircraft more tailored to a particular one. This is turn leads to the attractiveness of the variable cycle engine, for which the design choices may be less obviously constraining.

20.3 OVERVIEW

This short book has addressed a lot of different topics and has therefore needed to avoid too much detail. As a result specialists inside companies may find many areas where the book stops short of the level of detail they require. The emphasis has been towards understanding the basis of the design practice of aircraft engines and why the design choices are made the way they are. So far as possible, empiricism has been avoided, but some is essential: for example one has to take realistic values of turbine inlet temperature and component efficiencies. Where empiricism has been used it has often been possible to look back at how sensitive the calculated results are to these inputs and in all cases the trends are correctly predicted and the magnitudes are sensible.

Another choice has been made to keep the book manageable: it addresses only the aerodynamic and thermodynamic issues involved. It should hardly need saying that the mechanical and materials issues are every bit as important as aerodynamics and thermodynamics, but to have introduced these with a satisfactory level of detail would have made the book substantially longer. In any case it is the goals set by the thermodynamic and aerodynamic choices (such as pressure ratio) which tend to drive the overall performance of the engine (such as specific thrust and specific fuel consumption) with the mechanical and materials issues coming in as constraints; this book has therefore chosen to concentrate on the aerothermal aspect.

As the materials get stronger and their tolerance of high temperature increases, the specific thrust and specific fuel consumption of subsonic civil engines will decrease whilst the specific thrust of military engines will increase. In other words the civil transport engine and the combat engine will move ever further apart in performance specification. At present, and for some time in the future, the limit of operation of an engine is as likely to be provided by the compressor delivery temperature as by the turbine inlet temperature. Whereas the allowable turbine inlet temperature may be raised by increasing the amount of cooling air, or even cooling the cooling air in an external heat exchanger, the compressor delivery temperature alone defines the temperature which the discs must withstand. If overall pressure ratios are to increase there must be major materials changes or substantial increases in compressor efficiency.

For the military engine progress in producing better materials will, to a large extent, determine the progress towards higher thrust-weight ratio. For the civil engine better materials will reduce the size of the core, reducing the mass flow through it for a given size of fan, increasing the bypass ratio. As emphasised throughout the book, bypass ratio is not the best variable to describe the civil engine because of its combined dependence on fan pressure ratio and core power per unit mass flow.

The jet engine has been an important enabler of change since its introduction into military aviation in the 1940's and into commercial aviation in the 1950's. Not only has the specific fuel consumption dropped by nearly 50% for the civil engine, the maximum thrust has gone up by an order of magnitude. The higher thrust has allowed larger planes, whilst the lower *sfc* and greater thrust has contributed to providing longer range planes. Even more remarkable is the increase in reliability and longevity. These factors have led to a prolonged and continuing rise in public's appetite for travel and with it a rise in the carbon footprint. A large jet engine is expensive, several millions of dollars, and civil engine manufacture is confined to a handful of companies. The barriers to entry are huge, not least because of the need for extensive knowledge base from decades of experience. It is hoped that this book will have opened up some of the ideas which contribute to the aerodynamic and thermodynamic aspects of this knowledge. It is also hoped that the constraints on reducing fuel burn and carbon dioxide emissions from ever-expanding aviation will have been made more understandable.

APPENDIX NOISE AND ITS REGULATION

HISTORY AND REGULATION

As jet air transport increased in the 1960s the annoyance to people living and working around major airports was becoming intense. Regulations affecting international air transport are governed by the International Civil Aviation Organisation (ICAO), but this body was moving so slowly that in 1969 the US Federal Aviation Agency (FAA) made proposals for maximum permitted noise levels. After extensive discussions in the USA these were formally approved as Federal Aviation Regulation (FAR) Part 36 in 1971, retroactive with effect from 1969, but only for new aircraft. Shortly afterwards the ICAO Committee on Aircraft Noise published similar recommendations, to be known as Annex 16, a formal addendum to the 1944 Chicago Convention on Civil Aviation; each member state had then to accept the rules in Annex 16 and write them into their legal framework. The underlying principle for the noise certification of aircraft under FAR Part 36 and Annex 16 are similar and has remained unchanged ever since, with the levels under the US and ICAO rules subsequently becoming virtually identical.

The certification for noise relies on measurements at three positions, two for take-off (referred to as lateral and flyover) and one for landing (referred to as approach). The levels are expressed in decibels (EPNdB) using effective perceived noise level (EPNL), described in outline below. The layout for testing is shown in Figure A1.

The noise at the lateral position is the highest noise measured along a line parallel to the runway whilst the aircraft is departing at full power and the maximum usually occurs when the aircraft has climbed to about 1000 feet. Flyover noise is measured directly under the flight path after take-off and at an altitude where it is normal to cut-back the power to reduce the noise whilst still maintaining a safe rate of climb. The approach noise is also measured directly under the flight path as the aircraft prepares to land, with the glide slope carefully controlled. The flights are for the maximum allowed weight of the aircraft and correspond to standard day temperatures (which will generally require corrections to be made to the measurements since tests are rarely carried out at precisely the standard conditions). Needless to say, aircraft do not always operate as specified for the tests, but the tests do at least provide a standard way of comparing aircraft and thereby regulating airport operations.

Around 1977 noise levels for certification were lowered and these were known in the USA as FAR Part 36 Stage 3 and elsewhere as Chapter 3 of Annex 16. The levels of noise to qualify for certification increase with gross take-off weight up to 400 tonne for lateral noise, 385 tonne for

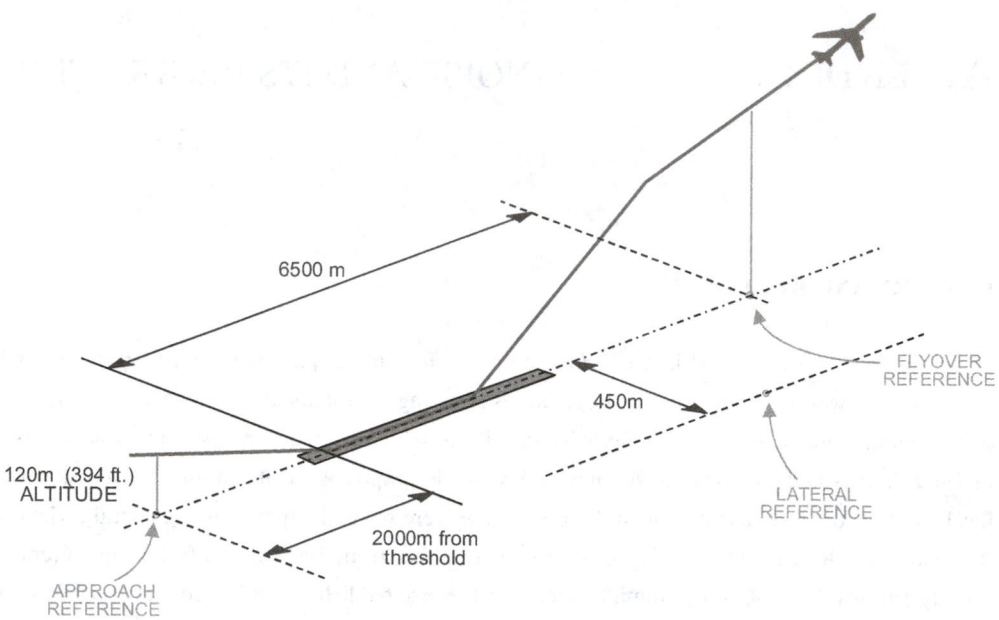

Figure A1. The three noise certification reference positions.

flyover and 280 tonne for approach; at greater weights the levels are constant. The allowable flyover noise is highest for four-engine planes and lowest for twins to make allowance for the slower climb rate with four engines. Figure A2 shows the noise levels permitted by ICAO for Chapter 3 of Annex 16 for the three conditions used. Shown on the figures are indications of the measured certification levels for a selection of modern aircraft, revealing very clearly that the recent aircraft types are well below the certification levels, notwithstanding the regulations failing to increase the allowed levels beyond maximum take-off weight of 400 tonne. The ICAO Chapter 3 levels are therefore a challenge only for older designs of aircraft and engine.

It may be noted that the absolute levels of the certification levels shown in Figure A2 are not equal for the different conditions. This is to reflect the different distances between the aircraft and measurement point for each condition.

In January 2001 Chapter 4 of Annex 16 was agreed by the Committee of Aviation Environment Protection, CAEP, part of ICAO, and was ratified by ICAO in the autumn of 2001. It affects aircraft for which a new-type certification was applied for from 2006. Chapter 4 requires a cumulative margin of 10 EPNdB from the levels in Chapter 3. (Cumulative means the numerical sum of the margins relative to Chapter 3 for the three noise measuring conditions: lateral, flyover and approach.) In addition at no condition must the level exceed that for Chapter 3 and there must be a cumulative margin of at least 2 EPNdB from Chapter 3 for any two conditions. In February 2014 Chapter 14 was agreed. This stipulates a cumulative reduction of 7 EPNdB relative to Chapter 4 (meaning 17 EPNdB relative to Chapter 3) and at least 1 EPNdB at each condition relative to

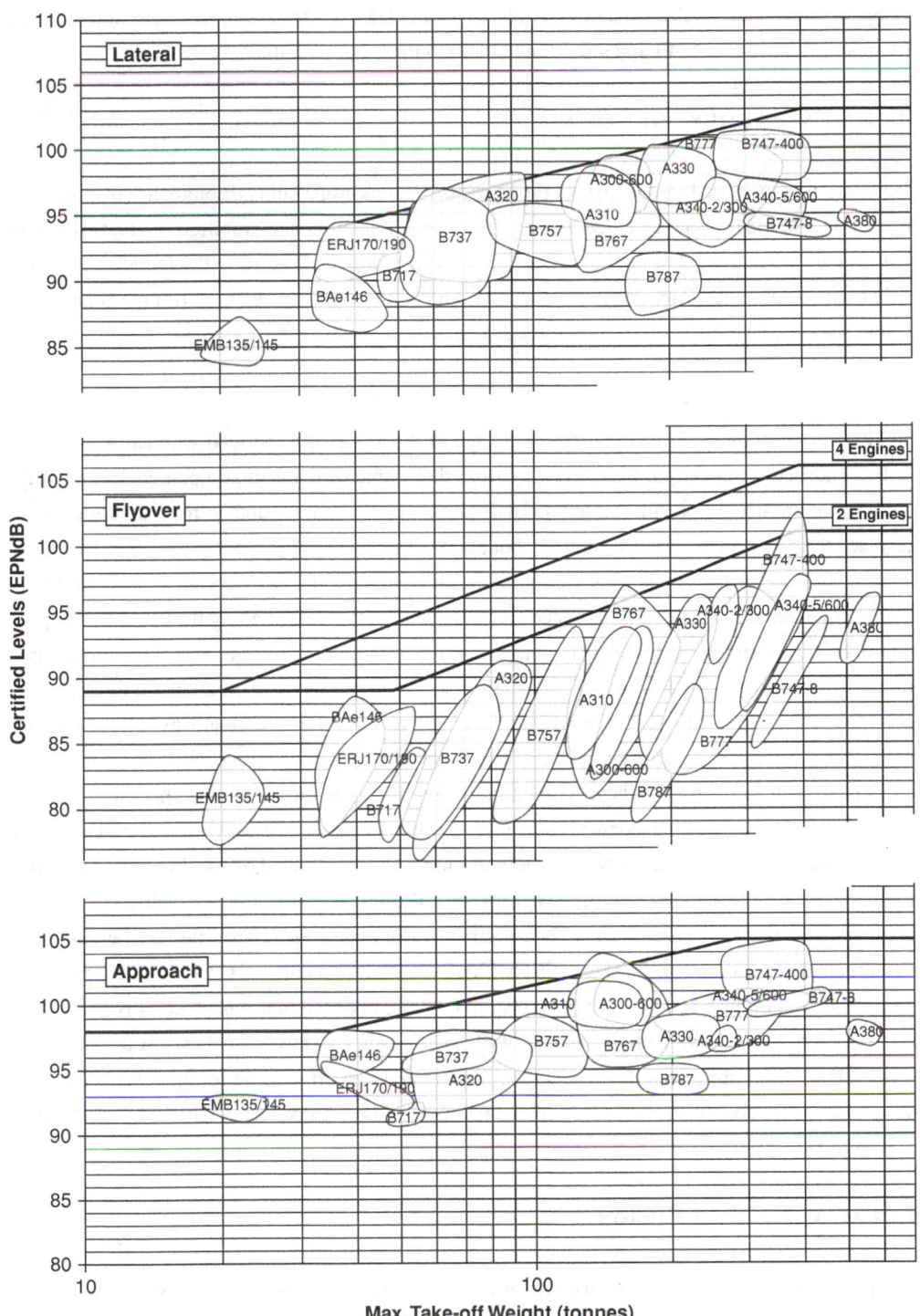

Figure A2. Certification noise levels for FAR Part 36 and ICAO Annex 16, Chapter 3.

Chapter 3. It is intended that Chapter 4 will be implemented after 31 December 2017 for new types, (that is, for new-design) of large aircraft and in 2020 for small aircraft.

THE MEASUREMENT OF NOISE

Noise is annoying sound and consists of pressure fluctuations propagating through the air as acoustic waves. The sound pressure level (SPL) relates the level of the pressure amplitude to that which the human ear can just detect at its most sensitive frequency, $p_{ref} = 20 \times 10^{-6}$ Pa. Because the ear is able to accommodate a wide range of amplitudes it is customary to use a logarithmic scale so the sound pressure level for a signal of amplitude p is written

$$SPL = 20 \times \log 10(p/p_{ref})$$

and expressed in units of decibel (dB). A change of 3 dB is readily noticed whereas a change of 1 dB is normally imperceptible. Well away from the source of noise in an open environment the sound pressure decreases with the square of distance from the source and so for a doubling of the distance the SPL can be shown to fall by about 6 dB.

The human ear is most sensitive to frequencies in the middle range (peaking at about 3 kHz) and to get a measure of annoyance it is necessary to correct the SPL to allow for this. The simplest approach is merely to adjust the levels in line with the sensitivity of the ear, and a commonly used measure based on this gives dBA, which can be read directly from a dial on an instrument. A more complex method, allowing for noise amplitude as well as frequency and requiring processing by a computer, can be used to give Perceived Noise Level (PNL), measured in PNdB. PNL forms the basis of the aircraft noise certification measurements, but it also turns out that humans are more annoyed by noise with a tonal content than noise of a broadband nature, and the measured noise can be corrected for this too to give PNLT. Finally the annoyance is affected by the duration over which the noise is present and allowance is made for this. The result is that a set of defined procedures are made to convert the instantaneous measurements of sound pressure level into a single number, the effective perceived noise level (EPNL), which is measured in units of EPNdB. The regulations adopted by the FAA, by ICAO, and by many airports, are couched in terms of EPNdB.

There is still quite widespread use of dBA, however, and this is the basis of the day-night level, LDN, which integrates the dBA and arbitrarily adds 10 dB at night to allow for the greater annoyance produced by aircraft noise during the night. LDN is strongly favoured by the FAA in the USA.

LOCAL NOISE REGULATION

Some residents living near busy airports have felt that the international agreements are moving too slowly to address their concerns and as a result local regulations have sprung up. The most important of these affecting the design of large aircraft, and currently the most onerous for large aircraft, are those for the London airports. Similar schemes have been adopted by some other European airports. In 1993 a new quota count system was introduced for London to affect night flights based on aircraft noise certification data. Each aircraft type is classified and awarded a quota count (QC) value depending on the amount of noise it generated under controlled certification conditions. The quieter is the aircraft the smaller is the QC value. Aircraft are classified separately for landing and take-off. Take-off quota count values are based on the average of the certificated flyover and side-line noise levels at maximum take-off weight, with 1.75 EPNdB added for Chapter 2 aircraft. Landing quota count values are based on the certificated approach noise level at maximum landing weight minus 9.0 EPNdB. Aircraft were originally divided into six QC bands from 0.5 to 16, but following a review QC 0.25 was added in 2007. Aircraft below 84 EPNdB are exempt. The bands are set out (in EPNdB) below.

exempt	QC0.25	QC0.5	QC1	QC2	QC4	QC8	QC16
<84	84–86.9	87–89.9	90–92.99	93–95.9	96–98.9	99–101.9	>101.9

The exact rules of operation of the quota count system have been varied over the years, but there are some general features which can be illustrated by the proposals becoming effective in 2016 and which serve to demonstrate how important noise has become to the airline operator. The total quota of night flights (i.e. product of the number of flights and the appropriate QC for the type of aircraft) for Heathrow for a whole year has been set at 9180. No scheduled departures for QC4 will be allowed between 11:30 PM and 6:00 AM, whilst for QC8 and QC16 this prohibition is from 11:00 PM to 7:00 AM. Delayed QC4 departures are allowed after 11:30 pm but QC8 and QC16 departures will not be permitted between 11:00 PM and 7:00 AM, even if the flight departure has been delayed. At Heathrow airport there is also a monitoring system, measured in dBA, to ensure that actual levels of noise are not exceeded, regardless of aircraft weight, with fines imposed in cases of infringement.

For the Airbus A380 operation in London was considered essential for many of the customers. Airbus and the engine manufacturers therefore agreed to target QC2 at departure, and QC1 at arrival, a cumulative margin from Chapter 3 of at least 20 EPNdb, much greater than the 10 EPNdB margin required for ICAO Chapter 4. The Boeing 787-8 achieves QC0.5 at take-off and QC0.25 at landing.

NOISE GENERATION

The understanding of the generation of noise in a quantitative way is one of the most challenging tasks in fluid mechanics. Since noise is merely propagating pressure fluctuations, any non-steady flow or any moving object, such as a fan blade, is capable of being a noise source. The problem is greatly increased for aircraft engines by the common tendency for the blade speeds and the flow velocities inside the engine to be close to sonic. Even in a qualitative sense there is not full agreement on what the noise sources are, nor what is their relative importance; this is most clearly

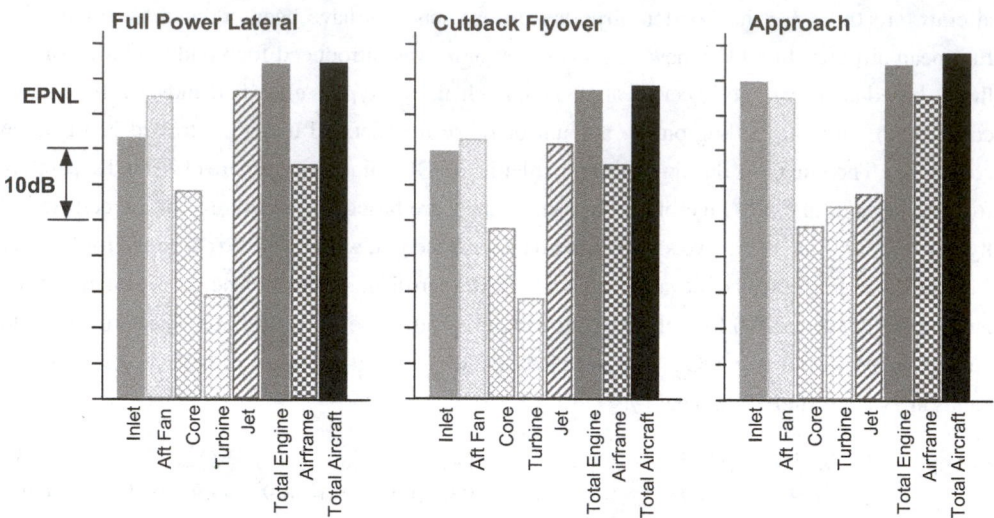

Figure A3. An approximate breakdown of major noise sources at the three certification conditions from NASA Contractor Report 198298, June 1996. (Bypass ratio ≈ 4, *fpr* ≈ 1.7.)

the case with broadband noise from the fan and from the LP turbine. The greatest broadband noise source is jet noise, attributable to the turbulent mixing which necessarily takes place. It was shown many years ago that jet noise is proportional to the eighth power of jet velocity; i.e. the acoustic power more than doubles (increases by 3 dB) for a 10% increase in jet velocity. This needs to be modified to allow for forward motion, high Mach numbers and the effects of separate core and bypass streams (each with different velocity and temperature), but the dependence on a high power of velocity remains.

Figure A3 shows the rough breakdown of the noise sources at the three conditions used for certification for a large engine in the mid 1990s. It will be seen that with a moderately high bypass ratio (and therefore relatively low jet velocity) the jet noise is still the largest source at the lateral position and it is comparable with the fan (inlet and aft radiated fan noise) at cut-back conditions during flyover. (It may be added that the cut-back manoeuvre, which is to reduce the engine thrust when a safe altitude has been reached, is precisely intended to reduce jet noise.) The fan noise from the rear (i.e. exhaust) is more important than from the intake at departure for this engine, a feature which does not concur with the subjective impression obtained standing near the runway during take-off. At approach the jet noise is negligible, but the fan is now the most important noise source from the engine. (Because the fan is rotating much more slowly during approach than take-off the *cause* of the fan noise will be different at the different conditions.) A more modern engine would have rather lower jet noise relative to fan noise than is shown in Figure A3, but with fan noise also lower in absolute terms.

An important feature shown in Figure A3 for the approach condition is the high level of noise from the airframe. In the case shown it is almost comparable to the noise from the engines.

With more modern engines, such as those for the Boeing 787, the Airbus A350 and the New Efficient Aircraft, the fan pressure ratio and rotational speed are lower so that the engine noise is lower and airframe noise is likely to dominate at approach. The high level of airframe noise is not altogether surprising when the configuration of the aircraft during the approach is considered. For stability reasons the flaps are lowered sufficiently far to give high drag as a result of large-scale separations, with the turbulence and noise that this entails. The high lift coefficient at approach, because of the low forward speed, gives strong trailing vortices which are turbulent and noisy. Lastly the undercarriage and the wheel wells are serious noise sources. Although the noise from the undercarriage and wells does seem amenable to reduction, it is not obvious how to reduce the other two sources which are the result of fundamental features of aircraft operation.

NOISE REDUCTION

Quite remarkable reductions have been achieved in aircraft noise over the last 40 or so years. A reduction at each condition of about 20 EPNdB has been achieved, notwithstanding the increase in aircraft size and weight which has occurred. Most of this has been achieved by the selection of the engine cycle to reduce jet noise, careful choice of the numbers of blades in the fan and LP turbine, and the use of acoustic treatment on solid walls of the inlet and outlet ducts. Whereas the design of blade shape to give good aerodynamic performance and maximum efficiency is well advanced, the similar selection of blade shape to minimise noise is currently in its infancy. The more sweeping steps that can and are taken to lower noise are listed below with some limitations and difficulties noted.

Lower the jet velocity, which means choosing a lower fan pressure ratio. This has the effect of increasing the propulsive efficiency to give a significant reduction in specific fuel consumption. For a fixed level of engine thrust, lowering the jet velocity requires a larger mass flow through the engine. It is possible that this limit has been crossed, so the minimising of noise is leading to higher fuel burn, as discussed in Chapter 7. Furthermore installation effects (can the engine fit under the wing? or will the engine be transportable in a 747 freighter?) come to limit the extent to which jet noise can be reduced.

Lower the speed of the rotating components, notably the fan. Lower blade speeds are generally, though not always, associated with lower noise. With the move towards lower jet velocity, the pressure ratio required across the fan is reduced and lower fan tip speeds can be used. The problems are now associated with the LP turbine which, as fan rotational speed is reduced, experiences higher non-dimensional stage loading. One solution is to add additional LP turbine stages, but these are heavy and expensive. There can also be a limit on the torque carrying capacity of the LP shaft, the diameter of which is fixed by the bore diameter of the compressor and turbine discs. A gearbox between the fan and LP turbine may provide the answer.

Avoid flow distortion into the fan. This is desirable, but some aspects are outside the engine designer's control. At take-off the intake is subject to natural cross winds and at both take-off and approach the flow into the intake is at incidence, which tends to cause flow non-uniformity. There has to be some structure or pylon arrangement downstream of the fan and this imposes a static pressure non-uniformity on the fan. Steps are taken to minimise all these effects, but some distortion seems inevitable at certain flight conditions.

Choose the numbers of rotor blades and stator blades to avoid interference patterns which rotate much faster than the blades and which produce strong unattenuated tones. This is inclined to make the turbomachinery designs less than optimal, either for cost or efficiency, but is already widely practised and there is little scope for further introduction. In practice this means having more than twice as many stator blades as rotor blades for fans and about 1.5 times as many stators as rotors in the LP turbine. The avoidance of resonance due to forced vibration also imposes constraints on blade number.

Design large axial gap between rotors and stators to weaken the interaction between the moving and stationary blades. This is most important for the bypass stream of the fan, but is also important for the core stream of the fan and in the LP turbine. Large axial gaps are also desirable for the reduction of forced vibration. However axial gaps come at a cost in terms of engine length, stiffness and weight, so there are limits on how much can be done to weaken the rotor-stator interaction.

Install acoustic liners in the intake, in the bypass duct and sometimes in the nozzle of the core. They are normally a perforated metal layer under which is a honeycomb structure and then a solid backing. The porosity of the perforated plate and the depth of the honeycomb are tuned for the application so as to maximise the impact on EPNL. For large civil engines liners are universally used in the intake, the bypass duct and sometimes in the duct downstream of the LP turbine. They substantially increase the cost and can be a cause of increased maintenance. Liners of improved performance, such as those which are in two layers (i.e. a perforated plate over a honeycomb over another perforated plate and honeycomb of different depth) are already in service. Recent designs have avoided discontinuities in the liner, such as axial joins, which were found to reduce line effectiveness. The amount of coverage by liners is also being increased but the scope for further noise benefits from the liners are now generally limited.

BIBLIOGRAPHY

ANDERSON J D. *Introduction to Flight*. McGraw-Hill, 7th edition, 2011.
    A useful and highly readable treatment of the aerodynamics of flight.

BAHR D W & DODDS W J. *Design of Modern Turbine Combustors*. Academic Press Inc., San Diego, CA, 1990.
    A more thorough and detailed book on the field of combustor technology.

BORGNAKKE C & SONNTAG R E. *Fundamentals of Thermodynamics*, John Wiley, 8th edition, 2008.

COHEN H, ROGERS G F C & SARAVANAMUTTOO H I H. *Gas Turbine Theory*. Longman, 6th edition, 2008.
    A wide-ranging book addressing many issues of relevance recently updated. A good basic introduction to turbomachinery aerodynamics and detailed treatment of off-design matching.

CUMPSTY N A. *Compressor Aerodynamics*. Longman, 1989. Reprinted with additions by Krieger of Malabar, CA, 2004.
    Directed at the compressor, it gives much of the basic theory of turbomachinery at a relatively high level. There is a solid treatment of dynamic scaling and dimensional analysis.

DIXON S L & HALL C A. *Fluid Mechanics, Thermodynamics of Turbomachinery*. Butterworth-Heinemann, 7th edition, 2013.
    A most useful introductory text to turbomachinery recently extensively updated.

GREEN J E. Greener by Design – the Technology Challenge. *The Aeronautical Journal*, Vol. 106, pp. 57–113, 2002.
    A stimulating and original recent look at aircraft and the potential to reduce emissions.

GUNSTON B. *The Development of Jet and Turbine Aero Engines*. Patrick Stephens, Sparkford, England, 3rd edition, 2002.
    Lots of interesting history and useful pictures of what the hardware looks like.

HILL P G & PETERSON C R. *Mechanics and Thermodynamics of Propulsion*. Addison-Wesley, 2nd edition, 1992. Paperback edition Pearson, 2009.
    A book which covers a very wide range of topics, including many beyond air-breathing jet propulsion. A good physical treatment of gas turbine engines, including the non-dimensional behaviour. A sound introduction to turbomachinery and combustion.

HÜNECKE K. *Jet Engines: Fundamentals of Theory, Design and Operation.* Airlife, Shrewsbury, England, 1997.

   Many good pictures and a particularly good treatment of intakes, nozzles and afterburners.

INTERGOVERNMENTAL PANEL ON CLIMATE CHANGE. *Aviation and the Global Atmosphere.* Special Report of IPCC Working Groups I & III. Cambridge University Press, 1999.

   A detailed, and very cautious, assessment of the current effect of aviation on the atmosphere by experts from around the world.

INTERNATIONAL STANDARDS ORGANIZATION. *Standard Atmosphere ISO 2533.* 1975.

   The details of the International Standard Atmosphere are given here. The significance of this standard is that it is widely used by engine makers, aircraft makers and aircraft operators.

KERREBROCK J L. *Aircraft Engines and Gas Turbines.* MIT Press, 2nd edition, 1992.

   A wide-ranging treatment touching on many subjects. It includes one of the few clear accounts, at an elementary level, of mechanical aspects of engine design. A good introduction to combustion emission control.

LEFEBVRE A H & BALLAL D R. *Gas Turbine Combustion: Alternative Fuels and Emissions.* CRC Press/Taylor & Francis, Philadelphia, 3rd edition, 2010.

   A detailed coverage of combustion.

MATTINGLY J D. *Elements of Gas Turbine Propulsion.* McGraw-Hill, 1996.

   A long book (960 pages) with a detailed coverage of many aspects.

MATTINGLY J D, HEISER W H & DALEY D H. *Aircraft Engine Design.* AIAA Education Series, 1987.

   More obviously a design-oriented book than most, this is directed mainly towards the military engine. A useful reference for information, but rather heavily algebraic in places.

MORAN M J, SHAPIRO H N, BOETTNER D D, & BAILEY M B. *Fundamentals of Engineering Thermodynamics*, John Wiley, 7th edition, 2010.

MUNSON B R, YOUNG D F, OKIISHI T H & HUEBSH W. *Fundamentals of Fluid Mechanics.* John Wiley, 6th edition, 2009.

   There are many fine textbooks on engineering fluid mechanics – this is a nice modern one.

ROLLS-Royce. *The Jet Engine.* 2005. ISBN: 0 902121 2 35

   A highly unusual book with exceptional illustrations of many aspects of engines and some highly relevant practical information. (See also the earlier edition, 1986, from which some figures were taken for this book.)

SHAPIRO A H. *Compressible Fluid Flow*, Vol 1. John Wiley, 1953.

   A classic text, written shortly after the topic became important and was put on a sound footing.

SHAW R L. *Fighter Combat*. United States Naval Institute, Annapolis, Maryland, 1985.

An interesting book from a different point of view, with good treatment in the appendix of military thrust requirements.

SMITH M J T. *Aircraft Noise*. Cambridge University Press, 1989.

Gives a good practical background to a very complicated subject, including subjective and regulatory aspects.

WALSH P B & FLETCHER P. *Gas Turbine Performance*. Blackwell Sciences, Oxford, 2nd edition, 2004.

This recent book contains a mine of information on gas turbines, including jet engines.

WHITFORD R. *Design for Air Combat*. Jane's Information Group, 1989.

A useful reference on aspects of military aircraft, engines, intakes and nozzles.

REFERENCES

BANES R, McINTYRE R W AND SIMS J A. Properties of air and combustion products with kerosine and hydrogen fuels. AGARD, 1967.

BREGUET, L. Calcul du poids de combustible consumme par un avion en vol ascendant. *Comptes Rendus de l'acadmie des sciences*, Vol 177, pp. 870–72, 1923.

BUCKNER J K AND WEBB J B. Selected results from the YF-16 wind tunnel test program. AIAA Paper 74-619, 1974.

CUMPSTY N A and MARQUIS A J. An approximate method to obtain thermodynamics gas properties for use in gas turbines. ASME Turbo Expo 2014, Düsseldorf. GT2014-26205.

DENNING R M AND MITCHELL N A. Trends in military aircraft propulsion. *Proceedings of Institution of Mechanical Engineers, Part G: Journal of Aerospace Engineering*, Vol 203, 1989.

GARWOOD K R, ROUND P AND HODGES G S. Advanced combat engines – tailoring the thrust to the critical flight regimes. AGARD Conference: Advanced Aero-Engine Concepts and Controls, AGARD cp-572, Paper 5, 1995.

KUNASAKA H A, MARTINEZ M M AND WEIR D S. *Definition of 1999 Technology Aircraft Noise Levels and Methodology for Assessing Airplane Noise Impact of Compared Noise Reductions Concepts*. NASA Contractor Report 198298, 1996.

The Northrop F-20 Tigershark, L'Aeronautique et L'Astronautique, Vol 102, 1983–85.

POLL, D I A. The optimum aeroplane and beyond. Lanchester Lecture of the Royal Aeronautical Society. *Aeronautical Journal,* Vol 113, 2009.

SMITH S F. A simple correlation of turbine efficiency. *Journal of Royal Aeronautical Society*, Vol 69, 1965.

Software

KURZKE, J. *Gasturb*. Available from http://www.gasturb.de/.
    This convenient and inexpensive package runs on a PC. It allows realistic on-design and off-design calculations to be run simply for a wide range of engines, including both engines for power production and jet engines.

INDEX

DESIGN SHEET FOR NEW EFFICIENT AIRCRAFT

This table gives quantities which are likely to be needed repeatedly. Numbers in parentheses indicate the exercise in which the data are evaluated.

Initial cruise altitude = 35,000 ft, Cruise Mach number = 0.78

Ambient temperature = 218.8 K, ambient pressure = 23.8 kPa   (Standard Atmosphere)

Speed of sound at cruise altitude $a =$ m/s; cruise speed of aircraft $V =$ m/s   (1.4b)

Aircraft mass at start of cruise = 171.5 tonne, aircraft lift-drag ratio at cruise = 21.6.

Thrust per engine at start of cruise at 35,000 ft (assume 2 engines) $F_N =$ kN   (2.4a)

Entering engine: stagnation temp. and press.	$T_{02} =$ K, $p_{02} =$	kPa (6.2)
Turbine inlet stagnation temperature at cruise,	$T_{04} = 1500$ K	
Engine pressure ratio at cruise for flow through core, $p_{03}/p_{02} = 45$		
Gas turbine cycle efficiency	$\eta_{cy} =$	(4.3e)

Entering HPC (after fan and booster): stagnation $T\&p$.	$T_{023} =$ K, $p_{023} =$	kPa (5.1a)
Leaving HP compressor: stagnation $T\&p$.	$T_{03} =$ K, $p_{03} =$	kPa (5.1a)
Leaving HP turbine: stagnation $T\&p$.	$T_{045} =$ K, $p_{045} =$	kPa (5.1b)
Leaving LP turbine: stagnation $T\&p$.	$T_{05} =$ K, $p_{05} =$	kPa (7.2)
Fan pressure ratio	$fpr = p_{013}/p_{02} = 1.5$	
Bypass stream leaving fan: stagnation $T\&p$.	$T_{013} =$ K, $p_{013} =$	kPa (7.2)
Bypass jet velocity at cruise	$V_{jb} =$	m/s (7.2)
Core jet velocity	$V_{jc} =$ $V_{jc} = V_{jb}$	m/s (7.2)
Bypass ratio	$bpr =$	(7.2)
Mass flow through the engine	$\dot{m}_{air} =$ kg/s	(7.4)
Core mass flow	$\dot{m}_c =$	
sfc of bare engine	$=$ kg h^{-1}kg^{-1}	(7.2)
Effective sfc (deducting drag of nacelle)	$=$ kg h^{-1}kg^{-1}	(7.5)
Fan tip radius	$=$ m	(7.6)
Rotational speed of LP shaft	$=$ radian/s	(9.1b)
Mean radius of booster = 0.5 times fan diameter	$=$ m	
Number of booster stages	$=$	(9.2c)
Mean radius of HP compressor	$=$ m	(9.3a)
HP compressor blade height at outlet	$=$ m	(9.4a)
Rotational speed of HP shaft	$=$ radian/s	(9.3b)
Core compressor mid-height blade speed	$=$ m/s	(9.3b)
Number of HP compressor stages	$=$	(9.4b)
HP turbine mean radius	$=$ m	(9.6)
HP turbine blade heights	$=$ mm	(9.8a & 9.8b)
LP turbine mean radius	$=$ m	(9.9b)
Number of LP turbine stages	$=$	(9.9b)

Design Sheet for New Fighter Aircraft

This table gives quantities which are likely to be needed repeatedly. Numbers in parentheses indicate the exercise in which the data are evaluated.

Aircraft mass at take-off $= 18 \times 10^3$ kg, aircraft mass at tropopause $= 15 \times 10^3$ kg.

Tropopause 11 km, $p_a = 22.7$ kPa, $T_a = 216.7$ K $\rho_a = 0.365$ kg/m³

			0.9	1.5	2.0	
Flight Mach number	M					
Flight speed	V	m/s				(14.1b)
Inlet stagn. temp.	T_{02}	K				(15.4)
Stagn. pressure relative to aircraft		kPa				(15.4)
Stagn. pressure into engine	p_{02}	kPa				(15.4)
Min. net engine thrust F_N for 1g	kN					(14.4a)
for 3g					$-$	(14.4c)

Sea level $p_a = 101.3$ kPa, $T_a = 288.15$ K $\rho_a = 1.223$ kg/m³

			0.9	1.5	2.0	
Flight Mach number	M					
Min. net engine thrust F_N for 5g turn	kN				$-$	(14.4b)
9g					$-$	(14.4b)

Design point for take-off, $p_a = 101.3$ kPa, $T_a = 288.15$ K,

HP turbine stator exit temp $T_{04} = 1850$ K; fan pr $= p_{013}/p_{02} = 4.5$; overall pr $= p_{03}/p_{02} = 30$, ·

Fan delivery temperature	$T_{013} = \quad T_{023} = \quad$ K	(16.6)
Core compressor delivery temperature	$T_{03} = \quad$ K	(16.6)
HP turbine exit pressure	$p_{045} = \quad$ kPa; temperature $T_{045} = \quad$ K	(16.6b)
Mixed out temperature at LP turbine inlet	$T_{045'} = \quad$ K; $c_{pm} = \quad$ J kg^{-1}K^{-1}; $\gamma_m = \quad$	(16.6c)
LP turbine exit pressure	$p_{05} = \quad$ kPa; temperature $T_{05} = \quad$ K	(16.6g)
Mixed out temperature at LP turbine exit	$T_{05'} = \quad$ K; $c_{pm} = \quad$ J kg^{-1}K^{-1}; $\gamma_m = \quad$	(16.6d)
Mixed out temperature at nozzle inlet (dry)	$T_{09} = \quad$ K; $c_{pm} = \quad$ J kg^{-1}K^{-1}; $\gamma_m = \quad$	(16.6f)
Mixed out temperature at nozzle inlet (a/b)	$T_{09} = 2200$ K; $c_{pm} = 1244$ J kg^{-1}K^{-1}; $\gamma_m = 1.30$	

		dry	a/b, $T_{09} = 2200$ K
Net thrust F_N	kN	(16.6b)	(16.8c)
Specific thrust	m/s	(16.6a)	(16.7b)
bypass ratio			
sfc	kg h^{-1}kg^{-1}	(16.6a)	(16.7b)
$\dot{m}_{air}$	kg/s	(16.6b)	(16.8c)
fan inlet dia.	m	(16.6b)	(16.8c)

For dry design

Fan first stage	$N = \quad$ rpm; tip speed $U_t = \quad$ m/s, tip Mach number $= \quad$	(18.1)
Fan stage temp rise	$= \quad$ K; first-stage pressure ratio $= \quad$	(18.2)
HP compressor	$N = \quad$ rpm; first-stage tip diameter $D = \quad$ m	(18.5)
HP turbine mean speed $U_m = \quad$	m/s; $\Delta h_0/U_m^2 = \quad$	(18.9)
LP turbine mean speed $U_m = \quad$	m/s; $\Delta h_0/U_m^2 = \quad$	(18.9)
HP and LP turbine blade heights at throat	mm; mm	(18.11)